DNA Replication

DNA Replication

Edited by **Tom Lee**

New York

Published by Callisto Reference,
106 Park Avenue, Suite 200,
New York, NY 10016, USA
www.callistoreference.com

DNA Replication
Edited by Tom Lee

International Standard Book Number: 978-1-63239-152-0 (Hardback)

Printed in the United States of America.

Contents

Preface

Every book is initially just a concept; it takes months of research and hard work to give it the final shape in which the readers receive it. In its early stages, this book also went through rigorous reviewing. The notable contributions made by experts from across the globe were first molded into patterned chapters and then arranged in a sensibly sequential manner to bring out the best results.

DNA replication is a basic part of the life cycle of all the organisms. Many characteristics of this process show thorough conservation across organisms in all domains of life. This book outlines and reviews the present condition of knowledge on numerous crucial aspects of the DNA replication procedure. DNA replication is a crucial process in both development and growth and in relation to a wide variety of pathological conditions including cancer. The book contains novel insights into the whole process of DNA replication covered under two broad sections, "Machines that Drive DNA Replication" and "Mechanisms that Protect Chromosome Integrity during DNA Replication". It is also a compilation of thought provoking questions and summaries to give way to future investigations.

It has been my immense pleasure to be a part of this project and to contribute my years of learning in such a meaningful form. I would like to take this opportunity to thank all the people who have been associated with the completion of this book at any step.

Editor

Machines that Drive DNA Replication

DNA Replication in Archaea, the Third Domain of Life

Yoshizumi Ishino and Sonoko Ishino

Additional information is available at the end of the chapter

1. Introduction

The accurate duplication and transmission of genetic information are essential and crucially important for living organisms. The molecular mechanism of DNA replication has been one of the central themes of molecular biology, and continuous efforts to elucidate the precise molecular mechanism of DNA replication have been made since the discovery of the double helix DNA structure in 1953 [1]. The protein factors that function in the DNA replication process, have been identified to date in the three domains of life (Figure 1).

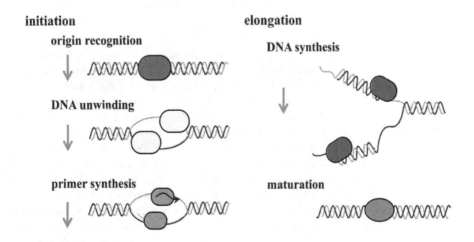

Figure 1. Stage of DNA replication

	Archaea	Eukaryota	Bacteria
initiation			
origin recognition	Cdc6/Orc1	ORC	DnaA
DNA unwinding	Cdc6/Orc1 MCM GINS	Cdc6 Cdt1 MCM GINS Cdc45	DnaC DnaB
primer synthesis	DNA primase	Pol α / primase	DnaG
elongation			
DNA synthesis	family B DNA polymerase (Pol B) family D DNA polymerase (Pol D)	family B DNA polymerase (Pol δ) (Pol ε)	family C DNA polymerase (Pol III)
	clamp loader (RFC) clamp (PCNA)	clamp loader (RFC) clamp (PCNA)	clamp loader (γ-complex) clamp (β-clamp)
maturation	Fen1 Dna2 DNA ligase	FEN1 DNA2 DNA ligase	Pol I RNaseH DNA ligase

Table 1. The proteins involved in DNA replication from the three domains of life

Archaea, the third domain of life, is a very interesting living organism to study from the aspects of molecular and evolutionary biology. Rapid progress of whole genome sequence analyses has allowed us to perform comparative genomic studies. In addition, recent microbial ecology has revealed that archaeal organisms inhabit not only extreme environments, but also more ordinary habitats. In these situations, archaeal biology is among the most exciting of research fields. Archaeal cells have a unicellular ultrastructure without a nucleus, resembling bacterial cells, but the proteins involved in the genetic information processing pathways, including DNA replication, transcription, and translation, share strong similarities with those of eukaryotes. Therefore, most of the archaeal proteins were identified as homologues of many eukaryotic replication proteins, including ORC (origin recognition complex), Cdc6, GINS (Sld5-Psf1-Psf2-Psf3), MCM (minichromosome maintenance), RPA (replication protein A), PCNA (proliferating cell nuclear antigen), RFC (replication factor C), FEN1 (flap endonuclease 1), in addition to the eukaryotic primase, DNA polymerase, and DNA ligase; these are obviously different from bacterial proteins (Table 1) and these proteins were biochemically characterized [2-4]. Their similarities indicate that the DNA replication machineries of Archaea and Eukaryota evolved from a common ancestor, which was different from that of Bacteria [5]. Therefore, the archaeal organisms are good models to elucidate the functions of each component of the eukaryotic type replication machinery complex. Genomic and comparative genomic research with archaea is made easier by the fact that the genome size and the number of genes of archaea are much smaller than those of eukaryotes. The archaeal replication machinery is probably a simplified form of that in eukaryotes. On the other hand, it is also interesting that the circular genome structure is conserved in Bacteria and Archaea and is different from the linear form of eukaryotic genomes. These features have encouraged us to study archaeal DNA replication, in the hopes of gaining fundamental insights into this molecular mechanism and its machinery from an evolutionary perspective. The study of bacterial DNA replication at a molecular level started in about 1960, and then eukaryotic studies followed since 1980. Because Archaea was recognized as the third domain of life later, the archaeal DNA replication research became active after 1990. With increasing the available total genome sequences, the progress of research on archaeal DNA replication has been rapid, and the depth of our knowledge of archaeal DNA replication has almost caught up with those of the bacterial and eukaryotic research fields. In this chapter, we will summarize the current knowledge of DNA replication in Archaea.

2. Replication origin

The basic mechanism of DNA replication was predicted as "replicon theory" by Jacob et al. [6]. They proposed that an initiation factor recognizes the replicator, now referred to as a replication origin, to start replication of the chromosomal DNA. Then, the replication origin of E. coli DNA was identified as oriC (origin of chromosome). The archaeal replication origin was identified in the Pyrococcus abyssi in 2001 as the first archaeal replication origin. The origin was located just upstream of the gene encoding the Cdc6 and Orc1-like sequences in the Pyrococcus genome [7]. We discovered a gene encoding an amino acid sequence that bore similarity to those of both eukaryotic Cdc6 and Orc1, which are the eukaryotic initiators. After confirming that this protein actually binds to the oriC region on the chromosomal DNA we named the gene product Cdc6/Orc1 due to its roughly equal homology with regions of eukaryotic Orc1 and Cdc6, [7]. The gene consists of an operon with the gene encoding DNA polymerase D (it was originally called Pol II, as the second DNA polymerase from Pyrococcus furiosus) in the genome [8]. A characteristic of the oriC is the conserved 13 bp repeats, as predicted earlier by bioinformatics [9], and two of the repeats are longer and surround a predicted DUE (DNA unwinding element) with an AT-rich sequence in Pyrococcus genomes (Figure 2) [10]. The longer repeated sequence was designated as an ORB (Origin Recognition Box), and it was actually recognized by Cdc6/Orc1 in a Sulfolobus solfataricus study [11]. The 13 base repeat is called a miniORB, as a minimal version of ORB. A whole genome microarray analysis of P. abyssi showed that the Cdc6/Orc1 binds to the oriC region with extreme specificity, and the specific binding of the highly purified P. furiosus Cdc6/Orc1 to ORB and miniORB was confirmed in vitro [12]. It has to be noted that multiple origins were identified in the Sulfolobus genomes. It is now well recognized that Sulfolobus has three origins and they work at the same time in the cell cycle [11, 13-16]. Analysis of the mechanism of how the multiple origins are utilized for genome replication is an interesting subject in the research field of archaeal DNA replication. The main questions are how the initiation of replication from multiple origins is regulated and how the replication forks progress after the collision of two forks from opposite directions.

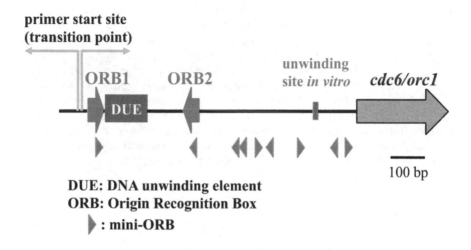

Figure 2. The *oriC* region in *Pyrococcus* genome. The region surrounding *oriC* is presented schematically. The ORB1 and ORB2 are indicated by large arrow, and the mini-ORB repeats are indicated by small arrowheads. DUE is indicated in red. The unwinding site, determined by *in vitro* analysis, is indicated in orange. The transition site is indicated by green arrows. The *cdc6/orc1* gene located in downstream is drawn by gray arrow.

3. How does Cdc6/Orc1 recognize *oriC*?

An important step in characterizing the initiation of DNA replication in Archaea is to understand how the Cdc6/Orc1 protein recognizes the *oriC* region. Based upon amino acid sequence alignments, the archaeal Cdc6/Orc1 proteins belong to the AAA⁺ family of proteins. The crystal structures of the Cdc6/Orc1 protein from *Pyrobaculum aerophilum* [17] and one of the two Cdc6/Orc1 proteins, ORC2 from *Aeropyrum pernix* (the two homologs in this organism are called ORC1 and ORC2 by the authors) [18] were determined. These Cdc6/Orc1 proteins consist of three structural domains. Domains I and II adopt a fold found in the AAA⁺ family proteins. A winged helix (WH) fold, which is present in a number of DNA binding proteins, is found in the domain III. There are four ORBs arranged in pairs on both sides of the DUE in the *oriC* region of *A. pernix*, and ORC1 binds to each ORB as a dimer. A mechanism was proposed in which ORC1 binds to all four ORBs to introduce a higher-order assembly for unwinding of the DUE with alterations in both topology and superhelicity [19]. Furthermore, the crystal structures of *S. solfataricus* Cdc6-1 and Cdc6-3 (two of the three

Cdc6/Orc1 proteins in this organism) forming a heterodimer bound to *ori2* DNA (one of the three origins in this organism) [20], and that of *A. pernix* ORC1 bound to an origin sequence [21] were determined. These studies revealed that both the N-terminal AAA+ ATPase domain (domain I+II) and C-terminal WH domain (domain III) contribute to origin DNA binding, and the structural information not only defined the polarity of initiator assembly on the origin but also indicated the induction of substantial distortion, which probably triggers the unwinding of the duplex DNA to start replication, into the DNA strands. These structural data also provided the detailed interaction mode between the initiator protein and the *oriC* DNA. Mutational analyses of the *Methanothermobactor thermautotrophicus* Cdc6-1 protein revealed the essential interaction between an arginine residue conserved in the archaeal Cdc6/Orc1 and an invariant guanine in the ORB sequence [22].

P. furiosus Cdc6/Orc1 is difficult to purify in a soluble form. A specific site in the *oriC* to start unwinding *in vitro*, was identified using the protein prepared by a denaturation-renaturation procedure recently [23]. As shown in Figure 2, the local unwinding site is about 670 bp away from the transition site between leading and lagging syntheses, which was determined earlier by an *in vivo* replication initiation point (RIP) assay [10]. Although the details of the replication machinery that must be established at the unwound site are not fully understood in Archaea, it is expected to minimally include MCM, GINS, primase, PCNA, DNA polymerase, and RPA, as described below. The following *P. furiosus* studies revealed that the ATPase activity of the Cdc6/Orc1 protein was completely suppressed by binding to DNA containing the ORB. Limited proteolysis and DNase I-footprint experiments suggested that the Cdc6/Orc1 protein changes its conformation on the ORB sequence in the presence of ATP. The physiological meaning of this conformational change has not been solved, but it should have an important function to start the initiation process [24] as in the case of bacterial DnaA protein. In addition, results from an *in vitro* recruiting assay indicated that MCM (Mcm protein complex), the replicative DNA helicase, is recruited onto the *oriC* region in a Cdc6/Orc1-dependent, but not ATP-dependent, manner [24], as described below. However, this recruitment is not sufficient for the unwinding function of MCM, and some other function remains to be identified for the functional loading of this helicase to promote the progression of the DNA replication fork.

4. MCM helicase

After unwinding of the *oriC* region, the replicative helicase needs to remain loaded to provide continuous unwinding of double stranded DNA (dsDNA) as the replication forks progress bidirectionally. The MCM protein complex, consisting of six subunits (Mcm2, 3, 4, 5, 6, and 7), is known to be the replicative helicase "core" in eukaryotic cells [25]. The MCM further interacts with Cdc45 and GINS, to form a ternary assembly referred to as the "CMG complex", that is believed to be the functional helicase in eukaryotic cells (Figure 3) [26]. However, this idea is still not universal for the eukaryotic replicative helicase.

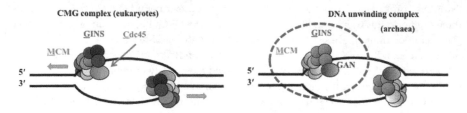

Figure 3. DNA-Unwinding complex in eukaryotes and archaea. The CMG complex is the replicative helicase for the template DNA unwinding reaction in eukaryotes. The archaeal genomes contain the homologs of the Mcm and Gins proteins, but a Cdc45 homolog has not been identified. Recent research suggests that a RecJ-like exonuclease GAN, which has weak sequence homology to that of Cdc45, may work as a helicase complex with MCM and GINS.

Most archaeal genomes appear to encode at least one Mcm homologue, and the helicase activities of these proteins from several archaeal organisms have been confirmed *in vitro* [27-31]. In contrast to the eukaryotic MCM, the archaeal MCMs, consist of a homohexamer or homo double hexamer, having distinct DNA helicase activity by themselves *in vitro*, and therefore, these MCMs on their own may function as the replicative helicase *in vivo*. The structure-function relationships of the archaeal Mcms have been aggressively studied using purified proteins and site-directed mutagenesis [32]. An early report using the ChIP method showed that the *P. abyssi* Mcm protein preferentially binds to the origin *in vivo* in exponentially growing cells [7, 12]. The *P. furiosus* MCM helicase does not display significant helicase activity *in vitro*. However, the DNA helicase activity was clearly stimulated by the addition of GINS (the Gins23-Gins51 complex), which is the homolog of the eukaryotic GINS complex (described below in more detail). This result suggests that MCM works with other accessory factors to form a core complex in *P. furiosus* similar to the eukaryotic CMG complex as described above [31].

Some archaeal organisms have more than two Cdc6/Orc1 homologs. It was found that the two Cdc6/Orc1 homologs, Cdc6-1 and Cdc6-2, both inhibit the helicase activity of MCM in *M. thermautotrophicus* [33. 34]. Similarly, Cdc6-1 inhibits MCM activity in *S. solfataricus* [35]. In contrast, the Cdc6-2 protein stimulates the helicase activity of MCM in *Thermoplasma acidophilum* [36]. Functional interactions between Cdc6/Orc1 and Mcm proteins need to be investigated in greater detail to achieve a more comprehensive understanding of the conservation and diversity of the initiation mechanism in archaeal DNA replication.

Another interesting feature of DNA replication initiation is that several archaea have multiple genes encoding Mcm homologs in their genomes. Based on the recent comprehensive genomic analyses, thirteen archaeal species have more than one *mcm* gene. However, many of the *mcm* genes in the archaeal genomes seem to reside within mobile elements, originating from viruses [37]. For example, two of the three genes in the *Thermococcus kodakarensis* genome are located in regions where genetic elements have presumably been integrated [38]. The establishment of a genetic manipulation system for *T. kodakarensis*, is the first for a hyperthermophilic euryarchaeon [39. 40], and is advantageous for investigating the function of these Mcm proteins. Two groups have recently performed gene disruption experiments for each *mcm* gene [41, 42]. These experiments revealed that the knock-out strains for *mcm1* and *mcm2* were easily isolated, but *mcm3* could not be disrupted. Mcm3 is relatively abundant in the *T. kodakarensis* cells. Furthermore, an *in vitro* experiment using purified Mcm proteins showed that only Mcm3 forms a stable hexameric structure in solution. These results support the contention that Mcm3 is the main helicase core protein in the normal DNA replication process in *T. kodakarensis*.

The functions of the other two Mcm proteins remain to be elucidated. The genes for Mcm1 and Mcm2 are stably inherited, and their gene products may perform some important functions in the DNA metabolism in *T. kodakarensis*. The DNA helicase activity of the recombinant Mcm1 protein is strong *in vitro*, and a distinct amount of the Mcm1 protein is present in *T. kodakarensis* cells. Moreover, Mcm1 functionally interacts with the GINS complex from *T. kodakarensis* [42]. These observations strongly suggest that Mcm1 does participate in some aspect of DNA transactions, and may be substituted with Mcm3. Our immunoprecipitation experiments showed that Mcm1 co-precipitated with Mcm3 and GINS, although they did not form a heterohexameric complex [42], suggesting that Mcm1 is involved in the replisome or repairsome and shares some function in *T. kodakarensis* cells. Although western blot analysis could not detect Mcm2 in the extract from exponentially growing *T. kodakarensis* cells [42], a RT-PCR experiment detected the transcript of the *mcm2* gene in the cells (Ishino et al., unpublished). The recombinant Mcm2 protein also has ATPase and helicase activities *in vitro*. [41] Therefore, the *mcm2* gene is expressed under normal growth conditions and may work in some process with a rapid turn over. Further experiments to measure the efficiency of *mcm2* gene transcription by quantitative PCR, as well as to assess the stability of the Mcm2 protein in the cell extract, are needed. Phenotypic analyses investigating the sensitivities of the Δ*mcm1* and Δ*mcm2* mutant strains to DNA damage caused by various mutagens, as reported for other DNA repair-related genes in *T. kodakarensis* [43], may provide a clue to elucidate the functions of these Mcm proteins.

Methanococcus maripaludis S2 harbors four *mcm* genes in its genome, three of which seem to be derived from phage, a shotgun proteomics study detected peptides originating from three out of the four *mcm* gene products [44]. Furthermore, the four gene products co-expressed in *E. coli* cells were co-purified in the same fraction [45]. These results suggest that multiple Mcm proteins are functional in the *M. maripaludis* cells.

5. Recruitment of Mcm to the *oriC* region

Another important question is how MCM is recruited onto the unwound region of *oriC*. The detailed loading mechanism of the MCM helicase has not been elucidated. It is believed that archaea utilize divergent mechanisms of MCM helicase assembly at the *oriC* [46].

An *in vitro* recruiting assay showed that *P. furiosus* MCM is recruited to the *oriC* DNA in a Cdc6/Orc1-dependent manner [24]. This assay revealed that preloading Cdc6/Orc1 onto the ORB DNA resulted in a clear reduction in MCM recruitment to the *oriC* region, suggesting that free Cdc6/Orc1 is preferable as a helicase recruiter, to associate with MCM and bring it to *oriC*. It would be interesting to understand how the two tasks, origin recognition and

MCM recruiting, are performed by the Cdc6/Orc1 protein, because the WH domain, which primarily recognizes and binds ORB, also has strong affinity for the Mcm protein. The assembly of the Mcm protein onto the ORB DNA by the Walker A-motif mutant of *P. furiosus* Cdc6/Orc1 occurred with the same efficiency as the wild type Cdc6/Orc1. The DNA binding of *P. furiosus* Cdc6/Orc1 was not drastically different in the presence and absence of ATP, as in the case of the initiator proteins from *Archaeoglobus fulgidus* [28], *S. solfataricus* [11], and *A. pernix* [19]. Therefore, it is still not known whether the ATP binding and hydrolysis activity of Cdc6/Orc1 regulates the Mcm protein recruitment onto *oriC* in the cells.

One more important issue is the very low efficiency of the Mcm protein recruitment in the reported *in vitro* assay [24]. Quantification of the recruited Mcm protein by the *in vitro* assay showed that less than one Mcm hexamer was recruited to the ORB. The linear DNA containing ORB1 and ORB2, used in the recruiting assay, may not be suitable to reconstitute the archaeal DNA replication machinery and a template that more closely mimics the chromosomal DNA may be required. Additionally, it may be that as yet unidentified proteins are required to achieve efficient *in vitro* helicase loading in the *P. furiosus* cells. Finally, it will ultimately be necessary to construct a more defined *in vitro* replication system to analyze the regulatory functions of Cdc6/Orc1 precisely during replication initiation.

In *M. thermautotrophicus*, the Cdc6-2 proteins can dissociate the Mcm multimers [47]. The activity of Cdc6-2 might be required as the MCM helicase loader in this organism. The interaction between Cdc6/Orc1 and Mcm is probably general. However, the effect of Cdc6/Orc1 on the MCM helicase activity differs among various organisms, as described above. Some other protein factors may function in various archaea, for example a protein that is distantly related to eukaryotic Cdt1, which plays a crucial role during MCM loading in Eukaryota, exists in some archaeal organisms, although its function has not been characterized yet [14].

6. GINS

The eukaryotic GINS complex was originally identified in *Saccharomyces cerevisiae* as essential protein factor for the initiation of DNA replication [48]. GINS consists of four different proteins, Sld5, Psf1, Psf2, and Psf3 (therefore, GINS is an acronym for Japanese go-ichi-ni-san, meaning 5-1-2-3, after these four subunits). The amino acid sequences of the four subunits in the GINS complex share some conservation, suggesting that they are ancestral paralogs [49]. However, most of the archaeal genomes have only one gene encoding this family protein, and more interestingly, the Crenarchaeota and Euryarchaeota (the two major subdomains of Archaea) characteristically have two genes with sequences similar to Psf2 and Psf3, and Sld5 and Psf1, respectively referred to as Gins23 and Gins51 [31, 49]. A Gins homolog, designated as Gins23, was biochemically detected in *S. solfataricus* as the first Gins protein in Archaea, in a yeast two-hybrid screening for interaction partners of the Mcm protein, and another subunit, designated as Gins15, was identified by mass-spectrometry analysis of an immunoaffinity-purified native GINS from an *S. solfataricus* cell extract. [50]. The *S. solfataricus* GINS, composed of two proteins, Gins23 and Gins15, forms a tetrameric structure with a 2:2 molar ratio [50]. The GINS from *P. furiosus*, a complex of Gins23 and Gins51 with a 2:2 ratio, was identified as the first euryarchaeal GINS [31]. Gins51 was preferred over Gins15 because of the order of the name of GINS.

The MCM2-7 hexamer was copurified in complex with Cdc45 and GINS from *Drosophila melanogaster* embryo extracts and *S. cerevisiae* lysates, and the "CMG (Cdc45-MCM2-7-GINS) complex" (Figure 3), as described above, should be important for the function of the replicative helicase. The CMG complex was also associated with the replication fork in *Xenopus laevis* egg extracts, and a large molecular machine, containing Cdc45, GINS, and MCM2-7, was proposed as the unwindosome to separate the DNA strands at the replication fork [51]. Therefore, GINS must be a critical factor for not only the initiation process, but also the elongation process in eukaryotic DNA replication. *S. solfataricus* GINS interacts with MCM and primase, suggesting that GINS is involved in the replisome. The concrete function of GINS in the replisome remains to be determined. No stimulation or inhibition of either the helicase or primase activity was observed by the interaction with *S. solfataricus* GINS *in vitro* [50]. On the other hand, the DNA helicase activity of *P. furiosus* MCM is clearly stimulated by the addition of the *P. furiosus* GINS complex, as described above [31].

In contrast to *S. solfataricus* and *P. furiosus*, which each express a Gins23 and Gins51, *Thermoplasma acidophilum* has a single Gins homolog, Gins51. The recombinant Gins51 protein from *T. acidophilum* was confirmed to form a homotetramer by gel filtration and electron microscopy analyses. Furthermore, a physical interaction between *T. acidophilum* Gins51 and Mcm was detected by a surface plasmon resonance analysis (SPR). Although the *T. acidophilum* Gins51 did not affect the helicase activity of its cognate MCM, when the equal ratio of each molecule was tested *in vitro* [52], an excess amount of Gins51 clearly stimulated the helicase activity (Ogino et al., unpublished). In the case of *T. kodakarensis*, the ATPase and helicase activities of MCM1 and MCM3 were clearly stimulated by *T. kodakarensis* GINS *in vitro*. It is interesting that the helicase activity of MCM1 was stimulated more than that of MCM3. Physical interactions between the *T. kodakarensis* Gins and Mcm proteins were also detected [53]. These reports suggested that the MCM-GINS complex is a common part of the replicative helicase in Archaea (Figure 3).

Recently, the crystal structure of the *T. kodakarensis* GINS tetramer, composed of Gins51 and Gins23 was determined, and the structure was conserved with the reported human GINS structures [53]. Each subunit of human GINS shares a similar fold, and assembles into the heterotetramer of a unique trapezoidal shape [54-56]. Sld5 and Psf1 possess the α-helical (A) domain at the N-terminus and the β-stranded domain (B) at the C-terminus (AB-type). On the other hand, Psf2 and Psf3 are the permuted version (BA-type). The backbone structure of each subunit and the tetrameric assembly of *T. kodakarensis* GINS are similar to those of human GINS. However, the location of the C-terminal B domain of Gins51 is remarkably different between the two GINS structures [53]. A homology model of the homotetrameric GINS from *T. acidophilum* was performed using the *T. kodakarensis* GINS crystal structure as a template. The Gins 51 protein has a long disordered region inserted between the A and B domains and this allows the conformation of the C-terminal domains to be more flexible.

This domain arrangement leads to the formation of an asymmetric homotetramer, rather than a symmetrical assembly, of the *T. kodakarensis* GINS [53].

The Cdc45 protein is ubiquitously distributed from yeast to human, supporting the notion that the formation of the CMG complex is universal in the eukaryotic DNA replication process. However, no archaeal homologue of Cdc45 has been identified. A recent report of bioinformatic analysis showed that the primary structure of eukaryotic Cdc45 and prokaryotic RecJ share a common ancestry [57]. Indeed, a homolog of the DNA binding domain of RecJ has been co-purified with GINS from *S. solfataricus* [50]. Our experiment detected the stimulation of the 5'-3' exonuclease activity of the RecJ homologs from *P. furiosus* and *T. kodakarensis* by the cognate GINS complexes (Ishino et al., unpublished). The RecJ homolog from *T. kodakarensis* forms a stable complex with the GINS, and the 5'-3' exonuclease activity is enhanced *in vitro*; therefore, the RecJ homolog was designated as GAN, from GINS-Associated Nuclease in a very recent paper [58]. Another related report found that the human Cdc45 structure obtained by the small angle X-ray scattering analysis (SAXS) is consistent with the crystallographic structure of the RecJ family members [59]. These current findings will promote further research on the structures and functions of the higher-order unwindosome in archaeal and eukaryotic cells (Figure 3).

7. Primase

To initiate DNA strand synthesis, a primase is required for the synthesis of a short oligonucleotide, as a primer. The DnaG and p48-p58 proteins are the primases in Bacteria and Eukaryota, respectively. The p48-p58 primase is further complexed with p180 and p70, to form DNA polymerase α-primase complex. The catalytic subunits of the eukaryotic (p48) and archaeal primases, share a little, but distinct sequence homology with those of the family X DNA polymerases [60]. The first archaeal primase was identified from *Methanococcus jannaschii*, as an ORF with a sequence similar to that of the eukaryotic p48. The gene product exhibited DNA polymerase activity and was able to synthesize oligonucleotides on the template DNA [61]. We characterized the p48-like protein (p41) from *P. furiosus*. Unexpectedly, the archaeal p41 protein did not synthesize short RNA by itself, but preferentially utilized deoxynucleotides to synthesize DNA strands up to several kilobases in length [62]. Furthermore, the gene neighboring the p41 gene encodes a protein with very weak similarity to the p58 subunit of the eukaryotic primase. The gene product, designated p46, actually forms a stable complex with p41, and the complex can synthesize a short RNA primer, as well as DNA strands of several hundred nucleotides *in vitro* [63]. The short RNA but not DNA primers were identified in *Pyrococcus* cells, and therefore, some mechanism to dominantly use RNA primers exists in the cells [10].

Further research on the primase homologs from *S. solfataricus* [64-66], *Pyrococcus horikoshii* [67-69], and *P. abyssi* [70] showed similar properties *in vitro*. Notably, p41 is the catalytic subunit, and the large one modulates the activity in the heterodimeric archaeal primases. The small and large subunits are also called PriS and PriL, respectively. The crystal structure of the N-terminal domain of PriL complexed with PriS of *S. solfataricus* primase revealed that PriL does not directly contact the active site of PriS, and therefore, the large subunit may interact with the synthesized primer, to adjust its length to a 7-14 mer. The structure of the catalytic center is similar to those of the family X DNA polymerases. The 3'-terminal nucleotidyl transferase activity, detected in the *S. solfataricus* primase [64, 66], and the gap-filling and strand-displacement activities in the *P. abyssi* primase [70] also support the structural similarity between PriS and the family X DNA polymerases.

A unique activity, named PADT (template-dependent Polymerization Across Discontinuous Template), in the *S. solfataricus* PriSL complex was published very recently [71]. The activity may be involved in double-strand break repair in Archaea.

The archaeal genomes also encode a sequence similar to the bacterial type DnaG primase. The DnaG homolog from the *P. furiosus* genome was expressed in *E. coli*, but the protein did not show any primer synthesis activity *in vitro*, and thus the archaeal DnaG-like protein may not act as a primase in *Pyrococcus* cells (Fujikane et al. unpublished). The DnaG-like protein was shown to participate in RNA degradation, as an exosome component [72, 73]. However, a recent paper reported that a DnaG homolog from *S. solfataricus* actually synthesizes primers with a 13 nucleotide length [74]. It would be interesting to investigate if the two different primases share the primer synthesis for leading and lagging strand replication, respectively, in the *Sulfolobus* cells, as the authors suggested [74]. A proposed hypothesis about the evolution of PriSL and DnaG from the last universal common ancestor (LUCA) is interesting [71].

The *Sulfolobus* PriSL protein was shown to interact with Mcm through Gins23 [50]. This primase-helicase interaction probably ensures the coupling of DNA unwinding and priming during the replication fork progression [50]. Furthermore, the direct interaction between PriSL and the clamp loader RFC (described below) in *S. solfataricus* may regulate the primer synthesis and its transfer to DNA polymerase in archaeal cells [75].

8. Single-stranded DNA binding protein

The single-stranded DNA binding protein, which is called SSB in Bacteria and RPA in Archaea and Eukaryota, is an important factor to protect the unwound single-stranded DNA from nuclease attack, chemical modification, and other disruptions during the DNA replication and repair processes. SSB and RPA have a structurally similar domain containing a common fold, called the OB (oligonucleotide/oligosaccharide binding)-fold, although there is little amino acid sequence similarity between them [76]. The common structure suggests that the mechanism of single-stranded DNA binding is conserved in living organisms despite the lack of sequence similarity. *E. coli* SSB is a homotetramer of a 20 kDa peptide with one OB-fold, and the SSBs from *Deinococcus radiodurans* and *Thermus aquaticus* consist of a homodimer of the peptide containing two OB-folds. The eukaryotic RPA is a stable heterotrimer, composed of 70, 32, and 14 kDa proteins. RPA70 contains two tandem repeats of an OB-fold, which are responsible for the major interaction with a single-stranded DNA in its central region. The N-terminal and C-terminal regions of RPA70 mediate interactions with RPA32 and also with many cellular or viral proteins [77, 78]. RPA32 contains an OB-fold in the central region [79-81], and the C-terminal region interacts with other RPA subunits and various cellular proteins [77, 78. 82, 83]. RPA14 also contains an OB-fold [77]. The eukaryotic RPA interacts with the SV40 T-antigen and the DNA polymerase α-primase complex, and thus forms part of the initiation complex at the replication origin [84]. The RPA also stimulates Polα-primase activity and PCNA-dependent Pol δ activity [85, 86].

The RPAs from *M. jannaschii* and *M. thermautotrophicus* were reported in 1998, as the first archaeal single-stranded DNA binding proteins [87-89]. These proteins share amino acid sequence similarity with the eukaryotic RPA70, and contain four or five repeated OB-fold and one zinc-finger motif. The *M. jannaschii* RPA exists as a monomer in solution, and has single-strand DNA binding activity. On the other hand, *P. furiosus* RPA forms a complex consisting of three distinct subunits, RPA41, RPA32, and RPA14, similar to the eukaryotic RPA [90]. The *P. furiosus* RPA strikingly stimulates the RadA-promoted strand-exchange reaction *in vitro* [90].

While the euryarchaeal organisms have a eukaryotic-type RPA homologue, the crenarchaeal SSB proteins appear to be much more related to the bacterial proteins, with a single OB fold and a flexible C-terminal tail. However, the crystal structure of the SSB protein from *S. solfataricus* showed that the OB-fold domain is more similar to that of the eukaryotic RPAs, supporting the close relationship between Archaea and Eukaryota [91].

The RPA from *Methanosarcina acetivorans* displays a unique property. Unlike the multiple RPA proteins found in other archaea and eukaryotes, each subunit of the *M. acetivorans* RPAs, RPA1, RPA2, and RPA3, have 4, 2, and 2 OB-folds, respectively, and can act as a distinct single-stranded DNA-binding proteins. Furthermore, each of the three RPA proteins, as well as their combinations, clearly stimulates the primer extension activity of *M. acetivorans* DNA polymerase BI *in vitro*, as shown previously for bacterial SSB and eukaryotic RPA [92]. Architectures of SSB and RPA suggested that they are composed of different combinations of the OB fold. Bacterial and eukaryotic organisms contain one type of SSB or RPA, respectively. In contrast, archaeal organisms have various RPAs, composed of different organizations of OB-folds. A hypothesis that homologous recombination might play an important role in generating this diversity of OB-folds in archaeal cells was proposed, based on experiments characterizing the engineered RPAs with various OB-folds [93].

9. DNA polymerase

DNA polymerase catalyzes phosphodiester bond formation between the terminal 3'-OH of the primer and the α-phosphate of the incoming triphosphate to extend the short primer, and is therefore the main player of the DNA replication process. Based on the amino acid sequence similarity, DNA polymerases have been classified into seven families, A, B, C, D, E, X, and Y (Table 2) [94-98].

The fundamental ability of DNA polymerases to synthesize a deoxyribonucleotide chain is widely conserved, but more specific properties, including processivity, synthesis accuracy, and substrate nucleotide selectivity, differ depending on the family. The enzymes within the same family have basically similar properties. *E. coli* has five DNA polymerases, and Pol I, Pol II, and Pol III belong to families A, B, and C, respectively. Pol IV and Pol V are classified in family Y, as the DNA polymerases for translesion synthesis (TLS). In eukaryotes, the replicative DNA polymerases, Pol α, Pol δ, and Pol ε, belong to family B, and the translesion DNA polymerases, η, ι, and κ, belong to family Y [99].

The most interesting feature discovered at the inception of this research area was that the archaea indeed have the eukaryotic Pol α-like (Family B) DNA polymerases [100-102]. Members of the Crenarchaeota have at least two family B DNA polymerases [103, 104]. On the other hand, there is only one family B DNA polymerase in the Euryarchaeota. Instead, the euryarchaeal genomes encode a family D DNA polymerase, proposed as Pol D, which seems to be specific for these archaeal organisms and has never been found in other domains [95, 105]. The genes for family Y-like DNA polymerases are conserved in several, but not all, archaeal genomes. The role of each DNA polymerase in the archaeal cells is still not known, although the distribution of the DNA polymerases is getting clearer (Table 2) [106].

	families of DNA polymerases						
	A	B	C	D	E	X	Y
Archaea							
Crenarchaeota		Pol BI, Pol BII Pol BIII			Pol E*		Pol Y
Euryarchaeota		Pol BI		Pol D	Pol E*		Pol Y
Korarchaeota		Pol BI, Pol BII		Pol D			
Aigarchaeota		Pol BI, Pol BII		Pol D			Pol Y
Thaumarchaeota		Pol BI		Pol D			Pol Y
Bacteria	Pol I	Pol II	Pol III				Pol IV Pol V
Eukaryota	Pol θ Pol γ**	Pol α, Pol δ Pol ε, Pol ζ				Pol β, Pol λ Pol μ, Pol σ	Pol η, Pol ι Pol κ

* plasmid-encoded

** mitochondrial

Table 2. Distribution of DNA polymerases from seven families in the three domains of life.

The first family D DNA polymerase was identified from *P. furiosus*, by screening for DNA polymerase activity in the cell extract [107]. The corresponding gene was cloned, revealing that this new DNA polymerase consists of two proteins, named DP1 and DP2, and that the deduced amino acid sequences of these proteins were not conserved in the DNA polymerase families [8]. *P. furiosus* Pol D exhibits efficient strand extension activity and strong proofreading activity [8, 108]. Other family D DNA polymerases were also characterized by several groups [109-115]. The Pol D genes had been found only in Euryarchaeota. However, recent environmental genomics and cultivation efforts revealed novel phyla in Archaea: Thaumarchaeota, Korarchaeota, and Aigarchaeota, and their genome sequences harbor the genes encoding Pol D.

A genetic study on *Halobacterium* sp. NRC-1 showed that both Pol B and Pol D are essential for viability [116]. An interesting issue is to elucidate whether Pol B and Pol D work together at the replication fork for the synthesis of the leading and lagging strands, respectively. According to the usage of an RNA primer and the presence of strand displacement activity, Pol D may catalyze lagging strand synthesis [106, 114].

Thaumarchaeota and Aigarchaeota harbor the genes encoding Pol D and crenarchaeal Pol BII [117, 118], while Korarchaeota encodes Pol BI, Pol BII and Pol D [119]. Biochemical characterization of these gene products will contribute to research on the evolution of DNA polymerases in living organisms. A hypothesis that the archaeal ancestor of eukaryotes encoded three DNA polymerases, two distinct family B DNA polymerases and a family D DNA polymerase, which all contributed to the evolution of the eukaryotic replication machinery, consisting of Pol α, δ, and ε, has been proposed [120].

A protein is encoded in the plasmid pRN1 isolated from a *Sulfolobus* strain [121]. This protein, ORF904 (named RepA), has primase and DNA polymerase activities in the N-terminal domain and helicase activity in the C-terminal domain, and is likely to be essential for the replication of pRN1 [122, 123]. The amino acid sequence of the N-terminal domain lacks homology to any known DNA polymerases or primases, and therefore, family E is proposed. Similar proteins are encoded by various archaeal and bacterial plasmids, as well as by some bacterial viruses [124]. Recently, one protein, tn2-12p, encoded in the plasmid pTN2 isolated from *Thermococcus nautilus*, was experimentally identified as a DNA polymerase in this family [125]. This enzyme is likely responsible for the replication of the plasmids. Further investigations of this family of DNA polymerases will be interesting from an evolutional perspective.

10. PCNA and RFC

The sliding clamp with the doughnut-shaped ring structure is conserved among living organisms, and functions as a platform or scaffold for proteins to work on the DNA strands. The eukaryotic and archaeal PCNAs form a homotrimeric ring structure [126, 127], which encircles the DNA strand and anchors many important proteins involved in DNA replication and repair (Figure 4). PCNA works as a processivity factor that retains the DNA polymerase on the DNA by binding it on one surface (front side) of the ring for continuous DNA strand synthesis in DNA replication (Figure 5). To introduce the DNA strand into the central hole of the clamp ring, a clamp loader is required to interact with the clamp and open its ring. The archaeal and eukaryotic clamp loader is called RFC (Figure 5). The most studied archaeal PCNA and RFC molecules to date are *P. furiosus* PCNA [128-132] and RFC [133-136]. The PCNA and RFC molecules are essential for DNA polymerase to perform processive DNA synthesis. The molecular mechanism of the clamp loading process has been actively investigated [137] (Figure 5). An intermediate PCNA-RFC-DNA complex, in which the PCNA ring is opened with out-of plane mode, was detected by a single particle analysis of electron microscopic images using *P. furiosus* proteins (Figure 6) [138]. The crystal structure of the complex, including the ATP-bound clamp loader, the ring-opened clamp, and the template-primer DNA, using proteins from bacteriophage T4, has recently been published [139], and our knowledge about the clamp loading mechanism is continuously progressing.

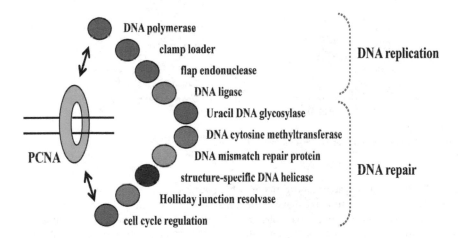

Figure 4. PCNA-interacting proteins

After clamp loading, DNA polymerase accesses the clamp and the polymerase-clamp complex performs processive DNA synthesis. Therefore, structural and functional analyses of the DNA polymerase-PCNA complex is the next target to elucidate the overall mechanisms of replication fork progression. The PCNA interacting proteins contain a small conserved sequence motif, called the PIP box, which binds to a common site on PCNA [140]. The PIP box consists of the sequence "Qxxhxxaa", where "x" represents any amino acid, "h" represents a hydrophobic residue (e.g. L, I or M), and "a" represents an aromatic residue (e.g. F, Y or W). Archaeal DNA polymerases have PIP box-like motifs in their sequences [141]). However, only a few studies have experimentally investigated the function of the motifs. The crystal structure of *P. furiosus* Pol B complexed with a monomeric PCNA mutant was determined, and a convincing model of the polymerase-PCNA ring interaction was constructed [142]. This study revealed that a novel interaction is formed between a stretched loop of PCNA and the thumb domain of Pol B, in addition to the authentic PIP box. A comparison of the model structure with the previously reported structures of a family B DNA polymerase from RB69 phage, complexed with DNA [143, 144], suggested that the second interaction site plays a crucial role in switching between the polymerase and exonuclease modes, by inducing a PCNA-polymerase complex configuration that favors synthesis over editing. This putative mechanism for the fidelity control of replicative DNA polymerases is supported by experiments, in which mutations at the second interaction site enhanced the exonuclease activity in the presence of PCNA [144]. Furthermore, the three-dimensional structure of the DNA polymerase-PCNA-DNA ternary complex was analyzed by electron microscopic (EM) single particle analysis. This structural view revealed the entire domain configuration of the trimeric ring of PCNA and DNA polymerase, including the protein-protein or protein-DNA contacts. This architecture provides clearer insights into the switching mechanism between the editing and synthesis modes [145].

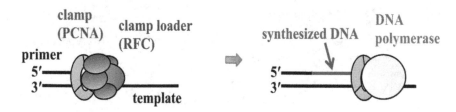

Figure 5. Mechanisms of processive DNA synthesis The clamp loader (RFC) tethers the clamp (PCNA) onto the primer terminus of the DNA strand. The clamp loader is then replaced by DNA polymerase, which can synthesize the DNA strand processively without falling off.

In contrast to most euryarchaeal organisms, which have a single PCNA homolog forming a homotrimeric ring structure, the majority of crenarchaea have multiple PCNA homologues, and they are capable of forming heterotrimeric rings for their functions [146, 147]. It is especially interesting that the three PCNAs, PCNA1, PCNA2, and PCNA3, specifically bind PCNA binding proteins, including DNA polymerases, DNA ligases, and FEN-1 endonuclease [147, 148]. Detailed structural studies of the heterologous PCNA from S. solfataricus revealed that the interaction modes between the subunits are conserved with those of the homotrimeric PCNAs [149, 150].

T. kodakarensis is the only euryarchaeal species that has two genes encoding PCNA homologs on the genome [38]. These two genes from the T. kodakarensis genome, and the highly purified gene products, PCNA1 and PCNA2, were characterized [151]. PCNA1 stimulated the DNA synthesis reactions of the two DNA polymerases, Pol B and Pol D, from T. kodakarensis in vitro. PCNA2 however only had an effect on Pol B. The T. kodakarensis strain with pcna2 disruption was isolated, whereas gene disruption for pcna1 was not possible. These results suggested that PCNA1 is essential for DNA replication, and PCNA2 may play a different role in T. kodakarensis cells. The sensitivities of the Δpcna2 mutant strain to ultraviolet irradiation (UV), methyl methanesulfonate (MMS) and mitomycin C (MMC) were indistinguishable to those of the wild type strain. Both PCNA1 and PCNA2 form a stable ring structure and work as a processivity factor for *T. kodakarensis* Pol B *in vitro*. The crystal structures of the two PCNAs revealed the different interactions at the subunit-subunit interfaces [152].

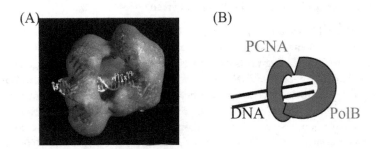

Figure 6. Electron Microscopic Analysis of *P. furious* DNA polymerase-PCNA-DNA complex. The complex in the editing mode of the DNA polymerase-PCNA DNA ternary complex was shown. (A) Electron microscopic (EM) map of the complex is depicted by gray surface. DNA polymerase and PCNA are shown in a ribbon representation colored purple and blue, respectively. The DNA is colored white. The exonuclease active site is shown in a green ribbon. (B) Schematic view of the complex.

The RFC molecule is conserved as a pentameric complex in Eukaryota and Archaea. However, the eukaryotic RFC is a heteropentameric complex, consisting of five different proteins, RFC1 to 5, in which RFC1 is larger than the other four RFCs. On the other hand, the archaeal RFC consists of two proteins, RFCS (small) and RFCL (large), in a 4 to 1 ratio. A different form of RFC, consisting of three subunits, RFCS1, RFCS2, and RFCL, in a 3 to 1 to 1 ratio, was also identified from *M. acetivorans* [153]. The three subunits of RFC may represent an intermediate stage in the evolution of the more complex RFC in Eukaryota from the less complex RFC in Archaea [153, 154]. The subunit organization and the spatial distribution of the subunits in the *M. acetivorans* RFC complex were analyzed and compared with those of the *E. coli* γ-complex, which is also a pentamer consisting of three different proteins. These two clamp loaders adopt similar subunit organizations and spatial distributions, but the functions of the individual subunits are likely to be diverse [154].

11. DNA ligase

DNA ligase is essential to connect the Okazaki fragments of the discontinuous strand synthesis during DNA replication, and therefore, it universally exists in all living organisms. This enzyme catalyzes phosphodiester bond formation via three nucleotidyl transfer steps [155, 156]. In the first step, DNA ligase forms a covalent enzyme-AMP intermediate, by reacting with ATP or NAD^+ as a cofactor. In the second step, DNA ligase recognizes the substrate DNA, and the AMP is subsequently transferred from the ligase to the 5'-phosphate terminus of the DNA, to form a DNA-adenylate intermediate (AppDNA). In the final step, the 5'-AppDNA is attacked by the adjacent 3'-hydroxy group of the DNA and a phospho-diester bond is formed. DNA ligases are grouped into two families, according to their re-quirement for ATP or NAD^+ as a nucleotide cofactor in the first step reaction. ATP-dependent DNA ligases are widely found in all three domains of life, whereas NAD^+-dependent DNA ligases exist mostly in Bacteria. Some halophilic archaea [157] and eukaryotic viruses [158] also have NAD^+-dependent enzymes.

Three genes (*LIG1*, *LIG3* and *LIG4*) encoding ATP-dependent DNA ligases have been identi-fied in the human genome to date and DNA ligase I (Lig I), encoded by *LIG1*, is a replicative enzyme that joins Okazaki fragments during DNA replication [156]. The first gene encoding a eukaryotic-like ATP-dependent DNA ligase was found in the thermophilic archaeon, *De-sulfolobus ambivalens* [159]. Subsequent identifications of the DNA ligases from archaeal or-ganisms revealed that these enzymes primarily use ATP as a cofactor. However, this classification may not be so strict. The utilization of NAD^+, as well as ATP, as a cofactor has been observed in several DNA ligases, including those from *T. kodakarensis* [160], *T. fumico-lans*, *P. abyssi* [161]), *Thermococcus* sp. NA1 [162], *T. acidophilum*, *Picrophilus torridus*, and *Fer-roplasma acidophilum*, although ATP is evidently preferable in all of the cases [163] (Table 3). The dual co-factor specificity (ATP/NAD^+) is an interesting feature of these DNA ligase en-zymes and it will be enlightening to investigate the structural basis for this. Another dual co-factor specificity exists in the archaeal DNA ligases, which use ADP as well as ATP, as found in the enzymes from *A. pernix* [164] and *Staphylothermus marinus* [165], and in the case of *Sulfobococcus zilligii*, GTP is also the functional cofactor [166]. The DNA ligases from *P. horikoshii* [167] and *P. furiosus* [168] have a strict ATP preference (Table 3). Sufficient bio-chemical data have not been obtained to resolve the issue of dual co-factor specificity, and further biochemical and structural analyses are required.

cofactor			
ATP	ATP and ADP	ATP and NAD+	ATP, ADP, and GTP
Acidithiobacillus ferrooxidans	*Aeropyrum pernix*	*Ferroplasma acidophilum*	*Sulfophobococcus zilligii*
Ferroplasma acidarmanus	*Staphylothermus marinus*	*Picrophilus torridus*	
Methanothermobacterium thermoautotrophicum		*Thermococcus fumicolans*	
Pyrococcus horikoshii		*Thermococcus kodakarensis*	
Pyrococcus furiosus		*Thermococcus sp.*	
Sulfolobus acidocaldarius		*Thermoplasma acidophilum*	
Sulfolobus shibatae			
Thermococcus sp. 1519			

Table 3. Cofactor dependency of the archaeal DNA ligases

The crystal structure of *P. furiosus* DNA ligase [169] was solved and the physical and functional interactions between the DNA ligase and PCNA was shown [168]. The detailed interaction mode between human Lig I and PCNA is somewhat unclear, because of several controversial reports [170-172]. The stimulatory effect of *P. furiosus* PCNA on the enzyme activity of the cognate DNA ligase was observed at a high salt concentration, at which a DNA ligase alone cannot bind to a nicked DNA substrate. Interestingly, the PCNA-binding site is located in the middle of the N-terminal DNA binding domain (DBD) of the *P. furiosus* DNA ligase, and the binding motif, QKSFF, which is proposed as a shorter version of the PIP box, is actually looped out from the protein surface [168]. Interestingly, this motif is located in the middle of the protein chain, rather than the N- or C-terminal region, where the PIP boxes are usually located. To confirm that this motif is conserved in the archaeal/eukaryotic DNA ligases, the physical and functional interactions between *A. pernix* DNA ligase and PCNA was analyzed and the interaction was shown to mainly depend on the phenylalanine 132 residue, which is located in the predicted region from the multiple sequence alignment of the ATP-dependent DNA ligases [173].

The crystal structure of the human Lig I, complexed with DNA, was solved as the first ATP-dependent mammalian DNA ligase, although the ligase was an N-terminal truncated form [174]. The structure comprises the N-terminal DNA binding domain, the middle adenylation domain, and the C-terminal OB-fold domain. The crystal structure of Lig I (residues 233 to 919) in complex with a nicked, 5'-adenylated DNA intermediate revealed that the enzyme redirects the path of the dsDNA, to expose the nick termini for the strand-joining reaction. The N-terminal DNA-binding domain works to encircle the DNA substrate like PCNA and to stabilize the DNA in a distorted structure, positioning the catalytic core on the nick. The crystal structure of the full length DNA ligase from *P. furiosus* revealed that the architecture of each domain resembles those of Lig I, but the domain arrangements strikingly differ between the two enzymes [168]. This domain rearrangement is probably derived from the "domain-connecting" role of the helical extension conserved at the C-termini in the archaeal and eukaryotic DNA ligases. The DNA substrate in the open form of Lig I is replaced by motif VI at the C-terminus, in the closed form of *P. furiosus* DNA ligase. Both the shapes and electrostatic distributions are similar between motif VI and the DNA substrate, suggesting that motif VI in the closed state mimics the incoming substrate DNA. The subsequently solved crystal structure of *S. solfataricus* DNA ligase is the fully open structure, in which the three domains are highly extended [175]. In this work, the *S. solfataricus* ligase-PCNA complex was also analyzed by SAXS. *S. solfataricus* DNA ligase bound to the PCNA ring still retains an open, extended conformation. The closed, ring-shaped conformation observed in the Lig I structure as described above is probably the active form to catalyze a DNA end-joining reaction, and therefore, it is proposed that the open-to-closed movement occurs for ligation, and the switch in the conformational change is accommodated by a malleable interface with PCNA, which serves as an efficient platform for DNA ligation [175]. After the publication of these crystal structures, the three-dimensional structure of the ternary complex, consisting of DNA ligase-PCNA-DNA, using the *P. furiosus* proteins was obtained by EM single particle analysis [176]. In the complex structure, the three domains of the crescent-shaped *P. furiosus* DNA ligase surround the central DNA duplex, encircled by the closed PCNA ring. The relative orientations of the ligase domains remarkably differ from those of the crystal structures, and therefore, a large domain rearrangement occurs upon ternary complex formation. In the EM image model, the DNA ligase contacts PCNA at two sites, the conventional PIP box and a novel second contact in the middle adenylation domain. It is also interesting that a substantial DNA tilt from the PCNA ring axis is observed. Based on these structural analyses, a mechanism in which the PCNA binding proteins are bound and released sequentially. In fact, most of the PCNA binding proteins share the same binding sites in the interdomain connecting loop (IDCL) and the C-terminal tail of the PCNA. The structural features exclude the possibility that the three proteins contact the single PCNA ring simultaneously, because DNA ligase occupies two of the three subunits of the PCNA trimer. In the case of the RFC-PCNA-DNA complex structure obtained by the same EM technique, RFC entirely covers the PCNA ring, thus blocking the access of other proteins [138]. These ternary complexes appear to favor a mechanism involving the sequential binding and release of replication factors.

12. Flap endonuclease 1 (FEN1)

Efficient processing of Okazaki fragments to make a continuous DNA strand is essential for the lagging strand synthesis in asymmetric DNA replication. The primase-synthesized RNA/DNA primers need to be removed to join the Okazaki fragments into an intact continuous strand DNA. Flap endonuclease 1 (FEN1) is mainly responsible for this task. Okazaki fragment maturation is highly coordinated with continuous DNA synthesis, and the interactions of DNA polymerase, FEN1, and DNA ligase with PCNA allow these enzymes to act sequentially during the maturation process, as described above.

FEN1, a structure-specific 5′-endonuclease, specifically recognizes a dsDNA with an unannealed 5′-flap [177, 178]. In the eukaryotic Okazaki fragment processing system, 5′-flap DNA structures are formed by the strand displacement activity of DNA polymerase δ. Lig I seals the nick after the flapped DNA is cleaved by FEN1. These processing steps are facilitated by PCNA [179]. The interactions between eukaryotic FEN1 and PCNA have been well characterized [140, 171], and the stimulatory effect of PCNA on the FEN1 activity was also shown [180]. The crystal structure of the human FEN1-PCNA complex revealed three FEN1 molecules bound to each PCNA subunit of the trimer ring in different configurations [181]. Based on these structural analyses together with the description in the DNA ligase section, a flip-flop transition mechanism, which enables proteins to internally switch for different functions on the same DNA clamp are currently being considered.

The eukaryotic homologs of FEN1 were found in Archaea [182]. The crystal structures of FEN1 from M. jannaschii [183], P. furiosus [184], P. horikoshii [185], A. fulgidus [186], and S. solfataricus [150] have been determined. In addition, detailed biochemical studies were performed on P. horikoshii FEN1 [187, 188]. Thus, studies of the archaeal FEN1 proteins have provided important insights into the structural basis of the cleavage reaction of the flapped DNA. Our recent research showed that the flap endonuclease activity of P. furiosus FEN1 was stimulated by PCNA. Furthermore, the stimulatory effect of PCNA on the sequential action of FEN1 and DNA ligase was observed *in vitro* (Kiyonari et al., unpublished). Based on these results, a model of the molecular switching mechanisms of the last steps of Okazaki-fragment maturation was constructed. The quaternary complex of FEN1-Lig-PCNA-DNA was also isolated for the EM single particle analysis. These studies will provide more concrete image of the molecular mechanism.

13. Summary and perspectives

Research on the molecular mechanism of DNA replication has been a central theme of molecular biology. Archaeal organisms became popular in the total genome sequencing age, as described above, and most of the DNA replication proteins are now equally understood by biochemical characterizations. In addition, the archaeal studies are especially interesting to understand the mechanisms by which cells live in extreme environmental conditions. Furthermore, it is also noteworthy that the proteins from the hyperthermophilic archaea are more stable than those from mesophilic organisms, and they are advantageous for the structural and functional analyses of higher-ordered complexes, such as the replisome. Studies on the higher-ordered complexes, rather than single proteins, are essential for understanding each of the events involved in DNA metabolism, and the archaeal research will continuously contribute to the development and advancement of the DNA replication research field, as summarized in part in a recent review [189, 190].

In addition to basic molecular biology research, DNA replication proteins from thermophiles have been quite useful reagents for gene manipulations, including genetic diagnosis, forensic DNA typing, and detection of bacterial and virus infections, as well as basic research. Numerous enzymes have been commercialized around the world, and are utilized daily. An example of the successful engineering of an archaeal DNA polymerase for PCR is the creation of the fusion protein between *P. furiosus* Pol B and a nonspecific dsDNA binding protein, Sso7d, from *S. solfataricus*, by genetic engineering techniques [191]. The fusion DNA polymerase overcame the low processivity of the wild type Pol B by the high affinity Sso7d to the DNA strand. As another example, we successfully developed a novel processive PCR method, using the archaeal Pol B with the help of a mutant PCNA [192, 193]. Several DNA sequencing technologies, referred to as "next-generation sequencing", have been developed [194, 195], and are now commercially available. Single-molecule detection, using dye-labeled modified nucleotides and longer read lengths, is now known as "third-generation DNA sequencing" [196]. These technologies apply DNA polymerases or DNA ligases from various sources, indicating that these DNA replication enzymes are indispensable for the development of DNA manipulation technology. These facts prove that the progress of the basic research on the molecular biology of archaeal DNA replication will promote the development of the new technologies for genetic engineering.

Author details

Yoshizumi Ishino and Sonoko Ishino

Department of Bioscience and Biotechnology, Graduate School of Bioresource and Bioenvironmental Sciences, Kyushu University, Japan

References

[1] Watson JD, Crick FH. Molecular structure of nucleic acids; a structure for deoxyribose nucleic acid. Nature, 1953;171(4356) 737-738.

[2] Kelman Z, White MF. Archaeal DNA replication and repair. Curr Opin Microbiol, 2005;(6) 669-676.

[3] Barry ER, Bell SD. DNA replication in the archaea. Microbiol Mol Biol Rev, 2006;70(4) 876-887.

[4] Wigley DB ORC proteins: marking the start. Curr Opin Struct Biol, 2009;19(1) 72-78.

[5] Leipe DD, Aravind L, Koonin EV. Did DNA replication evolve twice independently? Nucleic Acids Res, 1999;27(17) 3389-3401.

[6] Jacob F, Brenner SCR. Hebd Seances Acad Sci. 1963;256 298-300.

[7] Matsunaga F, Forterre P, Ishino Y, Myllykallio H. In vivo interactions of archaeal Cdc6/Orc1 and minichromosome maintenance proteins with the replication origin. Proc. Natl. Acad. Sci. USA, 2001;98(20) 11152-11157.

[8] Uemori T, Sato Y, Kato I, Doi H, Ishino Y. A novel DNA polymerase in the hyperthermophilic archaeon, *Pyrococcus furiosus*: gene cloning, expression, and characterization. Genes Cells, 1997;2(8) 499-512.

[9] Lopez P, Philippe H, Myllykallio H, Forterre P. Identification of putative chromosomal origins of replication in Archaea. Mol Microbiol, 1999;32(4) 883-886.

[10] Matsunaga F, Norais C, Forterre P, Myllykallio H. Identification of short 'eukaryotic' Okazaki fragments synthesized from a prokaryotic replication origin. EMBO Rep, 2003;4(2) 154-158.

[11] Robinson NP, Dionne I, Lundgren M, Marsh VL. Bernander R, Bell SD. Identification of two origins of replication in the single chromosome of the archaeon *Sulfolobus solfataricus* Cell, 2004;116(1) 25-38.

[12] Matsunaga F, Glatigny A, Mucchielli-Giorgi MH, Agier N, Delacroix H, Marisa L, et al. Genomewide and Biochemical Analyses of DNA-binding activity of Cdc6/Orc1 and Mcm proteins in *Pyrococcus* sp. Nucleic Acids Res, 2007;35(10)3214-3222.

[13] Lundgren M, Andersson A, Chen L, Nilsson P, Bernander R. Three replication origins in *Sulfolobus* species: synchronous initiation of chromosome replication and asynchronous termination. Proc Natl Acad Sci USA, 2004;101(18):7046-7051.

[14] Robinson NP, Bell SD. Extrachromosomal element capture and the evolution of multiple replication origins in archaeal chromosomes. Proc Natl Acad Sci USA, 2007;104(14) 5806-5811.

[15] Norais C, Hawkins M, Hartman AL, Eisen JA, Myllykallio H, Allers T. Genetic and physical mapping of DNA replication origins in *Haloferax volcanii*. PLoS Genet, 2007;3(5) e77.

[16] Capes MD, Coker JA, Gessler R, Grinblat-Huse V, DasSarma SL, Jacob CG, et al. The information transfer system of halophilic archaea. Plasmid, 2011;65(2) 77-101.

[17] Liu J, Smith CL, DeRyckere D, DeAngelis K, Martin GS, Berger JM. Structure and function of Cdc6/Cdc18: implications for origin recognition and checkpoint control. Mol Cell, 2000;6(3) 637-648.

[18] Singleton MR, Morales R, Grainge I., Cook N, Isupov MN, Wigley DB. Conformational changes induced by nucleotide binding in Cdc6/ORC from *Aeropyrum pernix*. J Mol Biol, 2004;343(3) 547-557.

[19] Grainge I, Gaudier M, Schuwirth BS, Westcott SL, Sandall J, Atanassova N, et al. Biochemical analysis of a DNA replication origin in the archaeon *Aeropyrum pernix*. J Mol Biol, 2006;363(2) 355-369.

[20] Dueber EL, Corn JE, Bell, SD, Berger JM. Replication origin recognition and deformation by a heterodimeric archaeal Orc1 complex. Science, 2007;317(5842) 1210-1213.

[21] Gaudier M, Schuwirth BS, Westcott SL, Wigley DB. Structural basis of DNA replication origin recognition by an ORC protein. Science, 2007;317(5842) 1213-1216.

[22] Majernik AI, Chong JP. A conserved mechanism for replication origin recognition and binding in archaea. Biochem J, 2008;409(2) 511-518.

[23] Matsunaga F, Takemura K, Akita M, Adachi A, Yamagami T, Ishino Y. Localized melting of duplex DNA by Cdc6/Orc1 at the DNA replication origin in the hyperthermophilic archaeon *Pyrococcus furiosus*. Extremophiles, 2010;14(1) 21-31.

[24] Akita M, Adachi A, Takemura K, Yamagami T, Matsunaga F, Ishino Y. Cdc6/Orc1 from *Pyrococcus furiosus* may act as the origin recognition protein and Mcm helicase recruiter. Gens Cells, 2010;15(5) 537-552.

[25] Masai H, You Z, Arai K. Control of DNA replication: regulation and activation of eukaryotic replicative helicase, MCM. IUBMB Life, 2005;57(4-5) 323-335.

[26] Ilves I, Petojevic T, Pesavento JJ, Botchan MR. Activation of the MCM2-7 helicase by association with Cdc45 and GINS proteins. Mol Cell, 2010;37(2) 247-258,

[27] Chong JP, Hayashi MK, Simon MN, Xu RM, Stillman B. A double-hexamer archaeal minichromosome maintenance protein is an ATP-dependent DNA helicase. Proc Natl Acad Sci USA, 2000;97(4) 1530-1535.

[28] Grainge I, Scaife S, Wigley DB. Biochemical analysis of components of the pre-replication complex of *Archaeoglobus fulgidus*. Nucleic Acids Res, 2003;31(16) 4888-4898.

[29] Kelman Z, Lee JK, Hurwitz J. The single minichromosome maintenance protein of *Methanobacterium thermoautotrophicum* ΔH contains DNA helicase activity. Proc Natl Acad Sci USA, 1999;96(26) 14783-14788.

[30] Shechter DF, Ying CY, Gautier J, The intrinsic DNA helicase activity of *Methanobacterium thermoautotrophicum* ΔH minichromosome maintenance protein. J Biol Chem, 2000;275(20) 15049-15059.

[31] Yoshimochi T, Fujikane R, Kawanami M, Matsunaga F, Ishino Y. The GINS complex from *Pyrococcus furiosus* stimulates the MCM helicase activity. J Biol Chem, 2008;283(3) 1601-1609.

[32] Sakakibara N, Kelman LM, Kelman Z. Unwinding the structure and function of the archaeal MCM helicase. Mol Microbiol, 2009;72(2) 286-896.

[33] Shin JH, Grabowski B, Kasiviswanathan R, Bell SD, Kelman Z. Regulation of minichromosome maintenance helicase activity by Cdc6. J Biol Chem, 2003;278(39) 38059-38067.

[34] Kasiviswanathan R, Shin JH, Kelman Z. Interactions between the archaeal Cdc6 and MCM proteins modulate their biochemical properties. Nucleic Acids Res, 2005;33(15) 4940-4950.

[35] De Felice M, Esposito L, Pucci B, Carpentieri F, De Falco M, Rossi M, et al. Biochemical characterization of a CDC6-like protein from the crenarchaeon *Sulfolobus solfataricus*. J Biol Chem, 2003;278(47) 46424-46431.

[36] Haugland GT, Shin JH, Birkeland NK, Kelman Z. Stimulation of MCM helicase activity by a Cdc6 protein in the archaeon *Thermoplasma acidophilum*. Nucleic Acids Res, 2006, 34(21) 6337-6344.

[37] Krupovic M, Gribaldo S, Bamford DH, Forterre P. The evolutionary history of archaeal MCM helicases: a case study of vertical evolution combined with hitchhiking of mobile genetic elements. Mol Biol Evol, 2010;27(12) 2716-2732.

[38] Fukui T, Atomi H, Kanai T, Matsumi R, Fujiwara S, Imanaka T.. Complete genome sequence of the hyperthermophilic archaeon *Thermococcus kodakarensis* KOD1 and comparison with *Pyrococcus* genomes. Genome Res, 2005;15(3) 352-363.

[39] Sato T, Fukui T, Atomi H, Imanaka T. Targeted gene disruption by homologous recombination in the hyperthermophilic archaeon *Thermococcus kodakaraensis* KOD1. J Bacteriol, 2003;185(1) 210-220.

[40] Sato T, Fukui T, Atomi H, Imanaka T. Improved and versatile transformation system
      allowing multiple genetic manipulations of the hyperthermophilic archaeon *Thermo-
      coccus kodakaraensis*. Appl Environ Microbiol, 2005;71(7) 3889-3899.

[41] Pan M, Santangelo TJ, Li Z, Reeve JN, Kelman Z. *Thermococcus kodakarensis* encodes
      three MCM homologs but only one is essential. Nucleic Acids Res, 2011;39(22)
      9671-9680.

[42] Ishino S, Fujino S, Tomita H, Ogino H, Takao K, Daiyasu H, et al. Biochemical and
      genetical analyses of the three mcm genes from the hyperthermophilic archaeon,
      Thermococcus kodakarensis. Genes Cells, 2011;16(12) 1176-1189.

[43] Fujikane R, Ishino S, Ishino Y. Forterre, P. Genetic analysis of DNA repair in the hy-
      perthermophilic archaeon, *Thermococcus kodakarensis*. Genes Genet Syst, 2010;
      85(4)243-257.

[44] Xia Q, Hendrickson E L, Zhang Y, Wang T, Taub F, Moore BC, et al. Quantitative
      proteomics of the archaeon *Methanococcus maripaludis* validated by microarray analy-
      sis and real time PCR. Mol Cell Proteomics, 2006;5(5) 868-881.

[45] Walters AD, Chong JP. An archaeal order with multiple minichromosome mainte-
      nance genes. Microbiology, 2010;156(Pt 5):1405-1414.

[46] Sakakibara N, Kelman LM, Kelman Z. How is the archaeal MCM helicase assembled
      at the origin? Possible mechanisms Biochem. Soc. Trans. 2007;37(Pt 1):7-11.

[47] Shin J-H, Heo GY, Kelman Z. The *Methanothermobactor thermautotrophicus* Cdc6-2 pro-
      tein, the putative helicase loader, dissociates the minichromosome maintenance heli-
      case. J Bacteriol, 2008;190(11) 4091-4094.

[48] Takayama Y, Kamimura Y, Okawa M, Muramatsu S, Sugino A, Araki H. et al. GINS,
      a novel multiprotein complex required for chromosomal DNA replication in bud-
      ding yeast. Genes Dev, 2003;17(9)1153-1165.

[49] Makarova KS, Wolf YI, Mekhedov SL, Mirkin BG, Koonin EV. Ancestral paralogs
      and pseudoparalogs and their role in the emergence of the eukaryotic cell. Nucleic
      Acids Res, 2005;33(14) 4626-4638.

[50] Marinsek N, Barry ER, Makarova KS, Dionne I, Koonin EV, Bell SD. GINS, a central
      nexus in the archaeal DNA replication fork. EMBO Rep, 2006;7(5) 539-545.

[51] Pacek M, Tutter AV, Kubota Y, Takisawa H, Walter JC. Localization of MCM2-7,
      Cdc45, and GINS to the site of DNA unwinding during eukaryotic DNA replication.
      Mol Cell, 2006;21(4) 581-587.

[52] Ogino H, Ishino S, Mayanagi K, Haugland GT, Birkeland NK, Yamagishi A, et al.
      The GINScomplex from the thermophilic archaeon, *Thermoplasma acidophilum* may
      function as a homotetramer in DNA replication. Extremophiles, 2011;15(4) 529-539.

[53] Oyama T, Ishino S, Fujino S, Ogino H, Shirai T, Mayanagi K, et al. Architectures of archaeal GINS complexes, essential DNA replication initiation factors. BMC Biol, 2011;9 28.

[54] Kamada K, Kubota Y, Arata T, Shindo Y, Hanaoka F. Structure of the human GINS complex and its assembly and functional interface in replication initiation. Nat Struct Mol Biol, 2007;14(5)388-396.

[55] Choi JM, Lim HS, Kim JJ, Song OK, Cho Y. Crystal structure of the human GINS complex. Genes Dev, 2007;21(11) 1316-1321.

[56] Chang, YP, Wang G, BermudezV, Hurwitz J, Chen XS, et al. Crystal structure of the GINS complex and functional insight into its role in DNA replication. Proc Natl Acad Sci USA, 2007;104(31) 12685-12690.

[57] Sanchez-Pulido L, Ponting CP. Cdc45: the missing RecJ ortholog in eukaryotes? Bioinformatics, 2011;27(14) 1885-1888.

[58] Li Z, Pan M, Santangelo T J, Chemnitz W, Yuan W, Edwards JL, et al. A novel DNA nuclease is stimulated by association with the GINS complex. Nucleic Acids Res, 2011;39(14) 6114-6123.

[59] Krastanova I, Sannino V, Amenitsch H, Gileadi O, Pisani FM, Onesti S., et al. Structural and functional insights into the DNA replication factor Cdc45 reveal an evolutionary relationship to the DHH family of phosphoesterases. J Biol Chem, 2011;287(6) 4121-4128.

[60] Kirk BW, Kuchta RD. Arg304 of human DNA primase is a key contributor to catalysis and NTP binding: primase and the family X polymerases share significant sequence homology. Biochemistry, 1999;38(31) 7727-7736

[61] Desogus G, Onesti S, Brick P, Rossi M, Pisani FM. Identification and characterization of a DNA primase from the hyperthermophilic archaeon *Methanococcus jannaschii*. Nucleic Acids Res, 1999;27(22) 4444-4450.

[62] Bocquier A, Liu L, Cann I, Komori K, Kohda D, Ishino Y. Archaeal primase: bridging the gap between RNA and DNA polymerase. Curr Biol, 2001;11(6) 452-456.

[63] Liu L, Komori K, Ishino S, Bocquier AA, Cann IK, Kohda D, et al. The archaeal DNA primase: biochemical characterization of the p41-p46 complex from *Pyrococcus furiosus*. J Biol Chem, 2001;276(48) 45484-45490.

[64] Lao-Sirieix SH, Bell SD. The heterodimeric primase of the hyperthermophilic archaeon *Sulfolobus solfataricus* possesses DNA and RNA primase, polymerase and 3'-terminal nucleotidyl transferase activities. J Mol Biol, 2004;344(5) 1251-1263.

[65] Lao-Sirieix SH, Nookala RK, Roversi P, Bell SD, Pellegrini L.et al. Structure of the heterodimeric core primase. Nat Struct Mol Biol, 2005;12(12) 1137-1144.

[66] De Falco M, Fusco A, DeFelice M, Rossi M, Pisani FM. The DNA primase is activated by substrates containing a thymine-rich bubble and has a 3'-terminal nucleotidyl-transferase activity. Nucleic Acids Res, 2004;32(17) 5223-5230.

[67] Ito N, Nureki O, Shirouzu M, Yokoyama S, Hanaoka F. Crystal structure of the *Pyrococcus horikoshii* DNA primase-UTP complex: implications for the mechanism of primer synthesis. Genes Cells, 2003;8(12) 913-923.

[68] Matsui E, Nishio M, Yokoyama H, Harata K, Damas S, Matsui I. Distinct domain functions regulating de novo DNA synthesis of thermostable DNA primase from hyperthermophile *Pyrococcus horikoshii*. Biochemistry. 2003;42(50) 14968-14976.

[69] Ito N, Matsui I, Matsui E. Molecular basis for the subunit assembly of the primase from an archaeon *Pyrococcus horikoshii*. FEBS J, 2007;274(5):1340-1351.

[70] Le Breton M, Henneke G, Norais C, Flament D, Myllykallio H, Querellou J,et al. The heterodimeric primase from the euryarchaeon *Pyrococcus abyssi*: a multifunctional enzyme for initiation and repair? J Mol Biol, 2007;374(5) 1172-1185.

[71] Hu J, Guo L, Wu K, Lang S, Huang L. Template-dependent polymerization across discontinuous templates by the heterodimeric primase from the hyperthermophilic archaeon *Sulfolobus solfataricus*. Nucleic Acids Res, 2011;40(8) 3470-3483.

[72] Evguenieva-Hackenberg E, Walter P, Hochleitner E, Lottspeich F, Klug G. An exosome-like complex in *Sulfolobus solfataricus*. EMBO Rep, 2003;4(9) 889-934.

[73] Walter P, Klein F, Lorentzen E, Ilchmann A, Klug G, Evguenieva-Hackenberg E. et al. Characterization of native and reconstituted exosome complexes from the hyperthermophilic archaeon *Sulfolobus solfataricus*. Mol Microbiol, 2006;62(4) 1076-1089.

[74] Zuo Z, Rodgers CJ, Mikheikin AL, Trakselis MA. Characterization of a functional DnaG-type primase in archaea: implications for a dual-primase system. J Mol Biol, 2010;397(3) 664-676.

[75] Wu K, Lai X, Guo X, Hu J, Xiang X, Huang L. Interplay between primase and replication factor C in the hyperthermophilic archaeon *Sulfolobus solfataricus*. Mol Microbiol, 2007;63(3) 826-837.

[76] Murzin AG. OB(oligonucleotide/oligosaccharide binding)-fold: common structural and functional solution for non-homologous sequences. EMBO J, 1993;12(3) 861-867.

[77] Braun KA, Lao Y, He Z, Ingles CJ, Wold MS. Wold MS Role of protein-protein interactions in the function of replication protein A (RPA): RPA modulates the activity of DNA polymerase alpha by multiple mechanisms. Biochemistry, 1997;36(28) 8443-8454.

[78] Lin YL, Chen C, Keshav KF, Winchester E, Dutta A. Dissection of functional domains of the human DNA replication protein complex replication protein A. J Biol Chem, 1996;271(29) 17190-17198.

[79] Bochkarev A, Bochkareva E, Frappier L, Edwards AM. The crystal structure of the complex of replication protein A subunits RPA32 and RPA14 reveals a mechanism for single-stranded DNA binding. EMBO J, 1999;18(16) 4498-4504.

[80] Bochkareva E, Frappier L, Edwards A M, Bochkarev A.The RPA32 subunit of human replication protein A contains a single-stranded DNA-binding domain. J Biol Chem, 1998;273(7) 3932-3936.

[81] Philipova D, Mullen JR, Maniar HS, Lu J, Gu C, Brill SJ, et al. A hierarchy of SSB pro-tomers in replication protein A. Genes Dev, 1996;10(17) 2222-2233.

[82] Gomes XV, Wold MS. Structural analysis of human replication protein A. Mapping functional domains of the 70-kDa subunit. J Biol Chem, 1995;270(9) 4534-4543.

[83] Mer G, Bochkarev A, Gupta R, Bochkareva E, Frappier L, Ingles CJ. Structural basis for the recognition of DNA repair proteins UNG2, XPA, and RAD52 by replication factor RPA. Cell, 2000;103(3) 449-456.

[84] Dornreiter I, Erdile LF, Gilbert IU, von Winkler D, Kelly TJ, Fanning E. Interaction of DNA polymerase alpha-primase with cellular replication protein A and SV40 T anti-gen. EMBO J, 1992;11(2) 769-776.

[85] Tsurimoto T, Stillman B. Multiple replication factors augment DNA synthesis by the two eukaryotic DNA polymerases, alpha and delta. EMBO J, 1989;8(12) 3883-3889.

[86] Kenny MK, Lee SH, Hurwitz J. Multiple functions of human single-stranded-DNA binding protein in simian virus 40 DNA replication: single-strand stabilization and stimulation of DNA polymerases alpha and delta. Proc Natl Acad Sci USA, 1989;86(24) 9757-9761.

[87] Chedin F, Seitz EM, Kowalczykowski SC. Novel homologs of replication protein A in archaea: implications for the evolution of ssDNA-binding proteins. Trends Biol Sci, 1998;23(8) 273-277.

[88] Kelly TJ, Simancek P, Brush GS. Identification and characterization of a single-stranded DNA-binding protein from the archaeon *Methanococcus jannaschii*. Proc Natl Acad Sci USA, 1998;95(25) 14634-14639.

[89] Kelman Z, Pietrokovski S, Hurwitz J. Isolation and characterization of a split B-type DNA polymerase from the archaeon *Methanobacterium thermoautotrophicum* ΔH. J Biol Chem, 1999;274(40) 28751-28761.

[90] Komori K, Ishino Y. Replication protein A in *Pyrococcus furiosus* is involved in homol-ogous DNA recombination. J Biol Chem, 2001;276(28) 25654-25660.

[91] Kerr ID, Wadsworth RI, Cubeddu L, Blankenfeldt W, Naismith JH, White MF In-sights into ssDNA recognition by the OB fold from a structural and thermodynamic study of *Sulfolobus* SSB protein. EMBO J, 2003;22(11) 2561-270.

[92] Robbins JB, Murphy MC, White BA, Mackie RI, Ha T, Cann IK.et al. Functional anal-ysis of multiple single-stranded DNA-binding proteins from *Methanosarcina acetivor-*

ans and their effects on DNA synthesis by DNA polymerase BI. J Biol Chem, 2004;279(8) 6315-6326.

[93] Lin Y, Lin LJ, Sriratana P, Coleman K, Ha T, Spies M, et al. Engineering of functional replication protein a homologs based on insights into the evolution of oligonucleotide/oligosaccharide-binding folds. J Bacteriol, 2008;190(17) 5766-5780.

[94] Braithwaite DK, Ito J. Compilation, alignment, and phylogenetic relationships of DNA polymerases. Nucleic Acids Res, 1993, 1993;21(4) 787-802.

[95] Cann I, Ishino Y. Archaeal DNA replication: Identifying the pieces to solve a puzzle. Genetics, 1999;152(4) 1249-1267.

[96] Ishino, Y., Cann, I. The euryarchaeotes, a subdomain of archaea, survive on a single DNA polymerase: fact or farce? Genes Genet Syst, 1998;73(6) 323-336.

[97] Lipps G, Röther S, Hart C, Krauss G. A novel type of replicative enzyme harbouring ATPase, primase and DNA polymerase activity. EMBO J, 2003;22(10) 2516-2525.

[98] Ohmori H, Friedberg EC, Fuchs RP, Goodman MF, Hanaoka F, Hinkle D, et al. The Y-family of DNA polymerases. Mol Cell;8(1) 7-8.

[99] Guo C, Kosarek-Stancel JN, Tang TS Friedberg EC. Y-family DNA polymerases in mammalian cells. Cell Mol Life Sci, 2009;66(14) 2363-2381.

[100] Pisani FM, De Martino C, Rossi M. A DNA polymerase from the archaeon *Sulfolobus solfataricus* shows sequence similarity to family B DNA polymerases. Nucleic Acids Res, 1992;20(11) 2711-2716.

[101] Perler, FB, Comb, DG, Jack, WE, Moran, LS, Qiang, B, Kucera, RB, et al. Intervening sequences in an Archaea DNA polymerase gene. Proc. Natl. Acad. Sci. USA 1992;89(12) 5577-5581.

[102] Uemori T, Ishino Y, Toh H, Asada K, Kato I. Organization and nucleotide sequence of the DNA polymerase gene from the archaeon *Pyrococcus furiosus*. Nucleic Acids Res, 1992;21(2) 259-265.

[103] Uemori T, Ishino Y, Doi H, Kato I. The hyperthermophilic archaeon *Pyrodictium occultum* has two alpha-like DNA polymerases. J Bacteriol, 1995;177(8) 2164-2177.

[104] Cann I, Ishino S, Nomura N, Sako Y, Ishino Y. Two family B DNA polymerases in *Aeropyrum pernix*, an obligate aerovic hyperthermophilic crenarchaeote. J Bacteriol, 1999;181(19) 5984-5992.

[105] Cann I, Komori K, Toh H, Kanai S, Ishino Y. A heterodimeric DNA polymerase: evidence that members of the euryarchaeota possess a novel DNA polymerase. Proc. Natl. Acad. Sci. USA, 1998;95(24) 14250-14255.

[106] Ishino S, Ishino Y. Conprehensive search for DNA polymerase in the hyperthermophilic archaeon, *Pyrococcus furiosus*. Nucleosides, Nucleotides, and Nucleic Acids. 2006;25(4-6):681-691.

[107] Imamura M, Uemori T, Kato I, Ishino Y.A non-α-like DNA polymerase from the hyperthermophilic archaeon *Pyrococcus furiosus*. Biol Pharm Bull, 1995;18(12) 1647-1652.

[108] Ishino Y, Ishino S. Novel DNA polymerases from Euryarchaeota. Meth Enzymol, 2001, 334 249-260

[109] Gueguen Y, Rolland JL, Lecompte O, Azam P, Le Romancer G, Flament D, et al. Characterization of two DNA polymerases from the hyperthermophilic euryarchaeon *Pyrococcus abyssi*. Eur J Biochem, 2001;268(22) 5961-5969.

[110] Shen Y, Musti K, Hiramoto M, Kikuchi H, Kawarabayashi Y, Matsui I. Invariant Asp-1122 and Asp-1124 are essential residues for polymerization catalysis of family D DNA polymerase from *Pyrococcus horikoshii*. J Biol Chem, 2001;276(29) 27376-27383.

[111] Shen Y, Tang X, Matsui I. Subunit interaction and regulation of activity through terminal domains of the family D DNA polymerase from *Pyrococcus horikoshii*. J Biol Chem, 2003;278(23) 21247-21257.

[112] Tang XF, Shen Y, Matsui E, Matsui I. Domain topology of the DNA polymerase D complex from a hyperthermophilic archaeon *Pyrococcus horikoshii*. Biochemistry, 2004;43(37) 11818-11827.

[113] Jokela M, Eskelinen A, Pospiech H, Rouvinen J, Syväoja JE. Characterization of the 3' exonuclease subunit DP1 of *Methanococcus jannaschii* replicative DNA polymerase D. Nucleic Acids Res, 2004;32(8)2430-2440.

[114] Henneke G, Flament D, Hübscher U, Querellou J, Raffin JP. The hyperthermophilic euryarchaeota *Pyrococcus abyssi* likely requires the two DNA polymerases D and B for DNA replication. J Mol Biol, 2005;350(1) 53-64.

[115] Castrec B, Laurent S, Henneke G, Flament D, Raffin JP. The glycine-rich motif of *Pyrococcus abyssi* DNA polymerase D is critical for protein stability. J Mol Biol, 2010;396(4) 840-848.

[116] Berquist BR, DasSarma P, DasSarma S. Essential and non-essential DNA replication genes in the model halophilic Archaeon, *Halobacterium* sp. NRC-1. BMC Genet, 2007;8 31.

[117] Brochier-Armanet C, Boussau B, Gribaldo S, Forterre P. Mesophilic Crenarchaeota: proposal for a third archaeal phylum, the Thaumarchaeota. Nat Rev Microbiol, 2008;6(3) 245-252.

[118] Nunoura T, Takaki Y, Kakuta J, Nishi S, Sugahara J, Kazama H, et al. Insights into the evolution of Archaea and eukaryotic protein modifier systems revealed by the genome of a novel archaeal group. Nucleic Acids Res, 2011;39(8) 3204-3223.

[119] Elkins JG, Podar M, Graham DE, Makarova KS, Wolf Y, Randau L, et al. A korarchaeal genome reveals insights into the evolution of the Archaea. Proc Natl Acad Sci USA, 2008;105(23) 8102-8107.

[120] Tahirov TH, Makarova KS, Rogozin IB, Pavlov YI, Koonin EV. Evolution of DNA polymerases: an inactivated polymerase-exonuclease module in Pol ε and a chimeric origin of eukaryotic polymerases from two classes of archaeal ancestors. Biol Direct, 2009;4 11.

[121] Zillig W, Kletzin A, Schleper C, et al. Screening for *Sulfolobales*, their plasmids and their viruses in Icelandic solfataras. Syst Appl Microbiol, 1993, 16: 609–628

[122] Lipps G, Röther S, Hart C, Krauss G. A novel type of replicative enzyme harbouring ATPase, primase and DNA polymerase activity. EMBO J, 2003;22(10) 2516-2525.

[123] Lipps G. Molecular biology of the pRN1 plasmid from *Sulfolobus islandicus*. Biochem Soc Trans, 2009;37(Pt 1) 42-45.

[124] Lipps G. The replication protein of the *Sulfolobus islandicus* plasmid pRN1. Biochem Soc Trans, 2004;32(Pt 2) 240-244.

[125] Soler N, Marguet E, Cortez D, Desnoues N, Keller J, van Tilbeurgh H. Two novel families of plasmids from hyperthermophilic archaea encoding new families of replication proteins. Nucleic Acids Res, 2010;38(15) 5088-5104.

[126] Moldovan G-L, Pfander B, Jentsch S. PCNA, the maestro of the replication fork. Cell, 2007;129(4)665-679.

[127] Pan M, Kelman L, Kelman Z. The archaeal PCNA proteins. Biochem Soc Trans, 2011;39(1) 20-24.

[128] Cann I, Ishino S, Hayashi I, Komori K, Toh H, Morikawa K, et al. Functional interactions of a homolog of proliferating cell nuclear antigen with DNA polymerases in Archaea. J. Bacteriol, 1999, Nov;181(21):6591-6599.

[129] Ishino Y, Tsurimoto T, Ishino S, Cann IK. Functional interactions of an archaeal sliding clamp with mammalian clamp loader and DNA polymerase δ. Genes Cells, 2001;6(8) 699-706.

[130] Matsumiya S, Ishino S, Ishino Y, Morikawa K. Physical interaction between proliferating cell nuclear antigen and replication factor C from *Pyrococcus furiosus*. Genes Cells, 2002;7(9) 911-922.

[131] Matsumiya S, Ishino S, Ishino Y, Morikaw K. Intermolecular ion pairs maintain toroidal structure of *Pyrococcus furiosus* PCNA. Prot Sci, 2003;12(4) 823-831.

[132] Tori K, Kimizu M, Ishino S, Ishino Y. Both DNA polymerase BI and D from the hyperthermophilic archaeon, *Pyrococcus furiosus* bind to PCNA at the C-terminal PIP box motifs. J Bacteriol, 2007;189(15) 5652-5657.

[133] Cann I, Ishino S, Yuasa M, Daiyasu H, Toh H, Ishino Y. Biochemical analysis of the replication factor C from the hyperthermophilic archaeon *Pyrococcus furiosus*. J Bacteriol, 2001;183(8) 2614-2623.

[134] Mayanagi K, Miyata T, Oyama T, Ishino Y, Morikawa K. Three-dimensional electron microscopy of clamp loader small subunit from *Pyrococcus furiosus*. J Struct Biol, 200;134(1) 35-45.

[135] Ishino S, Oyama T, Yuasa M, Morikawa K, Ishino Y. Mutational Analysis of *Pyrococcus furiosus* Replication Factor C based on the Three-Dimensional Structure. Extremophiles, 2003;7(3) 169-175.

[136] Oyama T, Ishino Y, Cann I, Ishino S, Morikawa K. Atomic structure of the clamp loader small subunit from *Pyrococcus furiosus*. Mol Cell, 2001;8(2) 455-463.

[137] Indiani C, O'Donnell M. The replication clamp-loading machine at work in the three domains of life. Nat Rev Mol Cell Biol, 2006;7(10) 751-761.

[138] Miyata T, Suzuki H, Oyama T, Mayanagi K, Ishino Y, Morikawa K. Open clamp structure in the clamp-loading complex visualized by electron microscopic image analysis. Proc Natl Acad Sci USA, 2005;102(39) 13795-13800.

[139] Kelch BA, Makino DL, O'Donnell M, Kuriyan J. The replication clamp-loading machine at work in the three domains of life. Science, 2011;334(6063) 1675-1680.

[140] Warbrick EM. The puzzle of PCNA's many partners. BioEssays, 2000;22(11) 997-1006.

[141] Vivona JB, Kelman Z. The diverse spectrum of sliding clamp interacting proteins. FEBS Lett, 2003;546(2-3) 167-172.

[142] Nishida H, Mayanagi K, Kiyonari S, Sato Y, Oyama T, Ishino Y, et al. Structural determinant for switching between the polymerase and exonuclease modes in the PCNA-replicative DNA polymerase complex. Proc Natl Acad Sci USA, 2009;106(49) 20693-20698.

[143] Shamoo Y, Steitz TA. Building a replisome from interacting pieces: sliding clamp complexed to a peptide from DNA polymerase and a polymerase editing complex. Cell, 1999;99(2) 155-166.

[144] Franklin MC, Wang J. Steitz TA. Structure of the replicating complex of a pol α family DNA polymerase. Cell, 2001;105(5) 657-667.

[145] Mayanagi K, Kiyonari S, Nishida H, Saito M, Kohda D, Ishino Y, et al. Architecture of the DNA polymerase B-proliferating cell nuclear antigen (PCNA)-DNA ternary complex. Proc Natl Acad Sci USA, 2011;108(5):1845-1849

[146] Daimon K, Kawarabayasi Y, Kikuchi H, Sako Y, Ishino Y. Three proliferating cell nuclear antigen-like proteins found in the hyperthermophilic archaeon *Aeropyrum pernix*: interactions with the two DNA polymerases. J Bacteriol, 2002;184(3) 687-694.

[147] Dionne I, Nookala RK, Jackson SP, Doherty AJ, Bell SD. A heterotrimeric PCNA in the hyperthermophilic archaeon *Sulfolobus solfataricus*. Mol Cell, 2003;11(1) 275-282.

[148] Imamura K, Fukunaga K, Kawarabayasi Y, Ishino Y. Specific interactions of three proliferating cell nuclear antigens with replication-related proteins in *Aeropyrum pernix*. Mol Microbiol, 2007;64(2) 308-318.

[149] Williams GJ, Johnson K, Rudolf J, McMahon SA, et al. Structure of the heterotrimeric PCNA from *Sulfolobus solfataricus*. Acta Crystallogr Sect F Struct Biol Cryst Commun, 2006;62(Pt 10) 944-948.

[150] Doré AS, Kilkenny ML, Jones SA, Oliver AW, Roe SM, Bell SD, et al. Structure of an archaeal PCNA1-PCNA2-FEN1 complex: elucidating PCNA subunit and client enzyme specificity. Nucleic Acids Res, 2006;34(16) 4515-4526.

[151] Kuba Y., Ishino, S., Yamagami, T., Tokuhara, M., Kanai, T., Fujikane, R., et al. Comparative analyses of the two PCNAs from the hyperthermophilic archaeon, *Thermococcus lodakarensis*. Genes Cells, 2012, 17, in press

[152] Ladner JE, Pan M, Hurwitz J, Kelman Z. Crystal structures of two active proliferating cell nuclear antigens (PCNAs) encoded by *Thermococcus kodakaraensis*. Proc Natl Acad Sci USA, 2011;108(7) 2711-2716.

[153] Chen YH, Kocherginskaya SA, Lin Y, Sriratana B, Lagunas AM, Robbins JB, et al. Biochemical and mutational analyses of a unique clamp loader complex in the archaeon *Methanosarcina acetivorans*. J Biol Chem, 2005;280(51) 41852-41863.

[154] Chen YH, Lin Y, Yoshinaga A, Chhotani B, Lorenzini JL, Crofts AA, et al. Molecular analyses of a three-subunit euryarchaeal clamp loader complex from *Methanosarcina acetivorans*. J Bacteriol, 2009;191(21) 6539-6549.

[155] Wilkinson A, Day J, Bowater R. Bacterial DNA ligases. Mol Microbiol, 2001;40(6) 1241-1248.

[156] Tomkinson AE, Vijayakumar S, Pascal JM, Ellenberger T. DNA ligases: structure, reaction mechanism, and function. Chem Rev, 2006;106(2) 687-699.

[157] Poidevin L, MacNeill SA. Biochemical characterisation of LigN, an NAD+-dependent DNA ligase from the halophilic euryarchaeon *Haloferax volcanii* that displays maximal in vitro activity at high salt concentrations. BMC Mol Biol, 2006;7 44.

[158] Benarroch D, Shuman S. Characterization of mimivirus NAD+-dependent DNA ligase. Virology, 2006;353(1) 133-143.

[159] Kletzin A. Molecular characterisation of a DNA ligase gene of the extremely thermophilic archaeon *Desulfurolobus ambivalens* shows close phylogenetic relationship to eukaryotic ligases. Nucleic Acids Res, 1992;20(20) 5389-5396.

[160] Nakatani M, Ezaki S, Atomi H, Imanaka T. A DNA ligase from a hyperthermophilic archaeon with unique cofactor specificity. J Bacteriol, 2000;182(22) 6424-6433.

[161] Rolland JL, Gueguen Y, Persillon C, Masson JM, Dietrich J. Characterization of a thermophilic DNA ligase from the archaeon *Thermococcus fumicolans*. FEMS Microbiol Lett, 2004;236(2)267-273.

[162] Kim YJ, Lee HS, Bae SS., Jeon JH, Yang SH, Lim JK, et al. Cloning, expression, and characterization of a DNA ligase fro3 a hyperthermophilic archaeon *Thermococcus* sp. Biotechnol Lett, 2006;28(6) 401-407.

[163] Ferrer M, Golyshina OV, Beloqui A, Böttger LH, Andreu JM, Polaina J, et al. A purple acidophilic di-ferric DNA ligase from *Ferroplasma*. Proc Natl Acad Sci USA, 2008;105(26) 8878-8883.

[164] Jeon SJ, Ishikawa K. A novel ADP-dependent DNA ligase from *Aeropyrum pernix* K1. FEBS Lett, 2003;550(1-3) 69-73.

[165] Seo MS, Kim YJ, Choi JJ, Lee MS, Kim JH, Lee JH, et al. Cloning and expression of a DNA ligase from the hyperthermophilic archaeon *Staphylothermus marinus* and properties of the enzyme. J Biotechnol, 2007;128(3) 519-530.

[166] Sun Y, Seo MS, Kim JH, Kim YJ, Kim GA, Lee JI, et al. Novel DNA ligase with broad nucleotide cofactor specificity from the hyperthermophilic crenarchaeon *Sulfophobococcus zilligii*: influence of ancestral DNA ligase on cofactor utilization. Environ Microbiol, 2008;10(12) 3212-3224.

[167] Keppetipola N, Shuman S. Characterization of a thermophilic ATP-dependent DNA ligase from the euryarchaeon *Pyrococcus horikoshii*. J Bacteriol, 2005;187(20) 6902-6908.

[168] Kiyonari S, Takayama K, Nishida H, Ishino Y. Identification of a novel binding motif in *Pyrococcus furiosus* DNA ligase for the functional interaction with proliferating cell nuclear antigen. J Biol Chem, 2006;281(38) 28023-28032.

[169] Nishida H, Kiyonari S, Ishino Y, Morikawa K. The closed structure of an archaeal DNA ligase from *Pyrococcus furiosus*. J Mol Biol, 2006;360(5) 956-967.

[170] Levin DS, Bai W, Yao N. An interaction between DNA ligase I and proliferating cell nuclear antigen: implications for Okazaki fragment synthesis and joining. Proc Natl Acad Sci USA, 1997;94(24) 12863-12868.

[171] Jónsson ZO, Hindges R, Hübscher U. Regulation of DNA replication and repair proteins through interaction with the front side of proliferating cell nuclear antigen. EMBO J, 1998;17(8) 2412-2425.

[172] Tom S, Henricksen LA, Park MS, Bambara RA. DNA ligase I and proliferating cell nuclear antigen form a functional complex. J Biol Chem, 2001;276(27) 24817-24825.

[173] Kiyonari S, Kamigochi T, Ishino Y. A single amino acid substitution in the DNA-binding domain of *Aeropyrum pernix* DNA ligase impairs its interaction with proliferating cell nuclear antigen, Extremophiles, 2007;11(5) 675-684.

[174] Pascal JM, O'Brien PJ, Tomkinson AE, Ellenberger T. Human DNA ligase I completely encircles and partially unwinds nicked DNA. Nature, 2004;432(7016) 473-478:

[175] Pascal JM, Tsodikov OV, Hura GL, Song W, Cotner EA, Classen S, et al. A flexible interface between DNA ligase and PCNA supports conformational switching and efficient ligation of DNA. Mol Cell, 2006;24(2) 279-291.

[176] Mayanagi K, Kiyonari S, Saito M, Shirai T, Ishino Y, Morikawa K. Mechanism of replication machinery assembly as revealed by the DNA ligase-PCNA-DNA complex architecture. Proc Natl Acad Sci USA, 2009;106(12) 4647-4652.

[177] Liu Y, Kao HI, Bambara RA. Flap endonuclease 1: a central component of DNA metabolism. Annu Rev Biochem, 2004, 2004;73 589-615.

[178] Zheng L, Jia J, Finger LD, Guo Z, Zer C, Shen B. Functional regulation of FEN1 nuclease and its link to cancer. Nucleic Acids Res, 2011;39(3) 781-794.

[179] Rossi ML, Bambara RA. Reconstituted Okazaki fragment processing indicates two pathways of primer removal. J Biol Chem, 2006;281(36):26051-26061.

[180] Tom S, Henricksen LA, Bambara RA. Mechanism whereby proliferating cell nuclear antigen stimulates flap endonuclease 1. J Biol Chem, 2000;275(14) 10498-10505.

[181] Sakurai S, Kitano K, Yamaguchi H, Hamada K, Okada K, Fukuda K, et al. Structural basis for recruitment of human flap endonuclease 1 to PCNA. EMBO J, 2005;24(4) 683-693.

[182] Shen B, Qiu J, Hosfield D, Tainer JA. Flap endonuclease homologs in archaebacteria exist as independent proteins. Trends Biochem Sci, 1998;23(5) 171-173.

[183] Hwang KY, Baek K, Kim HY, CHo Y. The crystal structure of flap endonuclease-1 from *Methanococcus jannaschii*. Nat Struct Biol, 1998;5(8) 707-713.

[184] Hosfield DJ, Mol CD, Shen B, Tainer JA. Structure of the DNA repair and replication endonuclease and exonuclease FEN-1: coupling DNA and PCNA binding to FEN-1 activity. Cell, 1998;95(1)135-146.

[185] Matsui E, Musti KV, Abe J, Yamasaki K, Matsui I, Harata K. Molecular structure and novel DNA binding sites located in loops of flap endonuclease-1 from *Pyrococcus horikoshii*. J Biol Chem, 2002;277(40) 37840-37847.

[186] Chapados BR, Hosfield DJ, Han S, Qiu J, Yelent B, Shen B, et al. Structural basis for FEN-1 substrate specificity and PCNA-mediated activation in DNA replication and repair. Cell, 2004;116(1) 39-50.

[187] Matsui E, Kawasaki S, Ishida H, Ishikawa K, Kosugi Y, Kikuchi H, et al. Thermostable flap endonuclease from the archaeon, *Pyrococcus horikoshii*, cleaves the replication fork-like structure endo/exonucleolytically. J Biol Chem, 1999;274(26) 18297-182309

[188] Matsui E, Abe J, Yokoyama H, Matsui I. Aromatic residues located close to the active center are essential for the catalytic reaction of flap endonuclease-1 from hyperthermophilic archaeon *Pyrococcus horikoshii*. J Biol Chem, 2004;279(16) 16687-16696.

[189] Beattie TR, Bell SD. Molecular machines in archaeal DNA replication. Curr Opin Chem Biol, 2011;15(5) 614-619.

[190] Ishino S and Ishino Y. Rapid progress of DNA replication studies in Archaea, the third domain of life. Sci China Life Sci, 2012;55(5) 386-403.

[191] Wang Y, Prosen DE, Mei L, Sullivan JC, Finney M, Vander Horn PB. A novel strategy to engineer DNA polymerases for enhanced processivity and improved performance in vitro. Nucleic Acids Res, 2004;32(3) 1197-1207

[192] Kawamura A, Ishino Y, Ishino S. Biophysical analysis of PCNA from *Pyrococcus furiosus*. J Jap Soc Extremophiles, 2012;11(1) 12-18

[193] Ishino S, Kawamura, A, Ishino Y. Application of PCNA to processive PCR by reducing the stability of its ring structure. J Jap Soc Extremophiles, 2012;11(1) 19-25

[194] Shendure J, Ji H. Next-generation DNA sequencing. Nat Biotechnol, 2008;26(10) 1135-1145.

[195] Ansorge WJ. Next-generation DNA sequencing techniques. Nat Biotechnol, 2009;25(4) 195-203

[196] Metzker ML. Sequencing technologies - the next generation. Nat Rev Genet, 2010;11(1) 31-46.

Chapter 2

Pulling the Trigger to Fire Origins of DNA Replication

David Stuart

Additional information is available at the end of the chapter

1. Introduction

DNA replication is a fundamental aspect of cell biology. The process is essential for chromosome doubling and segregation during cell division. Additionally, the DNA replication program can be manipulated to allow a reduction in ploidy as occurs during meiosis or an increase in ploidy as observed in endo-cycles during some developmental processes [1]. The importance of the integrity of the chromosome duplication process is inherently obvious. In somatic cells failure to replicate prevents cell division or leads to a catastrophic reductional division and cell death. Less drastic defects in DNA replication can appear as problems leading to gene amplification, chromosome breaks or chromosome missegregation [2]. These can manifest as birth defects or increased susceptibility to cancer [3]. The integrity of the DNA replication process is ensured partly by DNA repair mechanisms and checkpoint controls. However, the primary mechanism that safeguards the DNA replication process is the complex and multi-step process that leads to the assembly and activation of an active replication complex at chromosomal origins of DNA replication.

The assembly and activation of DNA replication complexes on eukaryotic chromosomes is critically dependent upon two cell cycle regulated protein kinase complexes; Cyclin Dependent Kinase (CDK) and Dbf4 Dependent Kinase (DDK). These protein kinases phosphorylate multiple protein substrates that play roles in assembling a replisome through promoting specific protein-protein interactions that recruit essential components to the complex and stabilize the assembled complex. Additionally, CDK and DDK play roles in the activation of the DNA replication complex and its helicase activity [4].

This chapter will review the key regulatory roles played by CDK and DDK activity in promoting timely assembly of DNA replication complexes. The focus of the article will be on the budding yeast *Saccharomyces cerevisae* where the assembly and activation of origins of DNA replication has been extensively studied. However, the yeast system will be compared and contrasted with other eukaryotes in order to emphasize universal features of the process and highlight unique characteristics of DNA replication in different organisms and cell types.

2. Origins of replication: Where it all starts

DNA replication is a fundamental aspect of cellular proliferation. Bacterial cells with relatively small chromosomes initiate DNA replication from a single well-defined site on each chromosome referred to as oriC [5]. Eukaryotic chromosomes can be from 10 to 1000 times larger than bacterial chromosomes. In order to completely replicate so much chromosomal DNA within a timely fashion that will allow proliferation, eukaryotic cells employ multiple sites on each chromosome that act as origins for the initiation of DNA replication. These sites are referred to as origins of DNA replication (ORIs). In most metazoans ORIs are poorly defined in the sense that they lack a specific consensus DNA sequence but appear to localize to large regions of a chromosome and are defined by the structure of the chromatin and modification state of the histones and chromatin proteins rather than by specific DNA sequences [6-8]. Indeed, even in the single celled fission yeast *Schizosaccharomyces pombe* DNA replication initiates from relatively broad chromosomal regions [9, 10]. The budding yeast and particularly *Saccharomyces cerevisiae* differs from other eukaryotes in this regard. Autonomously Replicating Sequences (ARS) were first identified in *S. cerevisiae* chromosomal DNA in 1979 [11]. When incorporated into plasmid DNA an ARS sequence allowed for efficient replication and maintenance of the extrachromosomal plasmid. Characterization of ARSs revealed specific DNA sequence elements that act as ORIs reviewed by [12]. These sequences are about 100 – 150 basepairs in length and are composed of elements referred to as A, B1, B2, other sequence elements referred to as B3 and C are sometimes present [13]. The A module harbors an AT-rich 11 basepair ARS Consensus Sequence (ACS). Together the A and B1 element contribute to the formation of a binding site for Origin Recognition Complex (ORC) proteins [14], discussed in the next section. The B2 sequence module contains a double stranded DNA unwinding element (DUE). This sequence is where unwinding of the double helical DNA initiates to create a replication bubble [15, 16]. The B3 element acts as a binding site for the transcription factor Abf1 and excludes nucleosome occupancy of the origin sites [17]. The C element has transcription factor binding sites that may stimulate the utilization of some ORIs but are not essential for ORI function [12, 18].

Although there are specific sequence determinants for *S. cerevisiae* origins of replication, even in this yeast not all ORIs are equal. Significant heterogeneity exists among ORIs in the frequency with which they are activated and utilized [19]. Indeed, there are some origin sequences in the *S. cerevisae* genome that are not utilized and appear to be dormant [20]. In addition to the frequency of activation there is a distinct temporal order to ORI activation with a subset of origins being activated at early times in S-phase and others being activated later in S-phase [21, 22]. DNA combing studies with *S. cerevisiae* have revealed that at the single molecule level origin activation is highly stochastic with different sets of ORIs being activated in each cell cycle [19, 23]. Indeed while there are approximately 700 potential ORIs in the *S. cerevisiae* nuclear genome only about 200 are activated in any given S-phase. Recent genome-wide studies investigating origin activation combined with mathematical modeling have suggested that replication timing can be explained by a stochastic mechanism [24-27]. The basis for the differential frequency of ORI activation and temporal regulation has been argued to be due to a limited availability of some essential activators [28-31]. In the case of S. cerevisiae over expression of Dbf4, the activating subunit of the Dbf4 dependent kinase (DDK) along with the Cdk substrates Sld2, Sld3 and their binding partner Dbp11 allow early activation of late firing ORIs [28]. Since Dbf4, Sld2, Sld3 do not remain associated with the replication complex once it has been activated, it has been proposed that once an origin fires, the limiting subunits are released from the complex and can then interact with another ORI and trigger its activation. In this scenario ORIs with the highest affinity for the rate limiting factors will have the highest probability of being activated and will have a high probability of being activated at early times in S-phase. ORIs with a lower affinity for the rate limiting factors will fire after those factors have been released from other ORIs. Hence a temporal order of ORI activation can be created. These models propose that the rate limiting activators of DNA replication have a higher affinity for some ORIs than others [28]. This differential affinity may be due to structural aspects of the chromatin in which the ORI is embedded as well as modification of the chromatin proteins by acetylation, methylation, and potentially other post-translational events [32-34]. Further, there is evidence that ORI usage can be influenced by the presence of nearby transcriptional units [35-37].

3. Assembly of the pre-RC: Orc marks the spot

The model of specific chromosomal locations acting as sequence specific sites for binding of protein complexes to initiate DNA replication is conserved across organisms from eukaryotes to prokaryotes and archaea. However, as already described there is no conservation of DNA sequences that act as ORIs across organisms. Indeed, even in S. cerevisiae, which has well defined ORIs the sequence of the origins of replication are rather degenerate with only the core ACS being well conserved. In other eukaryotic organisms ORIs display little similarity beyond being rich in AT sequences. Although the DNA sequences that act as sites for initiation of DNA replication are not conserved among eukaryotes the protein complex that binds to ORIs, the Origin Recognition Complex (ORC) is well conserved across eukaryotes and archaea [38-40]. The conserved ORC complex is a hetero-hexamer composed of six subunits Orc1 to Orc6. This complex binds directly to the chromosomal DNA. The S. cerevisiae Orc1-6 proteins bind as a hetero-hexamer to the ORI sequence constitutively throughout the cell cycle with Orc1, Orc2, Ocr4, and Orc5 making direct contact with the A and B1 sequence ORI DNA sequence [41-43]. In contrast metazoans and even the fission yeast S. pombe display regulated binding of the ORC complex to the chromosomal ORI sites. In particular the Orc1 subunit dissociates from the chromatin in G2-phase and re-associates with the complex in G1 [31, 44]. In D. melanogaster and human cells Orc1 is subject to degradation by the Anaphase Promoting Complex (APC) in G2-phase [44-48]. As Orc1-6 is required for DNA replication initiated at ORIs, the regulated binding of Orc in metazoans provides an additional layer of regulation that may be used to control the initiation of DNA replication.

The Orc1-6 proteins act as a marker of chromosomal ORI sites and a platform for the assembly of replication complexes. Orc1-6 does not perform this function in an entirely static fashion. Rather successful initiation of DNA replication requires that the Orc1-6 be capable of binding and hydrolyzing ATP, reviewed by [49]. The Orc1 and Orc5 subunits possess nucleotide-binding motifs, Orc1 has conserved Walker A and Walker B motifs and Orc5 has a Walker A motif and a questionable Walker B sequence [50]. Both Orc1 and Orc5 can bind DNA but only Orc1 displays ATPase activity and while mutations that inactivate the Orc1 Walker A sequence cause defects in DNA replication, mutations to the Orc5 Walker A sequence do not [50-52]. In yeast this activity is essential to allow Orc1-6 to bind specifically to chromosomal ORI DNA and to load other replication complex components on to the ORI [43, 50]. Site-specific binding of Orc1-6 to ORI DNA requires the ability to bind ATP; however ATP hydrolysis is not required, suggesting that ATP binding modulates Orc1 structure and its ability to complex with both DNA and other Orc subunits [50]. In contrast ATP hydrolysis is strictly required for the loading of other replication complex proteins and the formation of a functional DNA replication complex [50-52].

DNA replication is essential for developmental processes as well as for somatic cell prolifer-ation. It is frequently the case that the cell cycle is altered or modified from the canonical form it takes in mature cells to achieve specific developmental aims. Orc1-6 is essential for DNA replication in many developmental contexts. Mutations in human Orc1 and Orc4 proteins are responsible for Meier-Gorlin syndrome, a developmental disorder characterized by primary dwarfism, microcephaly, developmental abnormalities of ear and patella [53, 54]. Addition-ally, Orc3 is essential for neuronal development and maturation [55]. However, there is some diversity in the regulation of Orc1-6 during developmental. For example endo-reduplication in *D. melanogaster* does not require Orc1 [56, 57]. The developmental regulation of Orc binding to chromatin may be influenced by changes in chromatin modification that occur during development since changes in chromatin acetylation have been associated with and shown to regulate the transition to endo-reduplication and the redistribution of Orc proteins during development [58]. And, while Orc1-6 and DNA replication is essential for premeiotic DNA replication, the requirements for these proteins and the mechanism by which they are organ-ized to promote the initiation may differ between mitotic and meiotic S-phases [9].

4. Assembly of the pre-RC: Enter the helicase

The chromatin bound Orc1-6 acts as a nucleation site for the construction of a replication complex (RC). This begins with the assembly of a pre-Replicative Complex (pre-RC). The pre-RC is the multi-protein complex assembled on to ORIs in G1-phase prior to the initiation of DNA replication in S-phase. The base of the pre-RC is the chromatin bound Orc1-6, which acts as a landing pad for the assembly of a series of other protein factors required to assembly a replication fork and initiate bidirectional DNA synthesis. A key requirement for processive DNA synthesis is a dsDNA helicase that can unwind the chromosomal DNA. The Orc1-6 itself has no helicase activity but is essential for recruitment of the replicative helicase to origins of DNA replication. The replicative helicase in *S. cerevisiae* is the minichromosome maintenance complex (Mcm2-7). The Mcm complex is a hetero-hexamer composed of the subunits Mcm2 – Mcm7 [59-61]. The Mcm subunits interact with each other in a 1:1 ratio to form a ring-like structure that initially binds by wrapping around the DNA such that the double helix passes through the rings central channel. Extensive investigation using biochemical characterization and mutagenesis studies have revealed that the Mcm ring structure has a subunit assembly with the order Mcm5 – Mcm3 – Mcm7 – Mcm4 – Mcm6 – Mcm2 [62]. Sub-complexes of the full Mcm2-7 ring can exist in vivo and in vitro and indeed a trimer composed of Mcm4 – Mcm6 – Mcm7 has ATPas activity and can unwind duplex DNA in vitro [63, 64]. Multiple potential ATPase active sites are formed by interactions between the Mcm subunits: however, only the ATPase activity catalyzed by sites formed by Mcm3 – Mcm7 and Mcm7 – Mcm4 are essential for the helicase activity of the Mcm2-7 holo-complex [64, 65].

In G1 phase of the cell cycle the Mcm2-7 complex is recruited and loaded on to Orc1-6 bound ORI sequences. The helicase is loaded on to the B2 sequence element as a pair of hexamers arranged on the DNA in a head – to – head orientation [66, 67]. The helicase initially assembles on to the DNA as an open complex with a central channel; the ring can be closed around the DNA helix by an ATP dependent conformational change (Figure 1). This involves ATP binding to the Mcm2 – Mcm5 subunits and acting as a "switch" that closes the open gate around the duplex DNA [68].

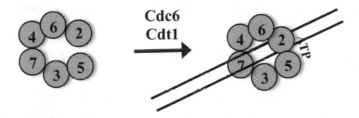

Figure 1. The Mcm2-7 hexamer assembles as an open complex that can be closed through ATP binding. The Mcm2-7 subunits can assemble with each other and in the presence of ATP the complex can assume a ring conformation. In vivo the hexamer is loaded on to Orc1-6 bound ORI duplex DNA. This loading is dependent upon the loading factors Cdc6 and Cdt1. The hexamer can be closed loosely around the duplex through binding to ATP.

Loading Mcm2-7 on to the Orc1-6 bound ORI DNA is accomplished through the combined action of the ATPase activity inherent to the chromatin bound Orc1-6 complex and interaction with the AAA[+] ATPase loading factor Cdc6. An additional protein required for loading of the Mcm complex is Cdt1, which was first identified in *S. pombe*, but subsequently functional homologs were discovered in *S. cerevisiae*, *X. laevis*, *D. melanogaster*, and mammalian cells [69-73]. The carboxyl-terminus of Cdt1 binds to the Mcm2 and Mcm6 subunits and these contacts are essential for recruitment of the functional Mcm2-7 helicase to Orc1-6 bound origins of DNA replication [74]. ATP hydrolysis catalyzed by both Orc1-6 and the Orc bound Cdc6 stimulate the recruitment of multiple Cdt1-Mcm2-7 complexes [75]. This allows two hexameric Mcm2-7 rings to bind the ORI in a head-to-head orientation, with the dsDNA running through a central channel in the complex [67, 76]. The double hexamers can slide on the duplex DNA creating the potential to load multimers of double hexamer structures at a single ORI. This may explain why the number of double hexamers loaded on to the DNA can greatly exceed the number of origins that are activated in the subsequent S-phase [77]. Following loading of the Mcm2-7 complexes Cdt1, and Cdc6 are released and do not remain at the ORI as the replication complex continues to assemble [78].

Association of the Mcm2-7 complex with Orc1-6 is a tightly regulated process. In *S. pombe*, Cdt1 mRNA accumulates in the G1 and early S-phase of the cell cycle and in both *S. pombe* and mammalian cells the abundance of the Cdt1 protein is regulated through its destruction by the ubiquitin-proteosome system [71, 73]. In contrast the abundance of Cdt1 protein in *S. cerevisiae* does not fluctuate throughout the cell cycle [69, 79]. In metazoans Cdt1 binding to Mcm2-7 and recruitment to Orc1-6 is negatively regulated by the protein geminin [80]. No protein with a similar function to geminin has been identified in yeast; however, recruitment of *S. cerevisae* Cdt1-Mcm2-7 complexes to Orc1-6 are negatively regulated by phosphorylation of Orc subunits by Cyclin Dependent Kinase (Cdk) activity [81]. This is an important mechanism to ensure that ORIs are loaded and licensed only once in each cell cycle. Additionally, the gene encoding the loader *CDC6* is transcriptionally regulated such that the mRNA accumulates exclusively during G1 and early S-phase [82]. The Cdc6 protein itself accumulates only in late G1 and early S-phase and is targeted for degradation outside of G1-phase by the Skip1-Cdc53-F box protein (SCF) mediated ubiqutin-proteosome complex [83]. The rigorous regulation applied to Cdc6 and Cdt1 ensures that the Orc1-6 complexes can only be loaded with the replicative DNA helicase machinery in G1 and early S-phase. This is essential to avoid the possibility of origin re-licensing during a cell cycle, which could lead to over replication of some segments of the genome, unscheduled changes in ploidy, the formation of structures that could be at risk for damage, and inappropriate recombination leading to chromosome damage and instability [2, 84].

5. Activating the pre-RC: DDK and CDK usher in the replication complex

Loading the Mcm2-7 helicase complex on to an Orc1-6 bound ORI creates a pre-RC, which licenses the origin and provides the potential for it to be activated or "fired" in S-phase. However, activation of the Mcm2-7 complex and unwinding of the DNA depends upon the further ordered addition of the protein factors Sld3, Cdc45, Sld2, Dpb11, the GINS complex (composed of Psf1, Psf2, Psf3, and Sld5], Mcm10, the replicative DNA polymerases Polε, Polδ, and Polα-primase, along with numerous accessory factors. The addition of these factors to the ORI bound Orc1-6 – Mcm2-7 is dependent upon the activity of two protein kinases DDK and CDK.

DDK (Dbf4 Dependant Kinase) is composed of a catalytic subunit, Cdc7 and an activating subunit, Dbf4 [4]. DDK is essential for the initiation of DNA replication and loss of function mutations in either subunit are lethal resulting in a G1 – S-phase arrest characterized by "dumbbell" morphology in *S. cerevisiae* [85, 86]. DDK is an acidiophilc protein kinase [87]. It phosphorylates serine/threonine residues and displays a preference for phosphorylating serine or threonine residues that are followed by an acidic aspartic acid or glutamic acid residue [88-90]. Additionally, DDK will phosphorylate serine or theronine residues that precede a serine or threonine that has been phosphorylated by another kinase. This is the case with the DDK substrate protein Mer2 where phosphorylation of a serine residue by Cdk1 acts as a priming event to allow phosphorylation by DDK [88, 91]. In *S. cerevisiae* the catalytic subunit Cdc7 does not fluctuate in abundance through the cell cycle; however the kinase activity associated with the protein significantly increases in late G1 and S-phase [92]. The kinase activity associated with Cdc7 is regulated primarily through the interaction of Cdc7 with its positively acting regulatory subunit Dbf4. While the abundance of Cdc7 is relatively constant through the cell cycle, Dbf4 displays a striking accumulation in late G1 and early S-phase and rapidly disappears following the completion of DNA replication [93]. The accumulation of Dbf4 in late G1 and S-phase is accounted for in part by transcriptional regulation; the gene is expressed exclusively in late G1 and S-phase [85], and by regulated destruction of Dbf4 by the ubiqutin-proteosome system [94]. Binding of Dbf4 to Cdc7 leads to a conformational shift in the structure of the inert Cdc7 monomer, that stabilizes the active state of the enzyme [95]. Dbf4 displays localization to ORIs [96]. This localization is driven by sequence motifs in Dbf4 that bind specifically to Orc2, Orc3, and to Mcm4 [97, 98]. Contacts with Mcm4 are particularly critical to achieve recruitment of DDK to the pre-RC. Thus, while Cdc7 possesses the catalytic kinase activity, Dbf4 is required to activate the enzyme and target its kinase activity to the appropriate substrates.

The second protein kinase required for conversion of the pre-RC into an active DNA replication complex is CDK. The enzyme is composed of a catalytic subunit Cdk1 (formerly known as Cdc28 in *S. cerevisiae*) that can be activated by association with a cyclin. Like Cdc7, the monomeric Cdk1 has little associated kinase activity [99]. Also similar to Cdc7 the abundance of Cdk1 does not vary appreciable through the cell cycle; however its associated kinase activity fluctuates from very low levels in early G1 to peak levels occurring in M-phase [100, 101]. Binding to an activating cyclin subunit triggers a conformational change in Cdk1 that reveals the active site and promotes the enzymes protein kinase activity [102]. *S. cerevisiae* expresses 9 Cdk1 activating cyclins that promote Cdk1 kinase activity in different phases of the cell cycle. Cln1, Cln2, and Cln3 are required for budding and events in G1 phase, Cln1 and Cln2 accumulate in late G1 and early S-phase while Cln3 is expressed throughout the cell cycle. Clb1, Clb2, Clb3, and Clb4 accumulate in G2 and M-phases, and promote events in G2 and mitosis [103]. Clb5-Cdk1, and Clb6-Cdk1 are the predominant Cdk complexes that promote the initiation of DNA replication during a normal cell cycle in *S. cerevisiae. CLB5* and *CLB6* are transcriptionally regulated such that their mRNAs accumulates in late G1 and S-phase. The Clb5 and Clb6 proteins begin to accumulate in late G1-phase [104-106]. Clb6 is targeted for destruction by the SCF and degraded early in S-phase whereas Clb5 persists into G2-phase [107]. Owing to its destruction early in S-phase Clb6-Cdk1 influences only early firing ORIs whereas Clb5-Cdk1 can regulate both early and later firing ORIs [107, 108]. Among the cyclin subunits Clb5 and Clb6 are the most effective at triggering ORI activation and henceforth I will refer to them as S-Cdk. Their effectiveness in activating DNA replication is in part due to the timing of their accumulation; however, even if other cyclins are expressed in late G1 and early S-phase they cannot activate DNA replication as effectively as S-Cdk [109-112]. Both Clb5 and Clb6 have a hydrophobic patch on their surfaces with an MRAIL sequence motif that allows them to interact with target proteins that have Arg–x–Leu or Lys-x-Leu sequences [111, 113, 114]. Whereas DDK physically interacts with the Mcm2-7 complex and this interaction is essential for conversion of a pre-RC to an active replicative complex, there is no evidence that Cdk must bind to the pre-RC in order to drive its conversion to an active complex. Clb5 can bind to Orc6 and does so following the initiation of DNA replication but this is a mechanism to prevent re-licensing and reactivation of ORIs rather than to promote their initial activation in S-phase [115].

6. Activating the licensed origins: All aboard the helicase train

The first additional components to interact with the loaded and licensed pre-RC are Sld3, its partner Sld7 and Cdc45 [116-118]. These factors associate with early firing ORIs and bind to the Mcm2-7 complex in G1 phase. Sld3 was originally identified in a genetic screen designed to isolate mutations that were synthetically lethal in an *S. cerevisiae* strain that harbored a temperature sensitive mutant allele of the DNA polymerase ε binding protein *DPB11* [119]. *CDC45* was discovered through its genetic interactions with *MCM5* and *MCM7* mutants [120]. Mutations in either *CDC45* or *SLD3* that cause loss of function prevent DNA replication and are thus lethal [116, 118]. Chromatin immunoprecipitation and in vitro reconstitution experiments indicate that the binding of Sld3 and Cdc45 to ORIs in G1-phase is relatively weak [121, 122]. DDK activity and binding of DDK to the pre-RC is required for the stable recruitment of Sld3 and Cdc45 both in vitro [121], and in vivo [116, 123, 124]. In addition, Sld3 and Cdc45 are required for each others interaction with the ORI bound Mcm2-7 complex.

Association of Cdc45, Sld3 and its partner Sld7 with ORIs is dependent upon DDK [29, 121]. Neither Sld3-Sld7 nor Cdc45 are directly phosphorylated by DDK rather Mcm2, Mcm4 and potentially Mcm6 are the critical S-phase substrates for DDK [89, 98, 125]. Indeed, modification of the structural architecture of the Mcm2-7 complex is likely the critical function for DDK in the activation of DNA replication since a mutation of Mcm5 that changes proline 83 to leucine alters the structure of the Mcm2-7 complex and allows cells lacking DDK to survive and replicate their DNA [122, 126, 127]. Additionally, DDK binds to the Mcm2-7 complex through interactions with a docking domain in Mcm4 and mutations in the Mcm can bypass the requirement for DDK [98, 125]. The initial interaction of DDK with the Mcm2-7 complex is dependent upon prior phosphorylation of at least Mcm4 and Mcm6 by yet to be identified protein kinases [89, 90].

The binding of Cdc45, Sld3 and Sld7 is a pre-requisite for the further assembly and conversion of the pre-RC to an active replication complex (RC). Following the loading of these factors Cdk activity is required. Accumulating S-Cdks interact with both Sld2 and Sld3 through RxL motifs in the substrate proteins [113-115, 128]. This leads to phosphorylation of Sld2 and Sld3 at multiple sites [129, 130]. The multi-site phosphorylation of Sld2 leads to a conformational change in the protein that allows the additional phosphorylation of threonine 84, which does not reside within a canonical Cdk recognition motif [131]. Phosphorylation of T[84] allows Sld2 to interact with Dpb11 a protein originally identified based upon its interactions with the replicative DNA polymerase, Polε [132]. Dpb11 has BRCT repeat domains at both its amino-terminal and carboxyl-terminal regions [133]. These sequence motifs function as phospho-peptide binding domains [134] allowing the phosphorylated Sld2 to bind the carboxyl-terminal BRCT phosphopeptide binding domain of Dpb11 [119, 129, 130]. Similarly phosphorylation of Sld3 allows Sld3 to bind the amino-terminal BRCT repeat of Dpb11 thus recruiting the Sld2-Sld3-Dpb11 complex to the Mcm2-7 complex and origin of replication [129, 130]. Dpb11 binds Polε, the leading strand replicative DNA polymerase in *S. cerevisiae* [132]. The interaction of Dpb11 with DNA Polε is not Cdk dependent but binding to phosphorylated Sld2 and Sld3 allows recruitment of the entire complex to the licensed ORI [135].

Although Sld2 and Sld3 are not the only components of the replication complex that can be phosphorylated by Cdk1 they are the critical substrates since phosphomimetic mutations in Sld2 and fusion of Sld3 with Dpb11 can bypass the need for Cdk1 activity to initiate DNA synthesis [129, 130].

The binding of Sld2 and Sld3 to the pre-RC allows the recruitment of GINS to the Mcm2-7 hexamer. GINS is a protein complex composed of Psf1, Psf2, Psf3 and Sld5 and is named after the number based names of its components Go, Ichi, Ni, San (Japanese for 5, 1, 2, 3]. Sld5 was identified in a genetic screen for mutants that displayed synthetic lethality when combined with a thermo-sensitive *dpb11* allele [116]. Subsequent investigations reveled partners of Sld5 (Psf1, Psf2, Psf3) that formed a complex required for initiation and DNA strand elongation during DNA replication [136]. GINS associates with Cdc45 at the ORI and its recruitment leads to stable engagement of Cdc45 with the Mcm2-7 complex. In vitro Cdc45 and GINS strongly stimulate the ATPase and DNA unwinding activity of Mcm2-7 complex [137]. There is evidence that Cdc45 makes specific contacts with Mcm2 while GINS binds to Mcm5, when GINS and Cdc45 bind one another this tightly closes the Mcm2-7 rings "gate" with DNA trapped within the central channel of the Mcm ring structure reviewed by [59]. There is no evidence that Cdk phosphorylates either Cdc45 or GINS or regulates their activity, the primary role played by the Cdk appears to be in promoting their recruitment to the chromatin bound Mcm2-7 complex. The binding of the additional components including GINS results in conversion of the pre-RC into the CMG (Cdc45/Mcm2-7/GINS) complex, this is also referred to as the pre-initiation complex (pre-IC) [138]. While Sld2, Sld3 and Sld7 are released from the complex following stable engagement of Cdc45 and GINS, both of the latter factors remain associated with the Mcm2-7 and are required for elongation of the nascent DNA strands following the initiation of DNA synthesis [136, 139].

Mcm10 is an additional factor required for assembly of a functional replisome and conversion of the pre-IC to an RC. Mcm10 was originally identified in a screen similar to that used for the identification of other *S. cerevisiae MCM* genes [140, 141]. Homologs of *MCM10* can be found from yeast to humans [142, 143]. Mcm10 is an abundant chromatin bound protein that interacts with all six subunits of the Mcm2-7 complex and localizes to origins of DNA replication [141, 142, 144]. Mcm10 has a critical role in conversion of the pre-RC to an active RC as it makes contacts with DNA Polα and the CMG complex components [145-147]. It is certain that Mcm10 plays a role in stabilizing the Mcm2-7 complex with DNA Polα [148]; however its precise role in the initial recruitment of DNA polymerases or their accessory factors to the replisome is not entirely clear.

The accumulation and action of DDK and CDK set in motion the assembly and conversion of the pre-RC to an activated RC. The use of two independent kinases to achieve this goal allows tight regulation over the assembly and activation process. Since both kinases are required to activate and "fire" the ORI it seems that there are in fact two triggers that can be pulled independently. For the initiation of DNA replication to take place both triggers must be pulled with the correct timing.

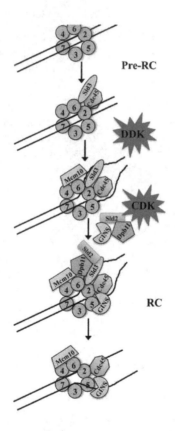

Figure 2. DDK and CDK promote assembly and activation of replication complexes at chromosomal origins of DNA replication. Sld3 and Cdc45 associate loosely to the ORI bound Mcm2-7 hexamer in G1-phase. Phosphorylation of the Mcm subunits by DDK promote tight binding by Cdc45 and Sld3, Mcm10 may associate with the complex at this time and plays an important role in unwinding of the ORI DNA duplex. CDK phosphorylation of Sld3, and Sld2 recruit Sld2, Dpb11, Pole and GINS to the Mcm2-7 complex. GINS binding increases the helicase activity of the Mcm2-7 hexamer allowing unwinding of duplex DNA. The association of GINS also marks a transition when Mcm2-7 binding to duplex DNA changes to binding such that a single strand is retained in the central channel, while the other strand is moved to the external surface of the complex.

7. The business end: Polymerases at the origin

The final critical steps of origin firing are the recruitment of the replicative polymerases, unwinding of the dsDNA and initiation of DNA synthesis. While all cells encode multiple different DNA polymerases the enzymes with the most well characterized roles in nuclear chromosomal DNA replication are DNA Polε, DNA Polδ, and DNA Polα – primase. DNA Polε acts as the leading strand DNA polymerase for nuclear DNA replication in *S. cerevisiae* [149]. Through its interaction with Dpb11 it is recruited to the pre-RC complex following Cdk1 mediated phosphorylation of Sld2, and Sld3. DNA Polδ is the major lagging strand DNA polymerase in *S. cerevisiae* [150]. Although DNA Polδ plays a key role in nuclear DNA replication it is currently unclear how this enzyme is recruited to the nascent RC. DNA Polα-primase is essential for the initiation of DNA replication as primase synthesizes RNA primers that Polα extends with short DNA oligonucleotides on the unwound ORI DNA providing primers for DNA Polε and DNA Polδ. [151, 152]. Mcm10 binds DNA Polα and this DNA polymerase may be initially recruited to the Mcm2-7 complex through these interactions. The primase polypeptide forms a complex with the carboxyl-terminus of Polα allowing the two to be incorporated into the growing replisome simultaneously [153]. Following or perhaps concurrent with recruitment of the replicative DNA polymerases there is a reorganization of the complex as it undergoes conversion from a pre-IC to RC. During this process Dpb11, Sld2 and Sld3 are ejected from the complex while Polε remains bound. Within the RC, DNA Polε makes contacts with Mrc1 that help to retain it within the complex [154]. It is currently unclear how Mrc1 is recruited to the complex upon conversion to a nascent RC or whether unwinding of the ORI DNA is required. Polα makes contacts initially with Mcm10 and once incorporated into the RC, it makes further contacts with Ctf4 a component of Replication Factor C (RFC), these contacts help stabilize the binding of Polα to the complex [155, 156]. During the remodeling of the pre-RC into an activated RC several accessory proteins: Replication Factor C (RFC), Proliferating Cell Nuclear Antigen (PCNA), and Replication Protein A (RPA) are added to the complex. The mechanism that leads to recruitment of these accessory proteins has not been determined. It may be that they simply recognize and bind to the protein-DNA structure formed by the initial unwinding of the ORI DNA. Owing to its ssDNA binding capability RPA associates with the RC once unwinding of the ORI DNA is underway; here it assists in stabilizing the nascent replication bubble and provides access for the replicative DNA polymerases [157]. All three subunits of DNA Polδ make contact with PCNA and these interactions are essential for processive lagging strand DNA synthesis [158]. These factors influence the processivity and integrity of DNA synthesis.

Unwinding the ORI DNA to provide ssDNA as template for the DNA polymerases and to construct bidirectional replication forks is accomplished by the activated Mcm2-7 hexamer in concert with associated proteins Cdc45, GINS, Mcm10 and the replicative DNA polymerases. In vitro the Mcm2-7 hexamer unwinds DNA by tracking along a single strand while displacing the other strand [65, 159]. Achieving this end requires that the dsDNA initially bound be melted and locally unwound allowing release of one strand to the outside surface of the complex and retaining the other within the central channel of the hexamer. Although the molecular details of this process remain unclear some of the current models to explain ORI unwinding by Mcm2-7 have been recently reviewed in detail [59].

Sld2, Sld3, and Mcm10 all display some ability to bind ssDNA and it has been speculated that they might participate in the initially melting of the dsDNA, allowing the Mcm2-7 rings to undergo conformational change such that they close around one of strands of the melted duplex. Mcm10 may be a real candidate for this role based upon its stable incorporation into the RC and ability to bind ssDNA [160]. Determining the precise mechanism and timing of ORI DNA unwinding will await higher resolution structural and biochemical analysis.

8. Who's on first? Ordered action of DDK and CDK in the activation of ORIs

The assembly of a preRC and its conversion first to an RC and then an active replication fork is a multistep process that requires the activity of both DDK and CDK. Multiple investigations have been performed to determine the order in which DDK and CDK act at the ORIs to trigger their activation. Genetic studies with *S. cerevisiae* have suggested that DDK cannot complete its function without prior S-Cdk activation implying either that Cdk must act before DDK or that DDK performs a multiple functions at the pre-RC and that some of them require Cdk activity for completion [161]. In *X. laevis* egg extracts DDK can complete its essential function in the absence of Cdk activity, however Cdk cannot perform its vital function in the absence of DDK [162, 163]. Recent investigations using an *S. cerevisiae* in vitro DNA replication system suggest that assembly and activation of origins of replication require that DDK act before Cdk but that completion of DDKs essential functions require Cdk activity [90, 121]. The apparent conflict in these results may reflect differences between DNA replication control in somatic cells and eggs. Additionally, some of the differences may stem from the limitations inherent to both genetic and in vitro biochemical experimental systems. Redundant systems and limits to the speed with which activities can be activated and inactivated in vivo place limits on genetic approaches to understanding the specific requirements for DDK and CDK. While in vitro it may be difficult to accurately recapitulate the in vivo environment. For example, Cdk activity increases during G1-phase in a graded fashion both in total kinase activity and kinase specificity. Relatively low levels of Cdk activity are sufficient to activate DNA replication and elevated levels of Cdk activity that accumulate in S, G2, and M-phases prevent licensing and activation of origins by promoting destruction of Cdc6, nuclear export of Mcm2-7 components and by binding to Orc6 and excluding recruitment of Mcm to ORIs [164, 165]. It is possible that low levels of Cdk activity are required prior to DDK initiating its function. Indeed phosphorylation of Mcm4 and Mcm6 is a prerequisite for DDK binding to the pre-RC and further inducing activation. It has been proposed that phosphorylation of Mcm4 by G1-Cdk activity may be required to allow DDK to bind to the Mcm2-7 complex [89].

9. Conclusion

DNA replication is a fundamental aspect of cellular proliferation and development. Many aspects of this process are well conserved not only within the domain of eukaryotes but also across bacteria and archaea. The multi-step assembly and activation of origins of DNA replication is more complicated and more rigorously regulated in eukayotes than it is in either prokaryotes or archaea. This complexity stems in part from the size of the eukaryotic genomes that necessitates multiple origins of replication on each chromosome. Additionally, multiple layers of regulation act as a safeguard that ensures each origin of DNA replication is activated only once in each cell cycle. This is crucial to prevent over replication, amplification of chromosomal segments and chromosome instability.

The initiation of DNA replication in *S. cerevisiae* has served as an exceptional model owing to the genetic and biochemical accessibility of this organism. Our current understanding of the steps leading to the initiation of DNA replication in *S. cerevisiae* can be summarized as follows. Orc1-6 bound ORI sequences act as a binding site for Cdc6, which in conjunction with Cdt1 recruits Mcm2-7 hexamers to the ORI. DDK is recruited to this structure by virtue of the affinity of Dbf4 for docking domains in Mcm4. DDK phosphorylates the Mcm2-7 helicase, promoting the recruitment of Sld3 and Cdc45. Next, S-CDK-dependent phosphorylation of Sld2 and Sld3 leads to their binding Dpb11 and recruitment of the complex, along with GINS and Polε to the pre-RC thus forming a CMG complex. These proteins then serve to both recruit Mcm10 and fully activate the Mcm2-7 helicase, which uses ATP hydrolysis to melt the origin DNA. Pola-primase and Polδ can then be loaded on to the ssDNA at the unwound ORI, leading to the formation of a complete replisome with accessory proteins such as PCNA, Mrc1, RFC, RPA, and topisomerase. The helicase activity of the Mcm2-7 hexamers then drives bidirectional dsDNA unwinding and replication fork movement along the chromosome allowing the synthesis of new DNA.

Initiating DNA replication is a serious event for a cell. The chromosomal DNA is rarely more at risk of damage than when it is being unwound and copied. During this processes single stranded DNA is revealed and the fork structures with the potential for breakage and recombination are formed. The requirement for two protein kinases, DDK and CDK, that perform non-redundant functions in the assembly and activation of replication complexes suggests that there are in fact two triggers that must be pulled to fire the origin. The requirement for two different kinases that are independently regulated and that each have distinct substrate specificity allows the initiation of DNA replication to be regulated with exquisite sensitivity. Perhaps rather than considering these two kinases as triggers they should really be though of as a double failsafe mechanism where each trigger must be pulled with the appropriate timing to allow DNA replication to proceed.

Despite our general understanding of this process many aspects of its molecular basis remain to be elucidated. How are Sld3 and Cdc45 initially recruited to the pre-RC? How does the Mcm2-7 helicase melt ORI DNA and what is the mechanism by which it is converted to a machine that directionally tracks along and unwinds dsDNA? Does DDK travel with the Mcm2-7 complex along the DNA? How are DNA Polδ and the accessory proteins RFC, and

PCNA recruited to the replication fork? It is likely that a combination of genetic analysis, biochemistry and high-resolution structure analysis will be required to answer these questions.

Acknowledgements

I thank all of the previous members of my lab who have participated in projects focused on understanding the mechanisms that control DNA replication. The work on DNA replication in my lab has been supported by operating grants from the Natural Sciences and Engineering Research Council (NSERC).

Author details

David Stuart

Department of Biochemistry, University of Alberta, Edmonton, Alberta, Canada

References

[1] Larkins BA, Dilkes BP, Dante RA, Coelho CM, Woo YM, Liu Y. Investigating the hows and whys of DNA endoreduplication. J Exp Bot. 2001;52(355):183-92.

[2] Green BM, Finn KJ, Li JJ. Loss of DNA replication control is a potent inducer of gene amplification. Science. 2010;329(5994):943-6.

[3] Pfau SJ, Amon A. Chromosomal instability and aneuploidy in cancer: from yeast to man. EMBO Rep. 2012;13(6):515-27.

[4] Johnston LH, Masai H, Sugino A. First the CDKs, now the DDKs. Trends Cell Biol. 1999;9(7):249-52.

[5] Baker TA, Wickner SH. Genetics and enzymology of DNA replication in *Escherichia coli*. Annu Rev Genet. 1992;26:447-77.

[6] Falaschi A, Giacca M, Zentilin L, Norio P, Diviacco S, Dimitrova D, et al. Searching for replication origins in mammalian DNA. Gene. 1993;135(1-2):125-35.

[7] Hamlin JL, Mesner LD, Lar O, Torres R, Chodaparambil SV, Wang L. A revisionist replicon model for higher eukaryotic genomes. J Cell Biochem. 2008;105(2):321-9.

[8] Norio P, Kosiyatrakul S, Yang Q, Guan Z, Brown NM, Thomas S, et al. Progressive activation of DNA replication initiation in large domains of the immunoglobulin heavy chain locus during B cell development. Mol Cell. 2005;20(4):575-87.

[9]   Heichinger C, Penkett CJ, Bahler J, Nurse P. Genome-wide characterization of fission yeast DNA replication origins. EMBO J. 2006;25(21):5171-9.

[10] Cotobal C, Segurado M, Antequera F. Structural diversity and dynamics of genomic replication origins in Schizosaccharomyces pombe. EMBO J. 2010;29(5):934-42.

[11] Stinchcomb DT, Struhl K, Davis RW. Isolation and characterisation of a yeast chromosomal replicator. Nature. 1979;282(5734):39-43.

[12] Dhar MK, Sehgal S, Kaul S. Structure, replication efficiency and fragility of yeast ARS elements. Res Microbiol. 2012;163(4):243-53.

[13] Marahrens Y, Stillman B. A yeast chromosomal origin of DNA replication defined by multiple functional elements. Science. 1992;255(5046):817-23.

[14] Bell SP, Stillman B. ATP-dependent recognition of eukaryotic origins of DNA replication by a multiprotein complex. Nature. 1992;357(6374):128-34.

[15] Natale DA, Schubert AE, Kowalski D. DNA helical stability accounts for mutational defects in a yeast replication origin. Proc Natl Acad Sci U S A. 1992;89(7):2654-8.

[16] Zou L, Stillman B. Assembly of a complex containing Cdc45p, replication protein A, and Mcm2p at replication origins controlled by S-phase cyclin-dependent kinases and Cdc7p-Dbf4p kinase. Mol Cell Biol. 2000;20(9):3086-96.

[17] Miyake T, Loch CM, Li R. Identification of a multifunctional domain in autonomously replicating sequence-binding factor 1 required for transcriptional activation, DNA replication, and gene silencing. Mol Cell Biol. 2002;22(2):505-16.

[18] Newlon CS, Theis JF. The structure and function of yeast ARS elements. Curr Opin Genet Dev. 1993;3(5):752-8.

[19] Czajkowsky DM, Liu J, Hamlin JL, Shao Z. DNA combing reveals intrinsic temporal disorder in the replication of yeast chromosome VI. J Mol Biol. 2008;375(1):12-9.

[20] Santocanale C, Sharma K, Diffley JF. Activation of dormant origins of DNA replication in budding yeast. Genes Dev. 1999;13(18):2360-4.

[21] Raghuraman MK, Winzeler EA, Collingwood D, Hunt S, Wodicka L, Conway A, et al. Replication dynamics of the yeast genome. Science. 2001;294(5540):115-21.

[22] Friedman KL, Brewer BJ, Fangman WL. Replication profile of Saccharomyces cerevisiae chromosome VI. Genes Cells. 1997;2(11):667-78.

[23] Patel PK, Arcangioli B, Baker SP, Bensimon A, Rhind N. DNA replication origins fire stochastically in fission yeast. Mol Biol Cell. 2006;17(1):308-16.

[24] Luo H, Li J, Eshaghi M, Liu J, Karuturi RK. Genome-wide estimation of firing efficiencies of origins of DNA replication from time-course copy number variation data. BMC Bioinformatics. 2010;11:247.

[25] Spiesser TW, Klipp E, Barberis M. A model for the spatiotemporal organization of DNA replication in *Saccharomyces cerevisiae*. Mol Genet Genomics. 2009;282(1):25-35.

[26] Barberis M, Spiesser TW, Klipp E. Replication origins and timing of temporal replication in budding yeast: how to solve the conundrum? Curr Genomics. 2010;11(3): 199-211.

[27] Yang SC, Rhind N, Bechhoefer J. Modeling genome-wide replication kinetics reveals a mechanism for regulation of replication timing. Mol Syst Biol. 2010;6:404.

[28] Mantiero D, Mackenzie A, Donaldson A, Zegerman P. Limiting replication initiation factors execute the temporal programme of origin firing in budding yeast. EMBO J. 2011;30(23):4805-14.

[29] Tanaka S, Nakato R, Katou Y, Shirahige K, Araki H. Origin association of Sld3, Sld7, and Cdc45 proteins is a key step for determination of origin-firing timing. Curr Biol. 2011;21(24):2055-63.

[30] Patel PK, Kommajosyula N, Rosebrock A, Bensimon A, Leatherwood J, Bechhoefer J, et al. The Hsk1(Cdc7) replication kinase regulates origin efficiency. Mol Biol Cell. 2008;19(12):5550-8.

[31] Wu PY, Nurse P. Establishing the program of origin firing during S phase in fission Yeast. Cell. 2009;136(5):852-64.

[32] Vogelauer M, Rubbi L, Lucas I, Brewer BJ, Grunstein M. Histone acetylation regulates the time of replication origin firing. Mol Cell. 2002;10(5):1223-33.

[33] Aparicio JG, Viggiani CJ, Gibson DG, Aparicio OM. The Rpd3-Sin3 histone deacetylase regulates replication timing and enables intra-S origin control in *Saccharomyces cerevisiae*. Mol Cell Biol. 2004;24(11):4769-80.

[34] Knott SR, Viggiani CJ, Tavare S, Aparicio OM. Genome-wide replication profiles indicate an expansive role for Rpd3L in regulating replication initiation timing or efficiency, and reveal genomic loci of Rpd3 function in *Saccharomyces cerevisiae*. Genes Dev. 2009;23(9):1077-90.

[35] Kohzaki H, Ito Y, Murakami Y. Context-dependent modulation of replication activity of *Saccharomyces cerevisiae* autonomously replicating sequences by transcription factors. Mol Cell Biol. 1999;19(11):7428-35.

[36] Murakami Y, Ito Y. Transcription factors in DNA replication. Front Biosci. 1999;4:D824-33.

[37] Sequeira-Mendes J, Diaz-Uriarte R, Apedaile A, Huntley D, Brockdorff N, Gomez M. Transcription initiation activity sets replication origin efficiency in mammalian cells. PLoS Genet. 2009;5(4):e1000446.

[38] Gavin KA, Hidaka M, Stillman B. Conserved initiator proteins in eukaryotes. Science. 1995;270(5242):1667-71.

[39] Duncker BP, Chesnokov IN, McConkey BJ. The origin recognition complex protein family. Genome Biol. 2009;10(3):214.

[40] Wigley DB. ORC proteins: marking the start. Curr Opin Struct Biol. 2009;19(1):72-8.

[41] Clarey MG, Erzberger JP, Grob P, Leschziner AE, Berger JM, Nogales E, et al. Nucleotide-dependent conformational changes in the DnaA-like core of the origin recognition complex. Nat Struct Mol Biol. 2006;13(8):684-90.

[42] Lee DG, Bell SP. Architecture of the yeast origin recognition complex bound to origins of DNA replication. Mol Cell Biol. 1997;17(12):7159-68.

[43] Speck C, Chen Z, Li H, Stillman B. ATPase-dependent cooperative binding of ORC and Cdc6 to origin DNA. Nat Struct Mol Biol. 2005;12(11):965-71.

[44] Araki M, Wharton RP, Tang Z, Yu H, Asano M. Degradation of origin recognition complex large subunit by the anaphase-promoting complex in Drosophila. EMBO J. 2003;22(22):6115-26.

[45] Mendez J, Zou-Yang XH, Kim SY, Hidaka M, Tansey WP, Stillman B. Human origin recognition complex large subunit is degraded by ubiquitin-mediated proteolysis after initiation of DNA replication. Mol Cell. 2002;9(3):481-91.

[46] Ohta S, Tatsumi Y, Fujita M, Tsurimoto T, Obuse C. The ORC1 cycle in human cells: II. Dynamic changes in the human ORC complex during the cell cycle. J Biol Chem. 2003;278(42):41535-40.

[47] Tatsumi Y, Ohta S, Kimura H, Tsurimoto T, Obuse C. The ORC1 cycle in human cells: I. cell cycle-regulated oscillation of human ORC1. J Biol Chem. 2003;278(42):41528-34.

[48] Li CJ, DePamphilis ML. Mammalian Orc1 protein is selectively released from chromatin and ubiquitinated during the S-to-M transition in the cell division cycle. Mol Cell Biol. 2002;22(1):105-16.

[49] Li H, Stillman B. The origin recognition complex: a biochemical and structural view. Subcell Biochem. 2012;62:37-58.

[50] Klemm RD, Austin RJ, Bell SP. Coordinate binding of ATP and origin DNA regulates the ATPase activity of the origin recognition complex. Cell. 1997;88(4):493-502.

[51] Lee DG, Bell SP. ATPase switches controlling DNA replication initiation. Curr Opin Cell Biol. 2000;12(3):280-5.

[52] Lee DG, Makhov AM, Klemm RD, Griffith JD, Bell SP. Regulation of origin recognition complex conformation and ATPase activity: differential effects of single-stranded and double-stranded DNA binding. EMBO J. 2000;19(17):4774-82.

[53] Hossain M, Stillman B. Meier-Gorlin syndrome mutations disrupt an Orc1 CDK inhibitory domain and cause centrosome reduplication. Genes Dev. 2012;26(16): 1797-810.

[54] Guernsey DL, Matsuoka M, Jiang H, Evans S, Macgillivray C, Nightingale M, et al. Mutations in origin recognition complex gene ORC4 cause Meier-Gorlin syndrome. Nat Genet. 2011;43(4):360-4.

[55] Cappuccio I, Colapicchioni C, Santangelo V, Sale P, Blandini F, Bonelli M, et al. The origin recognition complex subunit, ORC3, is developmentally regulated and supports the expression of biochemical markers of neuronal maturation in cultured cerebellar granule cells. Brain Res. 2010;1358:1-10.

[56] Asano M. Endoreplication: the advantage to initiating DNA replication without the ORC? Fly (Austin). 2009;3(2):173-5.

[57] Park SY, Asano M. The origin recognition complex is dispensable for endoreplication in Drosophila. Proc Natl Acad Sci U S A. 2008;105(34):12343-8.

[58] Aggarwal BD, Calvi BR. Chromatin regulates origin activity in Drosophila follicle cells. Nature. 2004;430(6997):372-6.

[59] Vijayraghavan S, Schwacha A. The eukaryotic mcm2-7 replicative helicase. Subcell Biochem. 2012;62:113-34.

[60] Forsburg SL. Eukaryotic MCM proteins: beyond replication initiation. Microbiol Mol Biol Rev. 2004;68(1):109-31.

[61] Tye BK. MCM proteins in DNA replication. Annu Rev Biochem. 1999;68:649-86.

[62] Bochman ML, Schwacha A. The Mcm complex: unwinding the mechanism of a replicative helicase. Microbiol Mol Biol Rev. 2009;73(4):652-83.

[63] Bochman ML, Schwacha A. Differences in the single-stranded DNA binding activities of MCM2-7 and MCM467: MCM2 and MCM5 define a slow ATP-dependent step. J Biol Chem. 2007;282(46):33795-804.

[64] Davey MJ, Indiani C, O'Donnell M. Reconstitution of the Mcm2-7p heterohexamer, subunit arrangement, and ATP site architecture. J Biol Chem. 2003;278(7):4491-9.

[65] Bochman ML, Bell SP, Schwacha A. Subunit organization of Mcm2-7 and the unequal role of active sites in ATP hydrolysis and viability. Mol Cell Biol. 2008;28(19):5865-73.

[66] Chang F, May CD, Hoggard T, Miller J, Fox CA, Weinreich M. High-resolution analysis of four efficient yeast replication origins reveals new insights into the ORC and putative MCM binding elements. Nucleic Acids Res. 2011;39(15):6523-35.

[67] Remus D, Beuron F, Tolun G, Griffith JD, Morris EP, Diffley JF. Concerted loading of Mcm2-7 double hexamers around DNA during DNA replication origin licensing. Cell. 2009;139(4):719-30.

[68] Bochman ML, Schwacha A. The *Saccharomyces cerevisiae* Mcm6/2 and Mcm5/3 ATPase active sites contribute to the function of the putative Mcm2-7 'gate'. Nucleic Acids Res. 2010;38(18):6078-88.

[69] Devault A, Vallen EA, Yuan T, Green S, Bensimon A, Schwob E. Identification of Tah11/ Sid2 as the ortholog of the replication licensing factor Cdt1 in *Saccharomyces cerevisiae*. Curr Biol. 2002;12(8):689-94.

[70] Whittaker AJ, Royzman I, Orr-Weaver TL. Drosophila double parked: a conserved, essential replication protein that colocalizes with the origin recognition complex and links DNA replication with mitosis and the down-regulation of S phase transcripts. Genes Dev. 2000;14(14):1765-76.

[71] Nishitani H, Lygerou Z, Nishimoto T, Nurse P. The Cdt1 protein is required to license DNA for replication in fission yeast. Nature. 2000;404(6778):625-8.

[72] Maiorano D, Moreau J, Mechali M. XCDT1 is required for the assembly of pre-replicative complexes in Xenopus laevis. Nature. 2000;404(6778):622-5.

[73] Nishitani H, Taraviras S, Lygerou Z, Nishimoto T. The human licensing factor for DNA replication Cdt1 accumulates in G1 and is destabilized after initiation of S-phase. J Biol Chem. 2001;276(48):44905-11.

[74] Liu C, Wu R, Zhou B, Wang J, Wei Z, Tye BK, et al. Structural insights into the Cdt1-mediated MCM2-7 chromatin loading. Nucleic Acids Res. 2012;40(7):3208-17.

[75] Takara TJ, Bell SP. Multiple Cdt1 molecules act at each origin to load replication-competent Mcm2-7 helicases. EMBO J. 2011;30(24):4885-96.

[76] Evrin C, Clarke P, Zech J, Lurz R, Sun J, Uhle S, et al. A double-hexameric MCM2-7 complex is loaded onto origin DNA during licensing of eukaryotic DNA replication. Proc Natl Acad Sci U S A. 2009;106(48):20240-5.

[77] Donovan S, Harwood J, Drury LS, Diffley JF. Cdc6p-dependent loading of Mcm proteins onto pre-replicative chromatin in budding yeast. Proc Natl Acad Sci U S A. 1997;94(11):5611-6.

[78] Randell JC, Bowers JL, Rodriguez HK, Bell SP. Sequential ATP hydrolysis by Cdc6 and ORC directs loading of the Mcm2-7 helicase. Mol Cell. 2006;21(1):29-39.

[79] Jacobson MD, Munoz CX, Knox KS, Williams BE, Lu LL, Cross FR, et al. Mutations in *SID2*, a novel gene in *Saccharomyces cerevisiae*, cause synthetic lethality with sic1 deletion and may cause a defect during S phase. Genetics. 2001;159(1):17-33.

[80] Yanagi K, Mizuno T, You Z, Hanaoka F. Mouse geminin inhibits not only Cdt1-MCM6 interactions but also a novel intrinsic Cdt1 DNA binding activity. J Biol Chem. 2002;277(43):40871-80.

[81] Chen S, Bell SP. CDK prevents Mcm2-7 helicase loading by inhibiting Cdt1 interaction with Orc6. Genes Dev. 2011;25(4):363-72.

[82] Zhou C, Jong AY. Mutation analysis of *Saccharomyces cerevisiae CDC6* promoter: defining its UAS domain and cell cycle regulating element. DNA Cell Biol. 1993;12(4): 363-70.

[83] Perkins G, Drury LS, Diffley JF. Separate SCF(*CDC4*) recognition elements target Cdc6 for proteolysis in S phase and mitosis. EMBO J. 2001;20(17):4836-45.

[84] Tanny RE, MacAlpine DM, Blitzblau HG, Bell SP. Genome-wide analysis of re-replication reveals inhibitory controls that target multiple stages of replication initiation. Mol Biol Cell. 2006;17(5):2415-23.

[85] Chapman JW, Johnston LH. The yeast gene, *DBF4*, essential for entry into S phase is cell cycle regulated. Exp Cell Res. 1989;180(2):419-28.

[86] Kitada K, Johnston LH, Sugino T, Sugino A. Temperature-sensitive *cdc7* mutations of *Saccharomyces cerevisiae* are suppressed by the *DBF4* gene, which is required for the G1/S cell cycle transition. Genetics. 1992;131(1):21-9.

[87] Mok J, Kim PM, Lam HY, Piccirillo S, Zhou X, Jeschke GR, et al. Deciphering protein kinase specificity through large-scale analysis of yeast phosphorylation site motifs. Sci Signal. 2010;3(109):ra12.

[88] Wan L, Niu H, Futcher B, Zhang C, Shokat KM, Boulton SJ, et al. Cdc28-Clb5 (CDK-S) and Cdc7-Dbf4 (DDK) collaborate to initiate meiotic recombination in yeast. Genes Dev. 2008;22(3):386-97.

[89] Randell JC, Fan A, Chan C, Francis LI, Heller RC, Galani K, et al. Mec1 is one of multiple kinases that prime the Mcm2-7 helicase for phosphorylation by Cdc7. Mol Cell. 2010;40(3):353-63.

[90] Francis LI, Randell JC, Takara TJ, Uchima L, Bell SP. Incorporation into the prereplicative complex activates the Mcm2-7 helicase for Cdc7-Dbf4 phosphorylation. Genes Dev. 2009;23(5):643-54.

[91] Sasanuma H, Hirota K, Fukuda T, Kakusho N, Kugou K, Kawasaki Y, et al. Cdc7-dependent phosphorylation of Mer2 facilitates initiation of yeast meiotic recombination. Genes Dev. 2008;22(3):398-410.

[92] Jackson AL, Pahl PM, Harrison K, Rosamond J, Sclafani RA. Cell cycle regulation of the yeast Cdc7 protein kinase by association with the Dbf4 protein. Mol Cell Biol. 1993;13(5):2899-908.

[93] Oshiro G, Owens JC, Shellman Y, Sclafani RA, Li JJ. Cell cycle control of Cdc7p kinase activity through regulation of Dbf4p stability. Mol Cell Biol. 1999;19(7):4888-96.

[94] Ferreira MF, Santocanale C, Drury LS, Diffley JF. Dbf4p, an essential S phase-promoting factor, is targeted for degradation by the anaphase-promoting complex. Mol Cell Biol. 2000;20(1):242-8.

[95] Hughes S, Elustondo F, Di Fonzo A, Leroux FG, Wong AC, Snijders AP, et al. Crystal structure of human CDC7 kinase in complex with its activator DBF4. Nat Struct Mol Biol. 2012;19(11):1101-7.

[96] Dowell SJ, Romanowski P, Diffley JF. Interaction of Dbf4, the Cdc7 protein kinase regulatory subunit, with yeast replication origins in vivo. Science. 1994;265(5176): 1243-6.

[97] Duncker BP, Shimada K, Tsai-Pflugfelder M, Pasero P, Gasser SM. An N-terminal domain of Dbf4p mediates interaction with both origin recognition complex (ORC) and Rad53p and can deregulate late origin firing. Proc Natl Acad Sci U S A. 2002;99(25): 16087-92.

[98] Sheu YJ, Stillman B. Cdc7-Dbf4 phosphorylates MCM proteins via a docking site-mediated mechanism to promote S phase progression. Mol Cell. 2006;24(1):101-13.

[99] Wittenberg C, Reed SI. Control of the yeast cell cycle is associated with assembly/disassembly of the Cdc28 protein kinase complex. Cell. 1988;54(7):1061-72.

[100] Mendenhall MD, Hodge AE. Regulation of Cdc28 cyclin-dependent protein kinase activity during the cell cycle of the yeast *Saccharomyces cerevisiae*. Microbiol Mol Biol Rev. 1998;62(4):1191-243.

[101] Nasmyth K. Control of the yeast cell cycle by the Cdc28 protein kinase. Curr Opin Cell Biol. 1993;5(2):166-79.

[102] Jeffrey PD, Russo AA, Polyak K, Gibbs E, Hurwitz J, Massague J, et al. Mechanism of CDK activation revealed by the structure of a cyclinA-CDK2 complex. Nature. 1995;376(6538):313-20.

[103] Andrews B, Measday V. The cyclin family of budding yeast: abundant use of a good idea. Trends Genet. 1998;14(2):66-72.

[104] Kuhne C, Linder P. A new pair of B-type cyclins from *Saccharomyces cerevisiae* that function early in the cell cycle. EMBO J. 1993;12(9):3437-47.

[105] Schwob E, Nasmyth K. *CLB5* and *CLB6*, a new pair of B cyclins involved in DNA replication in *Saccharomyces cerevisiae*. Genes Dev. 1993;7(7A):1160-75.

[106] Epstein CB, Cross FR. *CLB5*: a novel B cyclin from budding yeast with a role in S phase. Genes Dev. 1992;6(9):1695-706.

[107] Jackson LP, Reed SI, Haase SB. Distinct mechanisms control the stability of the related S-phase cyclins Clb5 and Clb6. Mol Cell Biol. 2006;26(6):2456-66.

[108] Donaldson AD, Raghuraman MK, Friedman KL, Cross FR, Brewer BJ, Fangman WL. *CLB5*-dependent activation of late replication origins in *S. cerevisiae*. Mol Cell. 1998;2(2): 173-82.

[109] Donaldson AD. The yeast mitotic cyclin Clb2 cannot substitute for S phase cyclins in replication origin firing. EMBO Rep. 2000;1(6):507-12.

[110] Hu F, Gan Y, Aparicio OM. Identification of Clb2 residues required for Swe1 regulation of Clb2-Cdc28 in *Saccharomyces cerevisiae*. Genetics. 2008;179(2):863-74.

[111] Cross FR, Jacobson MD. Conservation and function of a potential substrate-binding domain in the yeast Clb5 B-type cyclin. Mol Cell Biol. 2000;20(13):4782-90.

[112] DeCesare JM, Stuart DT. Among B-type cyclins only *CLB5* and *CLB6* promote premeiotic S phase in Saccharomyces cerevisiae. Genetics. 2012;190(3):1001-16.

[113] Koivomagi M, Valk E, Venta R, Iofik A, Lepiku M, Morgan DO, et al. Dynamics of Cdk1 substrate specificity during the cell cycle. Mol Cell. 2011;42(5):610-23.

[114] Loog M, Morgan DO. Cyclin specificity in the phosphorylation of cyclin-dependent kinase substrates. Nature. 2005;434(7029):104-8.

[115] Wilmes GM, Archambault V, Austin RJ, Jacobson MD, Bell SP, Cross FR. Interaction of the S-phase cyclin Clb5 with an "RXL" docking sequence in the initiator protein Orc6 provides an origin-localized replication control switch. Genes Dev. 2004;18(9):981-91.

[116] Kamimura Y, Tak YS, Sugino A, Araki H. Sld3, which interacts with Cdc45 (Sld4), functions for chromosomal DNA replication in *Saccharomyces cerevisiae*. EMBO J. 2001;20(8):2097-107.

[117] Tanaka T, Umemori T, Endo S, Muramatsu S, Kanemaki M, Kamimura Y, et al. Sld7, an Sld3-associated protein required for efficient chromosomal DNA replication in budding yeast. EMBO J. 2011;30(10):2019-30.

[118] Hopwood B, Dalton S. Cdc45p assembles into a complex with Cdc46p/Mcm5p, is required for minichromosome maintenance, and is essential for chromosomal DNA replication. Proc Natl Acad Sci U S A. 1996;93(22):12309-14.

[119] Kamimura Y, Masumoto H, Sugino A, Araki H. Sld2, which interacts with Dpb11 in *Saccharomyces cerevisiae*, is required for chromosomal DNA replication. Mol Cell Biol. 1998;18(10):6102-9.

[120] Hennessy KM, Lee A, Chen E, Botstein D. A group of interacting yeast DNA replication genes. Genes Dev. 1991;5(6):958-69.

[121] Heller RC, Kang S, Lam WM, Chen S, Chan CS, Bell SP. Eukaryotic origin-dependent DNA replication in vitro reveals sequential action of DDK and S-CDK kinases. Cell. 2011;146(1):80-91.

[122] Sclafani RA, Tecklenburg M, Pierce A. The *mcm5-bob1* bypass of Cdc7p/Dbf4p in DNA replication depends on both Cdk1-independent and Cdk1-dependent steps in Saccharomyces cerevisiae. Genetics. 2002;161(1):47-57.

[123] Aparicio OM, Stout AM, Bell SP. Differential assembly of Cdc45p and DNA polymerases at early and late origins of DNA replication. Proc Natl Acad Sci U S A. 1999;96(16):9130-5.

[124] Kanemaki M, Labib K. Distinct roles for Sld3 and GINS during establishment and progression of eukaryotic DNA replication forks. EMBO J. 2006;25(8):1753-63.

[125] Sheu YJ, Stillman B. The Dbf4-Cdc7 kinase promotes S phase by alleviating an inhibi-
        tory activity in Mcm4. Nature. 2010;463(7277):113-7.

[126] Hardy CF, Dryga O, Seematter S, Pahl PM, Sclafani RA. *mcm5/cdc46-bob1* bypasses the
        requirement for the S phase activator Cdc7p. Proc Natl Acad Sci U S A. 1997;94(7):
        3151-5.

[127] Hoang ML, Leon RP, Pessoa-Brandao L, Hunt S, Raghuraman MK, Fangman WL, et
        al. Structural changes in Mcm5 protein bypass Cdc7-Dbf4 function and reduce
        replication origin efficiency in *Saccharomyces cerevisiae*. Mol Cell Biol. 2007;27(21):
        7594-602.

[128] Masumoto H, Muramatsu S, Kamimura Y, Araki H. S-Cdk-dependent phosphorylation
        of Sld2 essential for chromosomal DNA replication in budding yeast. Nature.
        2002;415(6872):651-5.

[129] Tanaka S, Umemori T, Hirai K, Muramatsu S, Kamimura Y, Araki H. CDK-dependent
        phosphorylation of Sld2 and Sld3 initiates DNA replication in budding yeast. Nature.
        2007;445(7125):328-32.

[130] Zegerman P, Diffley JF. Phosphorylation of Sld2 and Sld3 by cyclin-dependent kinases
        promotes DNA replication in budding yeast. Nature. 2007;445(7125):281-5.

[131] Tak YS, Tanaka Y, Endo S, Kamimura Y, Araki H. A CDK-catalysed regulatory
        phosphorylation for formation of the DNA replication complex Sld2-Dpb11. EMBO J.
        2006;25(9):1987-96.

[132] Araki H, Leem SH, Phongdara A, Sugino A. Dpb11, which interacts with DNA
        polymerase II(epsilon) in *Saccharomyces cerevisiae*, has a dual role in S-phase progression
        and at a cell cycle checkpoint. Proc Natl Acad Sci U S A. 1995;92(25):11791-5.

[133] Bork P, Hofmann K, Bucher P, Neuwald AF, Altschul SF, Koonin EV. A superfamily
        of conserved domains in DNA damage-responsive cell cycle checkpoint proteins.
        FASEB J. 1997;11(1):68-76.

[134] Williams RS, Lee MS, Hau DD, Glover JN. Structural basis of phosphopeptide recog-
        nition by the BRCT domain of BRCA1. Nat Struct Mol Biol. 2004;11(6):519-25.

[135] Muramatsu S, Hirai K, Tak YS, Kamimura Y, Araki H. CDK-dependent complex
        formation between replication proteins Dpb11, Sld2, Pol (epsilon), and GINS in
        budding yeast. Genes Dev. 2010;24(6):602-12.

[136] Takayama Y, Kamimura Y, Okawa M, Muramatsu S, Sugino A, Araki H. GINS, a novel
        multiprotein complex required for chromosomal DNA replication in budding yeast.
        Genes Dev. 2003;17(9):1153-65.

[137] Ilves I, Petojevic T, Pesavento JJ, Botchan MR. Activation of the MCM2-7 helicase by
        association with Cdc45 and GINS proteins. Mol Cell. 2010;37(2):247-58.

[138] Sclafani RA, Holzen TM. Cell cycle regulation of DNA replication. Annu Rev Genet.
        2007;41:237-80.

[139] Tercero JA, Longhese MP, Diffley JF. A central role for DNA replication forks in checkpoint activation and response. Mol Cell. 2003;11(5):1323-36.

[140] Maine GT, Sinha P, Tye BK. Mutants of *S. cerevisiae* defective in the maintenance of minichromosomes. Genetics. 1984;106(3):365-85.

[141] Merchant AM, Kawasaki Y, Chen Y, Lei M, Tye BK. A lesion in the DNA replication initiation factor Mcm10 induces pausing of elongation forks through chromosomal replication origins in *Saccharomyces cerevisiae*. Mol Cell Biol. 1997;17(6):3261-71.

[142] Homesley L, Lei M, Kawasaki Y, Sawyer S, Christensen T, Tye BK. Mcm10 and the MCM2-7 complex interact to initiate DNA synthesis and to release replication factors from origins. Genes Dev. 2000;14(8):913-26.

[143] Izumi M, Yanagi K, Mizuno T, Yokoi M, Kawasaki Y, Moon KY, et al. The human homolog of Saccharomyces cerevisiae Mcm10 interacts with replication factors and dissociates from nuclease-resistant nuclear structures in G(2) phase. Nucleic Acids Res. 2000;28(23):4769-77.

[144] Gambus A, Jones RC, Sanchez-Diaz A, Kanemaki M, van Deursen F, Edmondson RD, et al. GINS maintains association of Cdc45 with MCM in replisome progression complexes at eukaryotic DNA replication forks. Nat Cell Biol. 2006;8(4):358-66.

[145] Warren EM, Huang H, Fanning E, Chazin WJ, Eichman BF. Physical interactions between Mcm10, DNA, and DNA polymerase alpha. J Biol Chem. 2009;284(36): 24662-72.

[146] van Deursen F, Sengupta S, De Piccoli G, Sanchez-Diaz A, Labib K. Mcm10 associates with the loaded DNA helicase at replication origins and defines a novel step in its activation. EMBO J. 2012;31(9):2195-206.

[147] Kanke M, Kodama Y, Takahashi TS, Nakagawa T, Masukata H. Mcm10 plays an essential role in origin DNA unwinding after loading of the CMG components. EMBO J. 2012;31(9):2182-94.

[148] Ricke RM, Bielinsky AK. Mcm10 regulates the stability and chromatin association of DNA polymerase-alpha. Mol Cell. 2004;16(2):173-85.

[149] Pursell ZF, Isoz I, Lundstrom EB, Johansson E, Kunkel TA. Yeast DNA polymerase epsilon participates in leading-strand DNA replication. Science. 2007;317(5834):127-30.

[150] Larrea AA, Lujan SA, Nick McElhinny SA, Mieczkowski PA, Resnick MA, Gordenin DA, et al. Genome-wide model for the normal eukaryotic DNA replication fork. Proc Natl Acad Sci U S A. 2010;107(41):17674-9.

[151] Santocanale C, Foiani M, Lucchini G, Plevani P. The isolated 48,000-dalton subunit of yeast DNA primase is sufficient for RNA primer synthesis. J Biol Chem. 1993;268(2): 1343-8.

[152] Plevani P, Foiani M, Valsasnini P, Badaracco G, Cheriathundam E, Chang LM. Poly-peptide structure of DNA primase from a yeast DNA polymerase-primase complex. J Biol Chem. 1985;260(11):7102-7.

[153] Kilkenny ML, De Piccoli G, Perera RL, Labib K, Pellegrini L. A conserved motif in the C-terminal tail of DNA polymerase alpha tethers primase to the eukaryotic replisome. J Biol Chem. 2012;287(28):23740-7.

[154] Lou H, Komata M, Katou Y, Guan Z, Reis CC, Budd M, et al. Mrc1 and DNA polymerase epsilon function together in linking DNA replication and the S phase checkpoint. Mol Cell. 2008;32(1):106-17.

[155] Gambus A, van Deursen F, Polychronopoulos D, Foltman M, Jones RC, Edmondson RD, et al. A key role for Ctf4 in coupling the MCM2-7 helicase to DNA polymerase alpha within the eukaryotic replisome. EMBO J. 2009;28(19):2992-3004.

[156] Lee C, Liachko I, Bouten R, Kelman Z, Tye BK. Alternative mechanisms for coordinating polymerase alpha and MCM helicase. Mol Cell Biol. 2010;30(2):423-35.

[157] Fanning E, Klimovich V, Nager AR. A dynamic model for replication protein A (RPA) function in DNA processing pathways. Nucleic Acids Res. 2006;34(15):4126-37.

[158] Acharya N, Klassen R, Johnson RE, Prakash L, Prakash S. PCNA binding domains in all three subunits of yeast DNA polymerase delta modulate its function in DNA replication. Proc Natl Acad Sci U S A. 2011;108(44):17927-32.

[159] Costa A, Ilves I, Tamberg N, Petojevic T, Nogales E, Botchan MR, et al. The structural basis for MCM2-7 helicase activation by GINS and Cdc45. Nat Struct Mol Biol. 2011;18(4):471-7.

[160] Watase G, Takisawa H, Kanemaki MT. Mcm10 plays a role in functioning of the eukaryotic replicative DNA helicase, Cdc45-Mcm-GINS. Curr Biol. 2012;22(4):343-9.

[161] Nougarede R, Della Seta F, Zarzov P, Schwob E. Hierarchy of S-phase-promoting factors: yeast Dbf4-Cdc7 kinase requires prior S-phase cyclin-dependent kinase activation. Mol Cell Biol. 2000;20(11):3795-806.

[162] Jares P, Blow JJ. Xenopus cdc7 function is dependent on licensing but not on XORC, XCdc6, or CDK activity and is required for XCdc45 loading. Genes Dev. 2000;14(12): 1528-40.

[163] Walter JC. Evidence for sequential action of cdc7 and cdk2 protein kinases during initiation of DNA replication in Xenopus egg extracts. J Biol Chem. 2000;275(50): 39773-8.

[164] Oikonomou C, Cross FR. Rising cyclin-CDK levels order cell cycle events. PLoS One. 2011;6(6):e20788.

[165] Ikui AE, Archambault V, Drapkin BJ, Campbell V, Cross FR. Cyclin and cyclin-dependent kinase substrate requirements for preventing rereplication reveal the need for concomitant activation and inhibition. Genetics. 2007;175(3):1011-22.

Chapter 3

Replicative Helicases as the Central Organizing Motor Proteins in the Molecular Machines of the Elongating Eukaryotic Replication Fork

John C. Fisk, Michaelle D. Chojnacki and
Thomas Melendy

Additional information is available at the end of the chapter

1. Introduction

Major processes in the cell often involve the coordinated and efficient assembly of macro-molecular complexes; such examples include: RNA transcription, DNA replication, translation, and cellular motion. These processes can be likened to miniature forms of machines, so-called "molecular machines" with multiple components and motors at their heart driving the systems. This term has been used by several researchers, which equate many of life's inner workings as homologous to machines; albeit much more efficient than their macro-type counterparts [104]. In 1998, Bruce Alberts wrote an elegant article for *Cell* noting the inherent beauty of molecular biology's machines, praising them and stating that as with all machines these macromolecular complexes must in turn contain an assortment of moving parts that act in a highly coordinated fashion with each other [1]. One such studied process is DNA replication, which has been extensively studied since the discovery of the DNA double helix. Due to the biological necessity for duplication of the genetic material, and the intricate link between the faithful replication of the genomic blueprint and its mismanagement leading to cancer, it is difficult to envision a process more important to human health than the study of DNA replication. The motor that drives the molecular machine that is DNA replication is the replicative DNA helicase. Replicative DNA helicases are well known as the motors that drive DNA replication forks along the DNA strands. But in more recent years it is becoming evident that replicative helicases also coordinate the necessary associations and dissociations of the various DNA replication complexes that need to act at the elongating replication fork. Here we will review the current knowledge of how the molecular motors, replicative DNA helicases, coordinate the actions of the molecular machines that are elongating eukaryotic DNA replication forks.

2. Phases of DNA replication

The replication of DNA during the Synthesis (S) Phase of the cell is generally differentiated into distinct stages. The first is the binding and **recognition of the origin** of replication by origin binding proteins. For cellular replication in eukaryotes, these proteins are the Origin Recognition Complex (ORC) proteins, many of which belong to AAA+ family of cellular ATPases [20, 97]. To begin activation of the origin (i.e. - **licensing**), two other proteins must act to make origins competent, Cdc6 and Cdt1 [5, 112]. These two proteins in turn are regulated by phosphorylation by Cdc7/Dbf4 as well as by geminin (in metazoans). The presence of ORC/Cdc6/Cdt1 are necessary for recruitment of the next set of vital DNA replication proteins, the minichromosome maintenance (MCM) proteins, which are components of the replicative DNA helicase [70, 115]. For many years, the MCM complex was proposed to be the replicative helicase; but both *in vitro* and *in vivo* studies could not verify that the MCM complex was in fact the DNA helicase necessary for eukaryotic replication [53, 68, 137]. However, it was well established that the six 'core' MCMs, MCM2-7, were essential for DNA replication and that their deletion was lethal in yeasts [125]. Additionally, MCMs appear to associate with chromatin just prior to S Phase, and dissociate from the chromatin as S Phase progresses, consistent with that of a DNA replication helicase [24, 117]. Only recently was it discovered that the MCM complex appears to be an incomplete DNA helicase, in that several additional proteins recruited during origin activation appear to be required to make up the DNA helicase holoenzyme. Cdc45 and the GINS (in Japanese Go-Ichi-Ni-San, which stands for the numbers 5-1-2-3 in the subunits Sld5, Psf1, Psf2, and Psf3) complex appear to make up the CMG (Cdc45-MCM-GINS), the complex multisubunit eukaryotic helicase [91], required for initiation of DNA replication. In spite of this elucidation of the CMG, the step-wise recruitment of these helicase components, and the complex nature of the post-translational modification steps required to reconstitute a functional CMG replicative DNA helicase, has severely constrained the ability to carry out detailed biochemical analyses of the eukaryotic DNA replication fork.

The formation of an active pre-replication complex at the origin, and the subsequent formation and activation of the CMG replicative DNA helicase allows for the recruitment of DNA polymerase α primase, which is necessary for the synthesis of RNA primers and a short DNA extension of those primers. Also recruited is RPA, the major ssDNA binding complex necessary to prevent the re-annealing of the DNA duplex [132], and topoisomerase I, which resolves the compression of the DNA helix caused by progression of the replication fork along the DNA duplex (**Initiation of DNA replication**). Following primer synthesis, the clamp loader, RFC, is loaded at the 5' end of the primers, and RFC in turn loads the DNA polymerase processivity factor, PCNA. Due to the 5'->3' nature of DNA replication, synthesis occurs continuously on the leading strand through the recruitment and activity of DNA polymerase ε [106] and discontinuously on the lagging strand by DNA polymerase δ extension of the repeated primers laid down by DNA polymerase α primase [98]. The components of polymerase δ are also often found, not surprisingly, associated with the proteins involved in "processing" the lagging strand Okazaki fragments, namely those proteins involved in removing the RNA primers (see below) [65, 103]. During this **elongation phase of DNA replication** is when the majority of DNA synthesis occurs. However, while the ORC complex and other components of the origin recognition/licensing machinery are dispensable following origin firing [26], the heart of the DNA replication apparatus remains associated with the replicative helicase, the molecular motor that is actively unwinding the DNA duplex. How the replicative helicase interacts with components of the elongation machinery is probably the least understood remaining aspect of DNA replication and is the focus of this review. Many other components are implicated in the elongation phase of eukaryotic DNA replication, such as Mrc1 (Claspin), which has been suggested to be involved in linking the helicase to the polymerases and has been found to be involved in the "uncoupling" of these two aspects of the fork during the DNA damage response [6, 56], and for regulating fork progression during uncompromised DNA synthesis [44, 78, 118, 122].

Following elongation, the RNA primers and the RNA-DNA linkages are removed through the actions of the flap endonuclease-1 (FEN1) nuclease and/or Pif1 helicase and Dna2 nuclease, assisted by RPA and DNA polymerase δ [74, 105, 108]. Following the removal of the primers, gaps are filled in, apparently by the action of the DNA polymerase δ and its cofactors, and the final DNA strands are ligated by DNA ligase I into long uninterrupted DNA chains. The removal of all the primers, filling of the subsequent gaps, and the final ligation of the products represent the completion of S-phase.

3. Model systems for elongation of DNA replication

As mentioned previously, eukaryotic cellular DNA replication is highly complicated, and only recently has the replicative DNA helicase finally been identified as MCM2-7 complexed with Cdc45 and the GINS complex (CMG) [91]; furthermore, the complex nature of assembly and regulation of this CMG replicative helicase has limited the ability to study the eukaryotic replication fork biochemically. However, early mechanistic studies of eukaryotic DNA replication were largely carried out using the small DNA tumor virus SV40 and to a lesser extent the papillomaviruses. What makes these viruses ideal models for the mechanistic study of eukaryotic DNA replication? One reason lies in their small genome size. To facilitate their duplication, these viruses make the most of their small number of ORFs by combining multiple replication functions into one or two proteins, and relying primarily on the host cell DNA replication machinery (see Table 1). In addition, the lack of these viruses utilizing the once-and-only-once per S Phase regulation of DNA replication means that their DNA replication systems were not subject to the complicated and constraining regulatory systems that control replication of cellular DNA.

SV40 DNA replication is driven by a single viral protein, SV40 large T-antigen (Tag), a protein that combines all the core DNA replication functions of the cellular initiation and origin activation proteins listed above for eukaryotic DNA replication. Tag recognizes and binds to the SV40 origin of replication, melts the DNA helix surrounding the origin, and establishes itself into a double hexameric structure. Tag then recruits the cellular DNA replication factors: RPA, topoisomerase I, and polymerase α primase. These four replication factors are all that is required for the initiation of SV40 DNA replication through the initial synthesis of RNA-DNA primers. Following these initiation events, the clamp loader, RFC, and the polymerase processivity factor, PCNA, are recruited and loaded, which leads to the binding and activity of DNA polymerase δ, which extends both lagging and leading strands in this viral DNA replication system. As in the mammalian system, Okazaki fragments are processed by FEN1, DNA helicase 2, and DNA ligase 1, completing synthesis of the viral DNA genomes. It was the early studies of this viral DNA synthesis system that elucidated these basic mechanisms of how eukaryotic DNA replication is carried out.

Replication Step/Function	Mammalian	SV40	Papillomavirus (PV)
Origin Recognition/Initiator	Orc complex (2-6)	T-antigen (Tag)	E2/E1
pre-RC	Cdc6, Cdt1, Cdc45, Geminin, MCM10, Sld2(RecQL4), Sld3, Dpb11(TopBP1)	Tag	E2/E1
Helicase	MCM 2-7, GINS, Cdc45	Tag	E1
SSB	RPA	RPA	RPA
Torsional relaxation	Topoisomerase I	Topoisomerase I	Topoisomerase I
Clamp loader	RFC	RFC	RFC
Processivity factor	PCNA	PCNA	PCNA
DNA polymerases	DNA pol α primase, DNA pol δ, DNA pol ε	DNA pol α primase, DNA pol δ	DNA pol α primase, DNA pol δ
Accessory factors	Mrc1(Claspin)	None	None?

Table 1. Known and Proposed Components of the DNA replication complex

Similar findings were also found for another virus family, the papillomaviruses, which have also proven to be an apt model for cellular DNA replication mechanisms due to their dependence on the host replication machinery. Initial studies were carried out in the bovine version BPV-1, and later corroborated with several human HPV isotypes. In general, papillomaviruses follow the same mode and progression of events found in SV40, except for the need for two viral proteins instead of the single Tag protein required for SV40. In addition, PV appears to require other cellular factors that SV40 does not [73, 80, 87], which to date remain unidentified. In papillomavirus DNA replication, the E2 protein assists and directs faithful viral origin recognition of E1 [79, 90, 110, 126], while E1 itself serves the role of the replicative DNA helicase, melting the DNA around the origin of replication and establishing itself as a double hexameric helicase. In a fashion similar to that of SV40 Tag, E1 also acts to recruit the cellular DNA replication proteins to the PV DNA replication fork [36, 113, 131]. E1 itself is a weak origin binding protein, but can bind to and unwind DNA even in the absence of E2 at high E1 concentrations, even on DNA without an apparent E1 binding sequence and is therefore relatively nonspecific without E2 [66]. Furthermore, following establishment of the double hexamer, the E2 protein is purportedly absent from subsequent steps of DNA replication, indicating E1 is the only viral protein implicated in the actual HPV elongating DNA fork [72]. Otherwise, these two small DNA viruses display very similar mechanisms of replication, especially during the elongation phase. So why rely on two very similar viruses as models and not just SV40? One reason is that by comparing and contrasting the DNA replication mechanisms in two subtly different systems, one gains further insight into the mechanisms of DNA replication. In specific aspects of DNA replication, one or the other virus might provide a more apt reflection of the mechanism of cellular DNA replication. Another reason lies in the diseases each virus causes and the implications for antiviral research. Although SV40 Tag is a potent transforming agent for cell culture due to its ability to inactivate p53, Rb protein, and many other components of the cell, SV40 itself does not appear to readily cause tumors in humans. Conversely, human papillomaviruses are the major cause of cervical, anogenital, and oral cancers and represent the major cause of infectious-agent-induced cancers in humans. These viruses represent historically important and still valuable models for DNA replication and can still be used to elucidate hitherto unknown mechanisms of mammalian DNA replication. Furthermore, the replicative DNA helicases of these viral DNA replication systems still provide the best biochemical system for investigating the role of DNA helicases in the elongation stage of eukaryotic DNA replication.

4. Replicative DNA helicases

When the structure of the DNA double helix was first proposed, one of the major questions concerning the replication of dsDNA was how the duplex would be opened to facilitate reading of the base sequence encoded by the DNA. The first such discovered protein that could carry out this function was the prokaryotic helicase of *E. coli*, discovered in 1976. All known helicases use the energy from NTPs to drive the remodeling of their substrate nucleic acids [75, 85]. Helicases are grouped into six superfamilies (SF1-SF6) and all possess typical Walker A and B motifs involved in NTP binding and hydrolysis. The motor proteins of the macromolecular machines at DNA replication forks are all AAA+ module-containing helicases, which function to unwind the DNA helix and to drive the replication machinery along the DNA template. Another common characteristic of replicative helicases is that most form higher order oligomeric structures to facilitate their functions as DNA helicases at DNA replication forks. The MCM complex of the CMG cellular helicase, SV40 Tag, and PV E1 all form hexamers. Both Tag and E1 have been recently crystalized in their hexameric forms, which has contributed significantly to elucidating how these helicases function in splitting the DNA helix [32, 38, 71]. Further, Tag and E1, and later MCMs, were shown to form dimers of two hexamers [34, 36], which are presumed to act in bridging the two DNA replication forks, holding them together during replication fork progression, and creating a system whereby the template DNA is threaded through the DNA replication machinery in both directions simultaneously.

Various models have been proposed for how DNA helicases unwind the DNA helix. Some early proposals included the monomers binding to the DNA backbone and essentially rolling one DNA strand away from the other using the circular nature of the hexamer. Other models included a hexamer 'embracing' ssDNA, excluding it from its partner, or two hexamers acting at a distance pumping dsDNA through their central pore. Some studies indicate the double hexamers stay associated during elongation, and this led to a double hexameric DNA pumping mechanism that pumps dsDNA through the central pore somehow splitting the helix [42]. The more recent structural studies of the BPV1 E1 helicase bound to DNA, ATP, and ADP indicate an intricate hybrid model whereby the E1 hexamer pumps ssDNA through each central pore in a staircase type mechanism as ATP is bound and hydrolyzed by each subsequent E1 monomer [32, 33, 109]. In this model E1 uses the ATP binding/hydrolysis-induced conformational changes of the individual monomers to drive each nucleotide base of the enclosed ssDNA template through the central pore, displacing the hybridized (lagging-strand template) DNA strand freeing it to be available as a template for lagging strand DNA synthesis [32]. Although the model for helicase action based on the SV40 Tag structure was not the same, the Tag structure was done in the absence of ssDNA, and the structural information on the Tag hexamer would be consistent with a helicase model similar to that of E1.

5. Helicase interactions with replication proteins that initiate elongation

As stated previously, DNA replication proteins commonly recruited by both of these viral replicative helicases are: RPA, topoisomerase I, and DNA polymerase α primase. In this section, we will look closer at the individual and combinatorial interactions between the helicase and these necessary DNA replication factors that are intimately involved in both the initiation and elongation stages of DNA replication. In many cases, studies have focused on specific interactions, often detailed down to specific amino acid residues required for recruitment of these factors. Various groups have used the powerful ability to investigate the interactions of these factors with the viral helicases both *in vitro* and *in vivo*, to elegantly

demonstrate the importance of these molecular contacts. For each of these three DNA replication factors, we will look into the extensive work that has been performed in the SV40 system with Tag, then in the PV system with E1. Following this, we will briefly touch on the
mammalian system, highlighting some of the similarities between the viral and the mammalian host systems.

6. Helicase interactions with replication proteins that initiate elongation: Topoisomerase I

The unwinding action of the DNA replication fork driving along the DNA helix creates torsional stress and overwound DNA that must be relieved to allow replication to proceed.
Topoisomerases are enzymes that help relieve this stress and aid in maintaining chromosome structure and integrity by modifying DNA topology, and resolving specific DNA
structures that arise from cellular processes such as DNA repair, replication, transcription,
recombination and chromosome compaction [13]. These processes result in compression
(positive supercoiling) of the DNA helix and the entanglement of DNA segments and chromosomal regions that can lead to cytotoxic or mutagenic breaks in the DNA if left unman-
aged [127]. Hence, topoisomerases play a vital role in living cells, particularly during DNA
replication.

Enzymatically, topoisomerases act through the action of a nucleophilic tyrosine; the enzyme
cleaves one or more DNA strands and generates an enzyme-DNA complex that serves to
prevent the release of nicked or broken DNA that could possibly result in chromosome
damage [127]. After passage of one or more DNA strands through this transient break(s), the
topoisomerase re-ligates the strands leaving the original DNA sequence intact. Though all
topoisomerases have this feature in common, topoisomerases are separated into two classes,
type I and type II, depending on whether they cleave one or two strands of DNA, respectively [127]. Type I topoisomerases act on one strand, and generally pass a single DNA
strand through the transient break, while type II topoisomerases break both DNA strands
and generally pass dsDNA through the transient break. Type I topoisomerases generally
work in front of replication or transcription forks, to relax positive supercoils in a highly
processive manner; while type II topoisomerases are involved in untangling intertwined duplex DNA such as that found in newly replicated molecules or during chromosome resolution during cell division [30].

Topoisomerases have roles in each of the major replicative phases: initiation, fork progression and termination. During DNA replication in eukaryotes, topoisomerases have been observed to bind directly to the replication origin to aid in activation in the initiation phase [45,
127]. During strand synthesis, topoisomerases are required to alleviate compression of the
DNA helix caused by positive supercoiling that results from DNA unwinding, which is
mediated by replicative helicases [127]. Topoisomerases are also required for daughter
strand resolution. Eukaryotes rely on topoisomerase I (topo I) to fulfill the initiation and
elongation functions during DNA replication [127].

Human topo I is an ATP-independent, 100-kDa monomeric protein capable of relaxing positive or negative superhelical twists by making a transient single-strand break, thus relieving the tension generated by the replicative helicases during the DNA-unwinding process [61, 127, 135]. Topo I can be divided into four domains: the highly charged NH_2 –terminal domain; the conserved core domain; a short, positively charged linker domain, which links the N-terminal domain to the core domain; and the highly conserved COOH-terminal domain, which contains the active-site tyrosine [116]. Due to the topologically constrained nature of a circular dsDNA molecule, it is no surprise that topo I is required for the replication of the genomes of small circular double-stranded DNA viruses. The role of topo I in DNA replication of the small DNA circular DNA viruses was first noted when it was observed that the extent of DNA replication in SV40 DNA replication *in vitro* was limited by the level of topoisomerase activity; addition of topo I to crude extracts stimulated SV40 DNA replication *in vitro* [51]. This effect could have been due to an enhanced rate of chain elongation resulting from an increased efficiency of unlinking of the parental DNA strands [135], or merely due to the presence of limited levels of topo I in the extracts used. Ultimately it was shown that the DNA replication of SV40 and PV both require topo I [134, 136].

While the role of topoisomerases in DNA replication had always been presumed to be due to their need to resolve topological constraint, more recent studies have indicated that topo I plays additional, highly specific, roles in DNA replication of the small DNA viruses, SV40 and PV. Topo I appears to be involved in the very earliest stages of DNA replication, namely origin recognition. It is evident that topo I is stably associated with the initiation complex and is one of the first cellular proteins to be recruited to the initiation machinery [11,45]. Topo I was shown to preferentially associate with the fully formed Tag double hexamer initiation complexes and to be recruited to the initiation complex prior to the beginning of unwinding [11]. This stable association of topo I with Tag results in an increased specificity of Tag for duplex unwinding at the origin by inhibiting unwinding at non-origin sites [39]. Perhaps for this reason, topo I was observed to be required at initiation to stimulate DNA replication *in vitro*, and was shown to have no effect on replication if introduced during the elongation phase, indicating it enhanced the synthesis of fully replicated DNA molecules by forming essential interactions with Tag and enabling initiation [45,11]. In contrast, topo I specifically enhances origin binding of PV E1 several-fold, but has no effect on non-origin binding [14]. After origin binding, E1 recruits topo I to the replication fork through direct protein interactions and the relaxation activity is strongly enhanced [14,4]. This enhancement of topo I is critical to relax the supercoiling created by the progressing replication fork during the elongation phase of DNA replication. Notably, although topo I plays a significant role in where Tag unwinds the DNA, topo I does not activate origin binding or unwinding and does not structurally distort the DNA [39]. Nonetheless, the similarities in these findings indicate that topo I plays an active role in origin recognition/specificity for the replication of both of these small DNA viral systems. Moreover, following initiation, the topo I-helicase complex remains stably associated and moves with the replication fork during DNA replication [45].

Topoisomerases have been proposed to act together with DNA helicases as "swivelases", tightly coordinating DNA duplex unwinding with the topoisomerase relaxing activity during DNA replication [15, 30, 61]. With the progression of the replication fork and unwinding of duplex DNA, topo I is needed to release the torsion created by the progressing replication fork [37]. Optimally topo I should be present and its activity regulated to suit the pace of the helicase [37]. This suggested that there might be direct interactions between the helicases and topo I, and that might be modulation of function due to these interactions. The early finding that topo I was localized at SV40 DNA replication forks supported this concept [4], as did evidence that topo I played an important role in the elongation phase of SV40 DNA replication. Reports of the interactions between SV40 TAg and E1 with topo I were also consistent with the swivelase model [15, 133]. The demonstration that E1 stimulates the enzymatic activity of topo I up to seven-fold and that SV40 TAg also stimulates topo I activity (R. Clower and T. Melendy, unpublished results) provided the first evidence of the cooperative nature of this interaction predicted by the swivelase model [15]. Based on these studies it is clear that the viral helicases interact productively with topo I at DNA replication forks forming active coordinated swivelase molecular machines.

The physical interactions between the viral helicases and topo I have been investigated. In 1996, it was found that two independent regions of Tag, one N-terminal and one C-terminal, bind to the cap region of topo I (see Fig. 1), and binding can take place while DNA-bound. Similarly, for PV E1 it was also observed that topo I binds two distinct regions within E1, within E1's DNA binding domain (DBD) and at the C-terminus [15, 45]. The E1 C-terminal region was shown to enhance topo I relaxation activity, and to a lesser extent, so did a truncation that included the DBD with additional sequence, flanking either side of the DBD [15]. More detailed studies identified mutants in the DNA binding domain of Tag that were unable to unwind the DNA and were partially defective in their association with topo I, suggesting that this interaction maybe important for proper unwinding of viral DNA at replication forks [114]. More recently, four specific amino acid residues within the C-terminal domain of Tag when mutated were shown to exhibit decreased topo I binding and to abolish SV40 DNA replication *in vitro* and to have dramatic effects on virus production *in vivo* [61]. These were the critical results that conclusively demonstrated the vital nature of the helicase-topo I interaction for DNA replication. Though first only believed to be involved in the relaxation of overwound DNA during replication fork progression, topo I has proven to be an integral part of the entire replication process in SV40 DNA replication, including critical roles in initiation and even in RNA-DNA primer synthesis in the elongation phase [37, 60, 61, 123]. Though less well-studied, topo I has been observed to be similarly important in these stages of PV DNA replication. These viral systems are vital models for eukaryotic DNA replication, and as of yet these biochemical studies cannot be recapitulated for cellular DNA replication. The only evidence to date of corroboration of these findings for chromosomal DNA replication is the co-purification of topo I with the GINS-MCM complex [39].

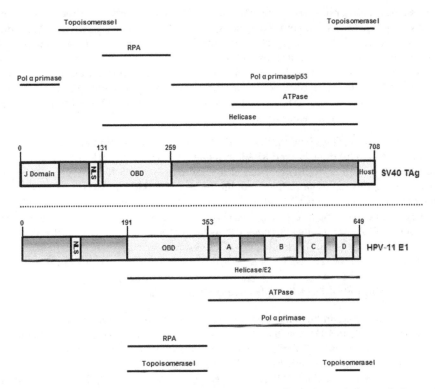

Figure 1. General replication domains of the SV40 T-antigen and papillomavirus E1 helicases. The known domains for SV40 Tag (upper) and HPV-11 E1 (lower) are indicated by horizontal lines. Four domains in E1 have limited homology with SV40 T-antigen (regions A D). A nuclear localization signal has only been elucidated for BPV-1 E1, therefore the HPV-11 NLS is currently only speculated to be in the analogous sequence area.

7. Helicase interactions with replication proteins that initiate elongation: Replication protein A

One of the first proteins identified as necessary for eukaryotic DNA replication is arguably also one of the most important DNA binding proteins in the cell, the ssDNA binding complex, Replication Protein A (RPA). RPA is a heterotrimeric complex conserved in all eukaryotes, and also shows strong homology to the ssDNA binding proteins of archaebacteria [57, 59]. The human RPA complex is comprised of three subunits, RPA70, RPA32, and RPA14, and the complex binds to ssDNA with extremely high affinity (approximately 10^{-9} to 10^{-10} M [62]), showing much lower affinity for dsDNA. RPA binds ssDNA with a distinct polarity, in a 5'->3' orientation [22, 51]. Like SSB [132], RPA is required for DNA replication *in vivo*; knockdown of the largest RPA subunit, RPA70, using siRNA results in inhibition of DNA synthesis [25]. The presence of RPA

as the ssDNA binding protein is critical in keeping the DNA double helix from reannealing during DNA replication, as well as protecting the exposed ssDNA from nuclease attack. And while
other non-related ssDNA binding proteins (such as *E. coli* or T4 SSB) can support some of these
functions (such as ssDNA stabilization and stimulation of the processive DNA polymerases)
RPA is specifically required for the early initiation steps of replication, including primer synthesis and stimulation of the DNA polymerase activity of DNA polymerase α primase [10, 81].
RPA is also involved in many DNA recombination and DNA repair pathways, acting as a central coordinator of DNA metabolism [52, 132].

RPA exhibits several DNA binding states. RPA70 has three ssDNA binding sites or oligonucleotide binding (OB) domains and RPA32 has one OB domain [8, 121]. When only RPA70
interacts, this is a lower affinity compacted state, binding to only 8-10 nts. When all four OB
domains bind, this represents a higher affinity extended mode that spans ~30 nts [7]. The
ability of other proteins to facilitate these binding modes in turn impact the binding of RPA
to ssDNA, either covering or exposing various stretches of ssDNA. Since several other proteins bind to RPA through its OB domains, this facilitates a model in which RPA cooperatively hands off and orients the binding of each DNA replication protein through increasing
affinity with the subsequent factor [64, 89, 138].

7.1. RPA loading onto ssDNA by replicative DNA helicases

RPA plays many roles in the initial steps of elongation as well as throughout DNA replication. Due to its role in ssDNA stabilization, RPA is one of first proteins required following
the unwinding of dsDNA. The critical question here is how this process is coordinated in
relation to the double hexameric helicase. The RPA heterotrimer itself makes direct contact
with the helicase, be it MCM, SV40 Tag, or PV E1 [3, 43, 77, 95, 101, 130]. The first such studied interaction was through Tag, which interacts with RPA through the helicase's origin
binding domain (OBD) (Figure 1). The importance of this interaction is implied by the absolute necessity for RPA for SV40 replication, RPA cannot be replaced by ssDNA binding proteins from *E. coli* or even RPA from *S. cerevisiae* [11, 58, 88]. In turn, RPA interacts with Tag
through both its RPA70 and possibly to a lesser degree its RPA32 subunits. In PV DNA replication, the E1 helicase interaction with RPA is also critical for viral DNA replication. E1 directly binds to RPA through its largest subunit, RPA70, but does not appear to bind to
RPA32 or RPA14 (unlike Tag which binds RPA70 and RPA32) [43, 52, 69, 77]. Similar to the
SV40 system, RPA binds to the PV E1 helicase through its major dsDNA binding domain
(Figure 1, Fisk JC and T. Melendy, unpublished data).

Evaluation of the multiple interactions between RPA, E1 and ssDNA in various combinations led to development of a novel model for how DNA helicases may 'load' ssDNA binding proteins onto ssDNA being displaced through helicase action [77]. RPA binds well to the
E1 protein, but only in the absence of free ssDNA. When RPA was prebound to short (~10
nt) stretches of ssDNA, thereby adopting the short compacted form of RPA, it still bound to
E1 as well as RPA not bound to DNA. However, when RPA was bound to longer ssDNA
templates (~30 nt or longer), consistent with RPA being in its fully-engaged extended form,
RPA would no longer bind to E1. This implied a 'releasing mechanism' by which the E1-

RPA interaction would be released upon RPA binding to ssDNA in RPA's extended form. Based on this data, a model was developed in which free, non-ssDNA-bound RPA is bound by E1. As the E1 helicase unwinds the dsDNA, producing ssDNA, it positions the RPA to bind to the newly exposed ssDNA, releasing RPA from the helicase complex (see Figure 2). As the helicase progresses, subsequent helicase monomers bring subsequent RPA molecules to the ssDNA continuously displaced by helicase action [77]. Very similar results were later shown for SV40 T-antigen, leading to a nearly identical model for RPA placement onto ssDNA during SV40 T-antigen helicase progression [9, 54]. Of course, this simplified model does not take into account topo I or polymerase α primase interactions, but it does suggest how the newly produced ssDNA can be rapidly coated with RPA to prevent reannealing or hairpin formation, and to protect from nuclease attack.

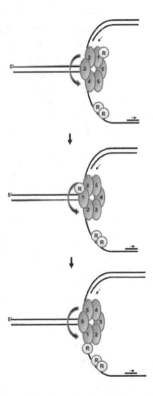

Figure 2. Generalized model for RPA deposition by replicative helicases. For simplicity, only a singular replication fork is shown. Free, unbound RPA interacts with a monomer of the helicase. As the helicase rotates relative to the DNA helix, the RPA bound monomer comes in juxtaposition to the freshly unwound ssDNA,. A 'hand off' occurs, whereby RPA binds to the ssDNA released by the helicase. Upon RPA binding ssDNA in the 'elongated' (~30 bp-bound) form, it can no longer be bound by the helicase monomer. This 'release' action leaves RPA bound to the newly exposed ssDNA, and allows the helicase to progress along the DNA template. As additional ssDNA is exposed this process is repeated, creating an array of RPA coating the ssDNA lagging strand template. (Adapted from [78].).

8. Helicase interactions with proteins that initiate elongation: DNA polymerase alpha-primase

In bacteria and the T4 bacteriophage, the importance of the primase is clear as they are linked physically to the helicase, which is necessary for efficient lagging strand synthesis [19, 92]. In the T7 bacteriophage, this is even more evident as the primase is actually fused to the functional hexameric helicase [31, 99, 102]. In a more complex fashion, in the mammalian system, GINS/ctf4 are required to link the helicase to the catalytic core of DNA polymerase (pol) alpha [40, 120, 140]. Clearly the interaction between primase and the helicase machinery is conserved throughout evolution.

Pol α primase was the first eukaryotic polymerase discovered in 1957 and was thought to be the only replicative DNA polymerase. The later discovery of the proofreading and highly processive polymerases δ and ε indicated that this was not the case [49, 50]. Pol α primase is a heterotetrameric complex comprised of a large p180 catalytic subunit, the regulatory p68 "B" subunit, and the two primase subunits of p55 and p49. Pol α primase is critical for first synthesizing an approximately 10 nt RNA primer, followed by a short ~20-30 nt DNA extension [23, 41, 119]. Polymerase switching then occurs on this RNA/DNA primer through the action of the eukaryotic clamp-loading complex, RFC, which loads the eukaryotic sliding clamp, PCNA, and then a processive DNA polymerase (DNA pol δ or ε) for synthesis of both leading and lagging DNA strands [124, 128]. RFC is integral here, as it competes with RPA for the end of the primer, disrupting the RPA-pol α interaction and allowing polymerase switching [138]. As with many of the core aspects of eukaryotic DNA replication, the functions of pol α primase were largely elucidated using the SV40 system. Pol α primase is absolutely essential for SV40 DNA replication *in vitro* [94]. Tag interacts with three subunits of pol α primase [17, 27-29, 48, 100, 129]. Recent work has demonstrated the importance of the Tag-p68 interaction for facilitating priming in both cell-free systems and in monkey cell culture [46, 139]. Mutations in Tag that abrogate the Tag-p68 interaction, but do not affect the interactions with p180 or primase, severely decrease priming (in the SV40 monopolymerase assay, which uses a plasmid with the SV40 ori, and purified Tag, pol α primase, RPA, and topo I, all of which are subjects of this review [86]). The amino residues in SV40 Tag shown to be critical for interaction with pol α p68 are H395, R548, K550, and K616, all of which are highly conserved between Tags from other polyomaviruses [139]. Interestingly, the helicase activity of Tag is dispensable for primosome activity (stimulation of priming by pol α primase), indicating that this effect on priming is likely due to the protein-protein interaction between the helicase and the pol, and not some indirect role of DNA helicase action [47]. In general, the Tag-pol α primase interaction mediates a process that allows the helicase to hand off the pol-primase to the ssDNA to enable primer synthesis [16, 35, 88]. Obviously there would also need to be interplay between the helicase and the two proteins competing for the exposed ssDNA, RPA and pol α primase.

The interactions between E1 and pol α primase show some differences between those found with SV40 Tag. Early studies indicated that the p180 catalytic subunit interacted with the N-terminal half of E1, while the p68 subunit interacted with the C-terminus of the helicase [18,

83]. A later study then looked closely at the role of the E1 interaction with pol α primase in regards to supporting HPV-11 DNA replication *in vitro* [2]. This study confirmed the earlier findings by indicating that E1 interacts with the pol p68 regulatory subunit through its C-terminal half (Figure 1). The presence of E2, whose trans-activation domain binds a similar region of E1, stimulates the E1-p68 interaction; but E2 and p68 nonetheless compete for [2, 83]. This is consistent with a step-wise mechanism whereby E2 helps E1 assemble into a functional helicase, which is then recognized by p68 of the pol α primase complex. No inter-action with the pol p180 subunit was detected in these latter studies. Whether this was due to subtle differences in the BPV-1 E1 used in the earlier study and HPV-11 E1 in the later studies has not been further investigated. Regardless, PV E1 appears to interact with pol α primase within the same E1 domain as the ATPase/helicase function. Further studies are necessary to determine if similar residues in E1 as those in SV40 Tag facilitate the binding to pol α primase; these studies may be beneficial as E1 may only use this subunit to bind and recruit pol α primase to the viral replication machinery.

9. Interactions between replication proteins that initiate elongation: coordination

While the earlier sections have alluded to interplay between the multiple cellular replication factors that interact with the viral helicases during DNA replication, the complexity of the interplay between these interactions is what truly epitomizes the term Molecular Machines.

9.1. RPA's involvement in de-repression of priming

While the interaction of the viral helicases with RPA has been shown to have a direct effect, apparently through the placement of RPA on the ssDNA being displaced by helicase action, this interaction has also been shown to play another vital role in DNA replication: de-repression of priming. RPA binds directly to pol α primase [10, 28, 96], and can stimulate the fidel-ity and processivity of pol α primase activity [10, 81]. However, when RPA is present in excess, which it is in human cell nuclei [76], RPA strongly represses synthesis of primers by pol α-primase, likely due to the high affinity of RPA out-competing pol α primase for the ssDNA template [16, 88]. While Tag and pol α primase are required for correct initiation of SV40 DNA replication [27, 130], and the interaction between Tag and pol α-primase is suffi-cient for stimulation of RNA/DNA primer synthesis by pol α-primase on ssDNA [16], these are insufficient for efficient primer synthesis when there is competition with ssDNA binding proteins. Tag can de-repress primer synthesis by pol α-primase, but only when the ssDNA template is coated by RPA, and not by other ssDNA binding proteins or evolutionarily di-vergent RPAs [88]. The interaction between Tag and RPA is vital for de-repression of pri-ming [88, 111]. E1 has similarly been shown to interact with RPA, and RPA is required for PV DNA replication (and RPA cannot be replaced by other ssDNA binding proteins in PV DNA replication). So while the E1-RPA interaction has not been shown to be essential for priming de-repression during PV DNA replication, this is nonetheless likely to be the case.

9.2. Topo I's involvement in priming

Similarly, in addition to its roles in origin recognition/specificity and release of DNA helix compression during elongation, another role for topo I was elucidated when it was observed that topo I induces pol α-primase to synthesize larger amounts of primers with
higher molecular weight [60]. In this study, Tag mutants that failed to bind topo I normally did not participate in the synthesis of expected amounts of primers or large molecular weight DNA molecules, indicating that the association of topo I with the C-terminal
Tag binding site is required for these processes. Whether this is due to a direct effect on
Tag function at the replication fork, or due to an indirect effect on pol α-primase through
Tag (analogous to the effect of the RPA-Tag effect on priming by pol α-primase described
above) is unclear. Additionally, topo I was shown to bind directly to RPA, and RPA
binds directly to pol α-primase, and can stimulate its DNA polymerase activity. It is unclear whether or not RPA may be influencing the interaction of Topo I with pol α-pri-
mase, or vice versa [60]. However these interactions are integrated, the binding of topo I
to the helicase domain of Tag significantly enhances the synthesis of DNA-RNA primers
and their extension by pol α-primase.

9.3. Helicase interactions with other proteins involved in elongation

What of helicase interaction with the other proteins involved in DNA replication elongation? In the model systems of SV40 and PV little has been elucidated about any direct interactions. Of the proteins involved in elongation, very little is known about the role of
helicase interaction with pol δ, RFC, PCNA, or the proteins involved in primer removal:
RNaseH, DNA2, Fen I, or DNA ligase I. In the accepted model of SV40 DNA replication,
the first primers synthesized by pol a primase on the two strands at the origin become
the primers for the leading strand of the opposite fork [124]. After recruitment of RFC,
PCNA and pol δ, the leading strand polymerase continuously tracks along behind the
helicase action. Since the helicase, in this case Tag, unwinds dsDNA at the relatively slow
rate of approximately 200 bp/min [93] while pol δ/RFC/PCNA polymerizes at about 80
nts/sec [12], it is reasonable to speculate that the slower speed of the helicase limits the
polymerase in such a way to coordinate the entire machinery mechanism. However, the
speeds of polymerases are often assayed on artificial templates, and this rate for pol
δ/RFC/PCNA is faster than the measured rate of eukaryotic replication forks (~ 2 kb/min).
Conversely the measured speed of Tag is far slower than the measured rate for eukaryotic replication forks. It is likely that coordination between the various factors and complexes involved in the replication fork lead to the final replication fork rate that is not
dependent on any one factor, but is a characteristic of the coordinated complex. Indeed, it
is critical that these machines are tightly regulated; without a tight molecular machine at
the fork, there would be wild exposure of ssDNA via the helicase leading to DNA damage signaling. It should also be noted here that DNA pol ε is not needed in SV40 DNA
replication [141]. This finding may be due to the lack of a need for two replicative helicases to duplicate small virus genomes. Alternately, DNA pol ε and TopBP1 (Dbp11) play
roles in initiation in mammalian replication; this role may be dispensable or even inter-

fere with the Tag/E1 initiator functions [82, 84]. In *E. coli* DNA replication it has been shown that the tau subunit actually links the leading strand DNA polymerase to the replicative helicase, dnaB [63] (which tracks along the lagging strand template, unlike the case for the SV40 and PV replicative helicases, that track along the leading strand template). It remains possible that these viral replicative helicases may have heretofore unobserved interactions with additional cellular factors involved in the elongation stages of replication that play important roles in DNA replication. This is a potential area for future study.

9.4. Extrapolation to the cellular chromosomal replication fork

The cellular 'replicative helicase' is still poorly defined. Some have designated the human CMG helicase (a large 11 subunit complex comprising Cdc45 and the MCMs and GINS sub-complexes [91]) to be the replicative helicase, while others have designated the RPC, the "replisome progression complex", comprised of the CMG in complex with Mrc1 (Claspin), Tof1 (Tim or Timeless), Csm3 (Swi3/Tipin), Ctf4 (And-1), and the FACT heterodimer (Spt16, and Pob3 (SSRP1) as the 'true replicative helicase' [39]. This study found that MCM10 and topo I associate weakly with this RPC complex, although it is unclear with which specific subunit. It is unknown if the MCM helicase itself interacts with topo I; however, considering the elaborate number of regulatory subunits now known in the eukaryotic helicase supercomplex, this may not be necessary, and may be unlikely. The GINS complex of CMG can bind to and directly stimulate the activity of pol α-primase [21]. A later study showed that the Ctf4 subunit couples the MCMs to pol α-primase and the Mrc1 subunit interacts with polymerase ε [40]. Other studies have found that both Mcm10 and Cdc45 interact with pol α-primase and also found that loss of Mcm10 in yeast led to uncoupling of the MCMs from pol α-primase and resulted in large stretches of ssDNA, a potent DNA damage signal [67, 107]. In human cells, Mcm10 has been suggested to interact with and regulate pol α-primase levels and prevent inappropriate induction of DNA damage [14]. RPA interacts with many components of the RPC, including Mcm3-7, Cdc45, and Claspin (Mrc1) and requires Mcm for chromatin localization [95]. It is intriguing that only RPA appears to directly interact with the Mcms in eukaryotes; this may be due to the intimate linkage with ssDNA and the helicase machine and the highest priority of multicellular organisms to prevent the aberrant signaling of DNA damage through ssDNA coating by RPA. Additionally, in the absence of the RPC interacting protein Mcm10 or in the presence of a mutant zinc finger bearing Mcm10, RPA is also prevented from loading [55]. In general, the major components of the elongation machinery interact with the replicative helicase in eukaryotes through multiple layers of regulation as the RPC complex, a feature that is nonexistent in the simplified machinery presented by these small DNA viral systems. These viral factories simplify the entire complex by using their own central multifunctional helicases. But this simplification has led to the ability to use these viral systems as models where the biochemical nature and functions of these important interactions that occur at the interface of initiation and elongation can be studied.

10. Conclusion

Replicative DNA helicases, modeled by the SV40 and PV DNA replication systems, play complex roles coordinating the multiple actions of multiple DNA replication factors at eukaryotic replication forks. Their interactions with topo I are involved in origin recognition/ specificity, DNA helix decompression function, and primer synthesis. Their interactions with pol α-primase are vital for primer synthesis. Their interactions with RPA are involved in loading of RPA onto ssDNA, and de-repression of priming on RPA-coated ssDNA. And the complex interplay between all these factors is intricate, highly-regulated, and appears to be coordinated at least in large part, through the action of the replicative helicases.

Using this wealth of knowledge about the viral replication forks, we have assembled a likely model of replication elongation using the viral helicases as the central molecular machine at the fork. For ease of the various steps of elongation, only a single helicase is pictured in this model (Figure 3). Following assembly of the replication machinery at the viral origin, there is a very intricate four-way interaction comprised of the helicase, topo I, RPA and pol α primase. Topo I has two interactions with helicase; one within the N-terminal half of the helicase and one within the C-terminus. Through these interactions the topo I-helicase interaction assists in helicase origin recognition and creates the swivelase; a machine that couples the unwinding of the DNA duplex with relaxation of torsional stress. During elongation, topo I is likely in front of the helicase to facilitate the easing of positive supercoiling, likely through interactions with the helicase N-terminus. The helicase encircles the leading strand of the newly unwound DNA, actively pumping the leading strand template through the central channel of the helicase and away from the lagging strand replication machinery. While the leading strand template is bound to the central channel and the helicase domain, the lagging strand template is therefore left relatively unprotected. To facilitate a protective role at this point, the OBD of the helicase binds to free RPA, which swings into place as the helicase turns, actively loading RPA onto the lagging strand template. This serves in the role of nuclease protection, as well as preventing aberrant ssDNA structures. However, this coating of the lagging strand template is counterproductive to the process of priming. Therefore, at regular intervals roughly equivalent to the length an Okazaki fragment, the helicase interacts with pol α primase and RPA to facilitate the placement of the pol α primase onto the template, possibly while simultaneously removing RPA in a localized fashion, so that pol α primase can synthesize the RNA-DNA primer. It is intriguing to speculate that it is through this regular placement that Okazaki fragments are placed and spaced; primarily through helicase action and its protein-protein interactions with the primase. Although given the size of eukaryotic Okazaki fragments, it is likely that interactions with histones may play a role as well. The coordinated and highly regulated roles of the multi-subunit DNA helicase in modulating the proteins and their protein-protein interactions involved in the late initiation and elongation stages of DNA replication clearly play a central organizing role in the molecular machine that is the eukaryotic DNA replication fork.

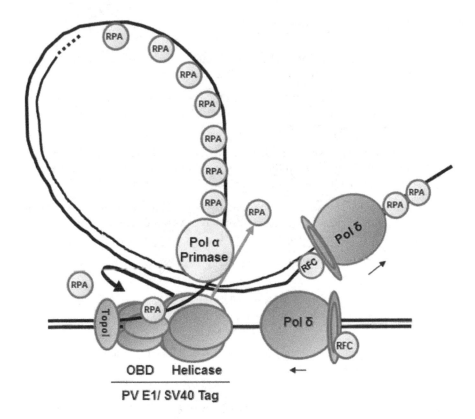

Figure 3. Proposed Model for the PV/SV40 DNA Replication Fork. Using the proposed helicase model presented in [33], the replicative helicase is shown oriented with the N-terminal OBD facing towards the unwound dsDNA. For simplicity, only one of the two hexamers is shown. The interaction of topo I with the OBD of the helicase both assists it in origin binding/specificity, and targets it to the incoming dsDNA, where topoisomerase action is vital for replication fork progression. The interaction of RPA within the OBD is involved in the process of directing loading RPA onto the ssDNA newly exposed by helicase action. The interaction of the helicase domain with pol α primase stimulates primer synthesis; and the interaction of the helicase with RPA allows for pol α primase to synthesize primers even in the presence of RPA, through localized RPA removal or 'priming de-repression'. As each primer is synthesized, RFC, in coordination with RPA, loads PCNA and DNA pol δ onto the 3' DNA end to allow for processive DNA synthesis. The various interactions of the helicase with topo I, RPA, and DNA pol α primase, as well as other interactions between the cellular factors themselves, coordinately the complex interplay necessary for replication fork function.

Author details

John C. Fisk, Michaelle D. Chojnacki and Thomas Melendy

Department of Microbiology & Immunology, University at Buffalo School of Medicine & Biomedical Sciences, Buffalo, NY, USA

References

[1] Alberts, B. 1998. The cell as a collection of protein machines: preparing the next generation of molecular biologists. Cell 92:291-294.

[2] Amin, A. A., S. Titolo, A. Pelletier, D. Fink, M. G. Cordingley, and J. Archambault. 2000. Identification of domains of the HPV11 E1 protein required for DNA replication in vitro. Virology 272:137-150.

[3] Arunkumar, A. I., V. Klimovich, X. Jiang, R. D. Ott, L. Mizoue, E. Fanning, and W. J. Chazin. 2005. Insights into hRPA32 C-terminal domain--mediated assembly of the simian virus 40 replisome. Nature Structural & Molecular Biology 12:332-339.

[4] Avemann, K., R. Knippers, T. Koller, and J. M. Sogo. 1988. Camptothecin, a specific inhibitor of type I DNA topoisomerase, induces DNA breakage at replication forks. Molecular and Cellular Biology 8:3026-3034.

[5] Bell, S. P., and A. Dutta. 2002. DNA replication in eukaryotic cells. Annual Review of Biochemistry 71:333-374.

[6] Bjergbaek, L., J. A. Cobb, M. Tsai-Pflugfelder, and S. M. Gasser. 2005. Mechanistically distinct roles for Sgs1p in checkpoint activation and replication fork maintenance. Embo J 24:405-417.

[7] Blackwell, L. J., J. A. Borowiec, and I. A. Mastrangelo. 1996. Single-stranded-DNA binding alters human replication protein A structure and facilitates interaction with DNA-dependent protein kinase. Molecular and Cellular Biology 16:4798-4807.

[8] Bochkarev, A., and E. Bochkareva. 2004. From RPA to BRCA2: lessons from single-stranded DNA binding by the OB-fold. Curr Opin Struct Biol 14:36-42.

[9] Bochkareva, E., D. Martynowski, A. Seitova, and A. Bochkarev. 2006. Structure of the origin-binding domain of simian virus 40 large T antigen bound to DNA. Embo J 25:5961-5969.

[10] Braun, K. A., Y. Lao, Z. He, C. J. Ingles, and M. S. Wold. 1997. Role of protein-protein interactions in the function of replication protein A (RPA): RPA modulates the activity of DNA polymerase alpha by multiple mechanisms. Biochemistry 36:8443-8454.

[11] Brill, S. J., and B. Stillman. 1989. Yeast replication factor-A functions in the unwinding of the SV40 origin of DNA replication. Nature 342:92-95.

[12] Burgers, P. M., and K. J. Gerik. 1998. Structure and processivity of two forms of Saccharomyces cerevisiae DNA polymerase delta. The Journal of Biological Chemistry 273:19756-19762.

[13] Champoux, J. J. 2001. DNA topoisomerases: structure, function, and mechanism. Annual Review of Biochemistry 70:369-413.

[14] Chattopadhyay, S., and A. K. Bielinsky. 2007. Human Mcm10 regulates the catalytic subunit of DNA polymerase-alpha and prevents DNA damage during replication. Molecular Biology of the Cell 18:4085-4095.

[15] Clower, R. V., J. C. Fisk, and T. Melendy. 2006. Papillomavirus E1 protein binds to and stimulates human topoisomerase I. Journal of Virology 80:1584-1587.

[16] Collins, K. L., and T. J. Kelly. 1991. Effects of T antigen and replication protein A on the initiation of DNA synthesis by DNA polymerase alpha-primase. Molecular and Cellular Biology 11:2108-2115.

[17] Collins, K. L., A. A. Russo, B. Y. Tseng, and T. J. Kelly. 1993. The role of the 70 kDa subunit of human DNA polymerase alpha in DNA replication. Embo J 12:4555-4566.

[18] Conger, K. L., J. S. Liu, S. R. Kuo, L. T. Chow, and T. S. Wang. 1999. Human papillomavirus DNA replication. Interactions between the viral E1 protein and two subunits of human dna polymerase alpha/primase. The Journal of Biological Chemistry 274:2696-2705.

[19] Corn, J. E., and J. M. Berger. 2006. Regulation of bacterial priming and daughter strand synthesis through helicase-primase interactions. Nucleic Acids Research 34:4082-4088.

[20] Davey, M. J., D. Jeruzalmi, J. Kuriyan, and M. O'Donnell. 2002. Motors and switches: AAA+ machines within the replisome. Nat Rev Mol Cell Biol 3:826-835.

[21] De Falco, M., E. Ferrari, M. De Felice, M. Rossi, U. Hubscher, and F. M. Pisani. 2007. The human GINS complex binds to and specifically stimulates human DNA polymerase alpha-primase. EMBO Reports 8:99-103.

[22] de Laat, W. L., E. Appeldoorn, K. Sugasawa, F. Weterings, N. G. Jaspers, and J. H. Hoeijmakers. 1998. DNA-binding polarity of human replication protein A positions nucleases in nucleotide excision repair. Genes & Development 12:2598-2609.

[23] Denis, D., and P. A. Bullock. 1993. Primer-DNA formation during simian virus 40 DNA replication in vitro. Molecular and Cellular Biology 13:2882-2890.

[24] Diffley, J. F., and K. Labib. 2002. The chromosome replication cycle. Journal of Cell Science 115:869-872.

[25] Dodson, G. E., Y. Shi, and R. S. Tibbetts. 2004. DNA replication defects, spontaneous DNA damage, and ATM-dependent checkpoint activation in replication protein A-deficient cells. The Journal of Biological Chemistry 279:34010-34014.

[26] Donovan, S., J. Harwood, L. S. Drury, and J. F. Diffley. 1997. Cdc6p-dependent loading of Mcm proteins onto pre-replicative chromatin in budding yeast. Proceedings of the National Academy of Sciences of the United States of America 94:5611-5616.

[27] Dornreiter, I., W. C. Copeland, and T. S. Wang. 1993. Initiation of simian virus 40 DNA replication requires the interaction of a specific domain of human DNA polymerase alpha with large T antigen. Molecular and Cellular Biology 13:809-820.

[28] Dornreiter, I., L. F. Erdile, I. U. Gilbert, D. von Winkler, T. J. Kelly, and E. Fanning. 1992. Interaction of DNA polymerase alpha-primase with cellular replication protein A and SV40 T antigen. Embo J 11:769-776.

[29] Dornreiter, I., A. Hoss, A. K. Arthur, and E. Fanning. 1990. SV40 T antigen binds directly to the large subunit of purified DNA polymerase alpha. Embo J 9:3329-3336.

[30] Duguet, M. 1997. When helicase and topoisomerase meet! Journal of Cell Science 110 (Pt 12):1345-1350.

[31] Egelman, E. H., X. Yu, R. Wild, M. M. Hingorani, and S. S. Patel. 1995. Bacteriophage T7 helicase/primase proteins form rings around single-stranded DNA that suggest a general structure for hexameric helicases. Proceedings of the National Academy of Sciences of the United States of America 92:3869-3873.

[32] Enemark, E. J., and L. Joshua-Tor. 2006. Mechanism of DNA translocation in a replicative hexameric helicase. Nature 442:270-275.

[33] Enemark, E. J., and L. Joshua-Tor. 2008. On helicases and other motor proteins. Curr Opin Struct Biol 18:243-257.

[34] Evrin, C., P. Clarke, J. Zech, R. Lurz, J. Sun, S. Uhle, H. Li, B. Stillman, and C. Speck. 2009. A double-hexameric MCM2-7 complex is loaded onto origin DNA during licensing of eukaryotic DNA replication. Proceedings of the National Academy of Sciences of the United States of America 106:20240-20245.

[35] Fanning, E., V. Klimovich, and A. R. Nager. 2006. A dynamic model for replication protein A (RPA) function in DNA processing pathways. Nucleic Acids Research 34:4126-4137.

[36] Fouts, E. T., X. Yu, E. H. Egelman, and M. R. Botchan. 1999. Biochemical and electron microscopic image analysis of the hexameric E1 helicase. The Journal of Biological Chemistry 274:4447-4458.

[37] Gai, D., R. Roy, C. Wu, and D. T. Simmons. 2000. Topoisomerase I associates specifically with simian virus 40 large-T-antigen double hexamer-origin complexes. Journal of Virology 74:5224-5232.

[38] Gai, D., R. Zhao, D. Li, C. V. Finkielstein, and X. S. Chen. 2004. Mechanisms of conformational change for a replicative hexameric helicase of SV40 large tumor antigen. Cell 119:47-60.

[39] Gambus, A., R. C. Jones, A. Sanchez-Diaz, M. Kanemaki, F. van Deursen, R. D. Edmondson, and K. Labib. 2006. GINS maintains association of Cdc45 with MCM in replisome progression complexes at eukaryotic DNA replication forks. Nat Cell Biol 8:358-366.

[40] Gambus, A., F. van Deursen, D. Polychronopoulos, M. Foltman, R. C. Jones, R. D. Edmondson, A. Calzada, and K. Labib. 2009. A key role for Ctf4 in coupling the

MCM2-7 helicase to DNA polymerase alpha within the eukaryotic replisome. Embo J 28:2992-3004.

[41] Garg, P., and P. M. Burgers. 2005. DNA polymerases that propagate the eukaryotic DNA replication fork. Crit Rev Biochem Mol Biol 40:115-128.

[42] Gomez-Lorenzo, M. G., M. Valle, J. Frank, C. Gruss, C. O. Sorzano, X. S. Chen, L. E. Donate, and J. M. Carazo. 2003. Large T antigen on the simian virus 40 origin of replication: a 3D snapshot prior to DNA replication. Embo J 22:6205-6213.

[43] Han, Y., Y. M. Loo, K. T. Militello, and T. Melendy. 1999. Interactions of the papovavirus DNA replication initiator proteins, bovine papillomavirus type 1 E1 and simian virus 40 large T antigen, with human replication protein A. Journal of Virology 73:4899-4907.

[44] Hodgson, B., A. Calzada, and K. Labib. 2007. Mrc1 and Tof1 regulate DNA replication forks in different ways during normal S phase. Molecular Biology of the Cell 18:3894-3902.

[45] Hu, Y., R. V. Clower, and T. Melendy. 2006. Cellular topoisomerase I modulates origin binding by bovine papillomavirus type 1 E1. Journal of Virology 80:4363-4371.

[46] Huang, H., B. E. Weiner, H. Zhang, B. E. Fuller, Y. Gao, B. M. Wile, K. Zhao, D. R. Arnett, W. J. Chazin, and E. Fanning. 2010. Structure of a DNA polymerase alpha-primase domain that docks on the SV40 helicase and activates the viral primosome. The Journal of Biological Chemistry 285:17112-17122.

[47] Huang, H., K. Zhao, D. R. Arnett, and E. Fanning. 2010. A specific docking site for DNA polymerase {alpha}-primase on the SV40 helicase is required for viral primosome activity, but helicase activity is dispensable. The Journal of Biological Chemistry 285:33475-33484.

[48] Huang, S. G., K. Weisshart, I. Gilbert, and E. Fanning. 1998. Stoichiometry and mechanism of assembly of SV40 T antigen complexes with the viral origin of DNA replication and DNA polymerase alpha-primase. Biochemistry 37:15345-15352.

[49] Hubscher, U., G. Maga, and S. Spadari. 2002. Eukaryotic DNA polymerases. Annual Review of Biochemistry 71:133-163.

[50] Hubscher, U., H. P. Nasheuer, and J. E. Syvaoja. 2000. Eukaryotic DNA polymerases, a growing family. Trends in Biochemical Sciences 25:143-147.

[51] Iftode, C., and J. A. Borowiec. 2000. 5' --> 3' molecular polarity of human replication protein A (hRPA) binding to pseudo-origin DNA substrates. Biochemistry 39:11970-11981.

[52] Iftode, C., Y. Daniely, and J. A. Borowiec. 1999. Replication protein A (RPA): the eukaryotic SSB. Crit Rev Biochem Mol Biol 34:141-180.

[53] Ishimi, Y. 1997. A DNA helicase activity is associated with an MCM4, -6, and -7 protein complex. The Journal of Biological Chemistry 272:24508-24513.

[54] Jiang, X., V. Klimovich, A. I. Arunkumar, E. B. Hysinger, Y. Wang, R. D. Ott, G. D. Guler, B. Weiner, W. J. Chazin, and E. Fanning. 2006. Structural mechanism of RPA loading on DNA during activation of a simple pre-replication complex. Embo J 25:5516-5526.

[55] Kanke, M., Y. Kodama, T. S. Takahashi, T. Nakagawa, and H. Masukata. 2012. Mcm10 plays an essential role in origin DNA unwinding after loading of the CMG components. Embo J 31:2182-2194.

[56] Katou, Y., Y. Kanoh, M. Bando, H. Noguchi, H. Tanaka, T. Ashikari, K. Sugimoto, and K. Shirahige. 2003. S-phase checkpoint proteins Tof1 and Mrc1 form a stable replication-pausing complex. Nature 424:1078-1083.

[57] Kelly, T. J., P. Simancek, and G. S. Brush. 1998. Identification and characterization of a single-stranded DNA-binding protein from the archaeon Methanococcus jannaschii. Proceedings of the National Academy of Sciences of the United States of America 95:14634-14639.

[58] Kenny, M. K., S. H. Lee, and J. Hurwitz. 1989. Multiple functions of human single-stranded-DNA binding protein in simian virus 40 DNA replication: single-strand stabilization and stimulation of DNA polymerases alpha and delta. Proceedings of the National Academy of Sciences of the United States of America 86:9757-9761.

[59] Kerr, I. D., R. I. Wadsworth, L. Cubeddu, W. Blankenfeldt, J. H. Naismith, and M. F. White. 2003. Insights into ssDNA recognition by the OB fold from a structural and thermodynamic study of Sulfolobus SSB protein. Embo J 22:2561-2570.

[60] Khopde, S., R. Roy, and D. T. Simmons. 2008. The binding of topoisomerase I to T antigen enhances the synthesis of RNA-DNA primers during simian virus 40 DNA replication. Biochemistry 47:9653-9660.

[61] Khopde, S., and D. T. Simmons. 2008. Simian virus 40 DNA replication is dependent on an interaction between topoisomerase I and the C-terminal end of T antigen. Journal of Virology 82:1136-1145.

[62] Kim, C., B. F. Paulus, and M. S. Wold. 1994. Interactions of human replication protein A with oligonucleotides. Biochemistry 33:14197-14206.

[63] Kim, S., H. G. Dallmann, C. S. McHenry, and K. J. Marians. 1996. Coupling of a replicative polymerase and helicase: a tau-DnaB interaction mediates rapid replication fork movement. Cell 84:643-650.

[64] Kowalczykowski, S. C. 2000. Initiation of genetic recombination and recombination-dependent replication. Trends in Biochemical Sciences 25:156-165.

[65] Kunkel, T. A., and P. M. Burgers. 2008. Dividing the workload at a eukaryotic replication fork. Trends Cell Biol 18:521-527.

[66] Kuo, S. R., J. S. Liu, T. R. Broker, and L. T. Chow. 1994. Cell-free replication of the human papillomavirus DNA with homologous viral E1 and E2 proteins and human cell extracts. The Journal of Biological Chemistry 269:24058-24065.

[67] Lee, C., I. Liachko, R. Bouten, Z. Kelman, and B. K. Tye. 2010. Alternative mechanisms for coordinating polymerase alpha and MCM helicase. Molecular and Cellular Biology 30:423-435.

[68] Lee, J. K., and J. Hurwitz. 2001. Processive DNA helicase activity of the minichromosome maintenance proteins 4, 6, and 7 complex requires forked DNA structures. Proceedings of the National Academy of Sciences of the United States of America 98:54-59.

[69] Lee, S. H., and D. K. Kim. 1995. The role of the 34-kDa subunit of human replication protein A in simian virus 40 DNA replication in vitro. The Journal of Biological Chemistry 270:12801-12807.

[70] Lei, M., and B. K. Tye. 2001. Initiating DNA synthesis: from recruiting to activating the MCM complex. Journal of Cell Science 114:1447-1454.

[71] Li, D., R. Zhao, W. Lilyestrom, D. Gai, R. Zhang, J. A. DeCaprio, E. Fanning, A. Jochimiak, G. Szakonyi, and X. S. Chen. 2003. Structure of the replicative helicase of the oncoprotein SV40 large tumour antigen. Nature 423:512-518.

[72] Lin, B. Y., A. M. Makhov, J. D. Griffith, T. R. Broker, and L. T. Chow. 2002. Chaperone proteins abrogate inhibition of the human papillomavirus (HPV) E1 replicative helicase by the HPV E2 protein. Molecular and Cellular Biology 22:6592-6604.

[73] Liu, J. S., S. R. Kuo, A. M. Makhov, D. M. Cyr, J. D. Griffith, T. R. Broker, and L. T. Chow. 1998. Human Hsp70 and Hsp40 chaperone proteins facilitate human papillomavirus-11 E1 protein binding to the origin and stimulate cell-free DNA replication. The Journal of Biological Chemistry 273:30704-30712.

[74] Liu, Y., H. I. Kao, and R. A. Bambara. 2004. Flap endonuclease 1: a central component of DNA metabolism. Annual Review of Biochemistry 73:589-615.

[75] Lohman, T. M. 1992. Escherichia coli DNA helicases: mechanisms of DNA unwinding. Molecular Microbiology 6:5-14.

[76] Loo, Y. M., and T. Melendy. 2000. The majority of human replication protein A remains complexed throughout the cell cycle. Nucleic Acids Research 28:3354-3360.

[77] Loo, Y. M., and T. Melendy. 2004. Recruitment of replication protein A by the papillomavirus E1 protein and modulation by single-stranded DNA. Journal of Virology 78:1605-1615.

[78] Lou, H., M. Komata, Y. Katou, Z. Guan, C. C. Reis, M. Budd, K. Shirahige, and J. L. Campbell. 2008. Mrc1 and DNA polymerase epsilon function together in linking DNA replication and the S phase checkpoint. Molecular Cell 32:106-117.

[79] Lusky, M., J. Hurwitz, and Y. S. Seo. 1993. Cooperative assembly of the bovine papilloma virus E1 and E2 proteins on the replication origin requires an intact E2 binding site. The Journal of Biological Chemistry 268:15795-15803.

[80] Ma, T., N. Zou, B. Y. Lin, L. T. Chow, and J. W. Harper. 1999. Interaction between cyclin-dependent kinases and human papillomavirus replication-initiation protein E1 is required for efficient viral replication. Proceedings of the National Academy of Sciences of the United States of America 96:382-387.

[81] Maga, G., I. Frouin, S. Spadari, and U. Hubscher. 2001. Replication protein A as a "fidelity clamp" for DNA polymerase alpha. The Journal of Biological Chemistry 276:18235-18242.

[82] Makiniemi, M., T. Hillukkala, J. Tuusa, K. Reini, M. Vaara, D. Huang, H. Pospiech, I. Majuri, T. Westerling, T. P. Makela, and J. E. Syvaoja. 2001. BRCT domain-containing protein TopBP1 functions in DNA replication and damage response. The Journal of Biological Chemistry 276:30399-30406.

[83] Masterson, P. J., M. A. Stanley, A. P. Lewis, and M. A. Romanos. 1998. A C-terminal helicase domain of the human papillomavirus E1 protein binds E2 and the DNA polymerase alpha-primase p68 subunit. Journal of Virology 72:7407-7419.

[84] Masumoto, H., A. Sugino, and H. Araki. 2000. Dpb11 controls the association between DNA polymerases alpha and epsilon and the autonomously replicating sequence region of budding yeast. Molecular and Cellular Biology 20:2809-2817.

[85] Matson, S. W., and K. A. Kaiser-Rogers. 1990. DNA helicases. Annual Review of Biochemistry 59:289-329.

[86] Matsumoto, T., T. Eki, and J. Hurwitz. 1990. Studies on the initiation and elongation reactions in the simian virus 40 DNA replication system. Proceedings of the National Academy of Sciences of the United States of America 87:9712-9716.

[87] Melendy, T., J. Sedman, and A. Stenlund. 1995. Cellular factors required for papillomavirus DNA replication. Journal of Virology 69:7857-7867.

[88] Melendy, T., and B. Stillman. 1993. An interaction between replication protein A and SV40 T antigen appears essential for primosome assembly during SV40 DNA replication. The Journal of Biological Chemistry 268:3389-3395.

[89] Mer, G., A. Bochkarev, W. J. Chazin, and A. M. Edwards. 2000. Three-dimensional structure and function of replication protein A. Cold Spring Harbor Symposia on Quantitative Biology 65:193-200.

[90] Mohr, I. J., R. Clark, S. Sun, E. J. Androphy, P. MacPherson, and M. R. Botchan. 1990. Targeting the E1 replication protein to the papillomavirus origin of replication by complex formation with the E2 transactivator. Science 250:1694-1699.

[91] Moyer, S. E., P. W. Lewis, and M. R. Botchan. 2006. Isolation of the Cdc45/Mcm2-7/ GINS (CMG) complex, a candidate for the eukaryotic DNA replication fork helicase.

Proceedings of the National Academy of Sciences of the United States of America 103:10236-10241.

[92] Mueser, T. C., J. M. Hinerman, J. M. Devos, R. A. Boyer, and K. J. Williams. 2010. Structural analysis of bacteriophage T4 DNA replication: a review in the Virology Journal series on bacteriophage T4 and its relatives. Virol J 7:359.

[93] Murakami, Y., and J. Hurwitz. 1993. DNA polymerase alpha stimulates the ATP-dependent binding of simian virus tumor T antigen to the SV40 origin of replication. The Journal of Biological Chemistry 268:11018-11027.

[94] Murakami, Y., C. R. Wobbe, L. Weissbach, F. B. Dean, and J. Hurwitz. 1986. Role of DNA polymerase alpha and DNA primase in simian virus 40 DNA replication in vitro. Proceedings of the National Academy of Sciences of the United States of America 83:2869-2873.

[95] Nakaya, R., J. Takaya, T. Onuki, M. Moritani, N. Nozaki, and Y. Ishimi. 2010. Identification of proteins that may directly interact with human RPA. Journal of Biochemistry 148:539-547.

[96] Nasheuer, H. P., D. von Winkler, C. Schneider, I. Dornreiter, I. Gilbert, and E. Fanning. 1992. Purification and functional characterization of bovine RP-A in an in vitro SV40 DNA replication system. Chromosoma 102:S52-59.

[97] Neuwald, A. F., L. Aravind, J. L. Spouge, and E. V. Koonin. 1999. AAA+: A class of chaperone-like ATPases associated with the assembly, operation, and disassembly of protein complexes. Genome Research 9:27-43.

[98] Nick McElhinny, S. A., D. A. Gordenin, C. M. Stith, P. M. Burgers, and T. A. Kunkel. 2008. Division of labor at the eukaryotic replication fork. Molecular Cell 30:137-144.

[99] Notarnicola, S. M., K. Park, J. D. Griffith, and C. C. Richardson. 1995. A domain of the gene 4 helicase/primase of bacteriophage T7 required for the formation of an active hexamer. The Journal of Biological Chemistry 270:20215-20224.

[100] Ott, R. D., C. Rehfuess, V. N. Podust, J. E. Clark, and E. Fanning. 2002. Role of the p68 subunit of human DNA polymerase alpha-primase in simian virus 40 DNA replication. Molecular and Cellular Biology 22:5669-5678.

[101] Park, C. J., J. H. Lee, and B. S. Choi. 2005. Solution structure of the DNA-binding domain of RPA from Saccharomyces cerevisiae and its interaction with single-stranded DNA and SV40 T antigen. Nucleic Acids Research 33:4172-4181.

[102] Patel, S. S., and M. M. Hingorani. 1993. Oligomeric structure of bacteriophage T7 DNA primase/helicase proteins. The Journal of Biological Chemistry 268:10668-10675.

[103] Pavlov, Y. I., and P. V. Shcherbakova. 2010. DNA polymerases at the eukaryotic fork-20 years later. Mutation Research 685:45-53.

[104] Piccolino, M. 2000. Biological machines: from mills to molecules. Nat Rev Mol Cell Biol 1:149-153.

[105] Pike, J. E., R. A. Henry, P. M. Burgers, J. L. Campbell, and R. A. Bambara. 2010. An alternative pathway for Okazaki fragment processing: resolution of fold-back flaps by Pif1 helicase. The Journal of Biological Chemistry 285:41712-41723.

[106] Pursell, Z. F., I. Isoz, E. B. Lundstrom, E. Johansson, and T. A. Kunkel. 2007. Yeast DNA polymerase epsilon participates in leading-strand DNA replication. Science 317:127-130.

[107] Ricke, R. M., and A. K. Bielinsky. 2004. Mcm10 regulates the stability and chromatin association of DNA polymerase-alpha. Molecular Cell 16:173-185.

[108] Rossi, M. L., V. Purohit, P. D. Brandt, and R. A. Bambara. 2006. Lagging strand replication proteins in genome stability and DNA repair. Chem Rev 106:453-473.

[109] Sanders, C. M., O. V. Kovalevskiy, D. Sizov, A. A. Lebedev, M. N. Isupov, and A. A. Antson. 2007. Papillomavirus E1 helicase assembly maintains an asymmetric state in the absence of DNA and nucleotide cofactors. Nucleic Acids Research 35:6451-6457.

[110] Sanders, C. M., and A. Stenlund. 1998. Recruitment and loading of the E1 initiator protein: an ATP-dependent process catalysed by a transcription factor. Embo J 17:7044-7055.

[111] Schneider, C., K. Weisshart, L. A. Guarino, I. Dornreiter, and E. Fanning. 1994. Species-specific functional interactions of DNA polymerase alpha-primase with simian virus 40 (SV40) T antigen require SV40 origin DNA. Molecular and Cellular Biology 14:3176-3185.

[112] Sclafani, R. A., and T. M. Holzen. 2007. Cell cycle regulation of DNA replication. Annu Rev Genet 41:237-280.

[113] Sedman, J., and A. Stenlund. 1998. The papillomavirus E1 protein forms a DNA-dependent hexameric complex with ATPase and DNA helicase activities. J Virol 72:6893-6897.

[114] Simmons, D. T., T. Melendy, D. Usher, and B. Stillman. 1996. Simian virus 40 large T antigen binds to topoisomerase I. Virology 222:365-374.

[115] Speck, C., Z. Chen, H. Li, and B. Stillman. 2005. ATPase-dependent cooperative binding of ORC and Cdc6 to origin DNA. Nature Structural & Molecular Biology 12:965-971.

[116] Stewart, L., G. C. Ireton, and J. J. Champoux. 1996. The domain organization of human topoisomerase I. The Journal of Biological Chemistry 271:7602-7608.

[117] Stillman, B. 2005. Origin recognition and the chromosome cycle. FEBS Letters 579:877-884.

[118] Szyjka, S. J., C. J. Viggiani, and O. M. Aparicio. 2005. Mrc1 is required for normal progression of replication forks throughout chromatin in S. cerevisiae. Molecular Cell
19:691-697.

[119] Taljanidisz, J., R. S. Decker, Z. S. Guo, M. L. DePamphilis, and N. Sarkar. 1987. Initiation of simian virus 40 DNA replication in vitro: identification of RNA-primed nascent DNA chains. Nucleic Acids Research 15:7877-7888.

[120] Tanaka, H., Y. Katou, M. Yagura, K. Saitoh, T. Itoh, H. Araki, M. Bando, and K. Shirahige. 2009. Ctf4 coordinates the progression of helicase and DNA polymerase alpha.
Genes Cells 14:807-820.

[121] Theobald, D. L., R. M. Mitton-Fry, and D. S. Wuttke. 2003. Nucleic acid recognition
by OB-fold proteins. Annu Rev Biophys Biomol Struct 32:115-133.

[122] Tourriere, H., G. Versini, V. Cordon-Preciado, C. Alabert, and P. Pasero. 2005. Mrc1
and Tof1 promote replication fork progression and recovery independently of Rad53.
Molecular Cell 19:699-706.

[123] Trowbridge, P. W., R. Roy, and D. T. Simmons. 1999. Human topoisomerase I promotes initiation of simian virus 40 DNA replication in vitro. Mol Cell Biol
19:1686-1694.

[124] Tsurimoto, T., T. Melendy, and B. Stillman. 1990. Sequential initiation of lagging and
leading strand synthesis by two different polymerase complexes at the SV40 DNA
replication origin. Nature 346:534-539.

[125] Tye, B. K. 1999. MCM proteins in DNA replication. Annual Review of Biochemistry
68:649-686.

[126] Ustav, M., and A. Stenlund. 1991. Transient replication of BPV-1 requires two viral
polypeptides encoded by the E1 and E2 open reading frames. Embo J 10:449-457.

[127] Vos, S. M., E. M. Tretter, B. H. Schmidt, and J. M. Berger. 2011. All tangled up: how
cells direct, manage and exploit topoisomerase function. Nat Rev Mol Cell Biol
12:827-841.

[128] Waga, S., and B. Stillman. 1994. Anatomy of a DNA replication fork revealed by reconstitution of SV40 DNA replication in vitro. Nature 369:207-212.

[129] Weisshart, K., H. Forster, E. Kremmer, B. Schlott, F. Grosse, and H. P. Nasheuer.
2000. Protein-protein interactions of the primase subunits p58 and p48 with simian
virus 40 T antigen are required for efficient primer synthesis in a cell-free system.
The Journal of Biological Chemistry 275:17328-17337.

[130] Weisshart, K., P. Taneja, and E. Fanning. 1998. The replication protein A binding site
in simian virus 40 (SV40) T antigen and its role in the initial steps of SV40 DNA replication. Journal of Virology 72:9771-9781.

[131] Wilson, V. G., M. West, K. Woytek, and D. Rangasamy. 2002. Papillomavirus E1 proteins: form, function, and features. Virus Genes 24:275-290.

[132] Wold, M. S. 1997. Replication protein A: a heterotrimeric, single-stranded DNA-bind-
       ing protein required for eukaryotic DNA metabolism. Annual Review of Biochemis-
       try 66:61-92.

[133] Wun-Kim, K., R. Upson, W. Young, T. Melendy, B. Stillman, and D. T. Simmons.
       1993. The DNA-binding domain of simian virus 40 tumor antigen has multiple func-
       tions. Journal of Virology 67:7608-7611.

[134] Yang, L., R. Li, I. J. Mohr, R. Clark, and M. R. Botchan. 1991. Activation of BPV-1 rep-
       lication in vitro by the transcription factor E2. Nature 353:628-632.

[135] Yang, L., M. S. Wold, J. J. Li, T. J. Kelly, and L. F. Liu. 1987. Roles of DNA topoiso-
       merases in simian virus 40 DNA replication in vitro. Proceedings of the National
       Academy of Sciences of the United States of America 84:950-954.

[136] Yang, L., M. S. Wold, J. J. Li, T. J. Kelly, and L. F. Liu. 1987. Roles of DNA topoiso-
       merases in simian virus 40 DNA replication in vitro. Proceedings of the National
       Academy of Sciences of the United States of America 84:950-954.

[137] You, Z., Y. Komamura, and Y. Ishimi. 1999. Biochemical analysis of the intrinsic
       Mcm4-Mcm6-mcm7 DNA helicase activity. Molecular and Cellular Biology
       19:8003-8015.

[138] Yuzhakov, A., Z. Kelman, J. Hurwitz, and M. O'Donnell. 1999. Multiple competition
       reactions for RPA order the assembly of the DNA polymerase delta holoenzyme.
       Embo J 18:6189-6199.

[139] Zhou, B., D. R. Arnett, X. Yu, A. Brewster, G. A. Sowd, C. L. Xie, S. Vila, D. Gai, E.
       Fanning, and X. S. Chen. 2012. Structural basis for the interaction of a hexameric rep-
       licative helicase with the regulatory subunit of human DNA polymerase alpha-pri-
       mase. The Journal of Biological Chemistry.

[140] Zhu, W., C. Ukomadu, S. Jha, T. Senga, S. K. Dhar, J. A. Wohlschlegel, L. K. Nutt, S.
       Kornbluth, and A. Dutta. 2007. Mcm10 and And-1/CTF4 recruit DNA polymerase al-
       pha to chromatin for initiation of DNA replication. Genes & Development
       21:2288-2299.

[141] 141.Zlotkin, T., G. Kaufmann, Y. Jiang, M. Y. Lee, L. Uitto, J. Syvaoja, I. Dornreiter, E.
       Fanning, and T. Nethanel. 1996. DNA polymerase epsilon may be dispensable for
       SV40- but not cellular-DNA replication. Embo J 15:2298-2305.

Chapter 4

The MCM and RecQ Helicase Families: Ancient Roles in DNA Replication and Genomic Stability Lead to Distinct Roles in Human Disease

Dianne C. Daniel*, Ayuna V. Dagdanova and
Edward M. Johnson

Additional information is available at the end of the chapter

1. Introduction

1.1. Rationale for comparison of MCM and RecQ helicase families

DNA helicases are currently organized into superfamilies based on their sequence structures and 3-D conformations. Within each superfamily, there are members that have further evolved for specialized functions [1]. There is conservation of RecQ proteins from bacteria to humans. Whereas bacteria have one RecQ helicase, humans have evolved at least five different proteins [2]. The RecQ members belong to the helicase Superfamily II, and as such have the characteristic Rec fold [1]. In this chapter, we will focus on RecQ family members WRN, BLM and RECQL4 (RecQ protein-like 4), which is also referred to in the literature as RECQ4. Eukaryotic and archaeal MCMs belong to the helicase Superfamily VI, and have the AAA+ (ATPases associated with diverse cellular activities) fold [1, 3, 4, 5]. Both Rec and AAA+ folds are based on the ancestral ASCE (additional strand conserved E) fold or an alpha-beta-alpha domain necessary for nucleoside triphosphate binding and catalysis [1, 6, 7]. A rationale for comparison of the RecQ and MCM family members relates to the importance of their activities for genomic integrity. The WRN and BLM proteins as well as other members of the RecQ family are characterized by this feature [8]. Both WRN and BLM are involved in DNA repair and a role for WRN in telomere homeostasis in humans is well established [2, 9]. MCM2-7 proteins, along with cofactors, are thought to function as the eukaryotic replicative helicase [10]. MCM8 [11, 12] and MCM9 [13, 14] are more recently discovered and their roles are less well defined. Although data point to a role for MCM8 in DNA replication, that role may be specialized in higher organisms. In human cells, MCM10 is recruited to chromosomal domains before they replicate and studies in yeast suggest a role in DNA replication, but not as a helicase [15-18]. Members of each family are essential for chromosome homeostasis. When replication forks stall, there may be involvement of members of each family. These proteins have interlocking functions since, for example, a stalled replication fork with attendant MCM proteins can lead to a DNA double-strand break (DSB), which requires RecQ proteins for repair [19].

2. RecQ and MCM family structures

Breaks in double-strand DNA can occur during DNA replication, at specific loci (e.g., at telomeres) and during meiosis. RecQ helicases are implicated in DNA repair based on their involvement in such processes as DNA end resection, branch migration, D-loop processing, Holliday Junction (HJ) and double Holliday junction (dHJ) resolution [2]. RecQ proteins function at multiple steps, both early and late, during repair of DSB [19]. RecQ helicases travel in a 3' to 5' direction on ssDNA [20]. The RecQ proteins have an ancient lineage based on an ancestral ASCE fold ($\alpha\beta\alpha$ domain) of distant relation to P-loop NTPase folds. RecQ structural domains include a conserved core helicase domain for binding and hydrolysis of nucleoside triphosphate that is equivalent to the Walker A and Walker B boxes seen in MCM proteins [21]. They have a helicase and RNAase D C-terminal (HRDC) domain thought to mediate structure-specific nucleic acid binding, double HJ dissolution and protein-protein interactions. They also have a RecQ C-terminal (RQC) domain thought to mediate interactions with other proteins, structure-specific nucleic acid binding and metal cofactor binding [22]. Acidic regions present in many RecQ proteins aid in protein-protein interactions. There are also nuclear localization signals in some RecQ proteins (e.g., *H. sapiens* WRN and BLM). There are two RecQ members with an exonuclease domain, one of which is *H. sapiens* WRN. Two members have been functionally characterized as having an N-terminal strand exchange domain, one of which is *H. sapiens* BLM [2].

RECQL4 has been reported to have ssDNA binding and DNA strand-annealing activities. In this single study, RECQL4 did not display substrate unwinding or resolution of substrates resembling replication or recombination intermediates [23]. Recognizable HRDC and RQC domains that are important in BLM activity are missing in RECQL4 [22, 24]. As observed, the ssDNA annealing activity would allow RECQL4 to function during synthesis-dependent strand annealing (SDSA) along with another helicase. RECQL4 could help direct pathway choice during HR in DSB repair through aiding ssDNA annealing activity in non-homologous end joining (NHEJ) [19]. Thus, RECQL4 is similar to other RecQ helicases in its core helicase domain, but its function as a helicase is unclear [23, 25, 26].

AAA+ proteins, including MCMs, have a core molecular motor. Like the RecQ proteins, the AAA+ fold is also based on the ancestral ASCE fold. Acquisition of a catalytic glutamate (Fig. 1) to initiate efficient hydrolysis of ATP marked the emergence of the ASCE division from the ancestral P-loop fold [7]. For further discussion of the glutamate "switch" in AAA+ proteins, see reference [27]. Mechanisms of action are diverse, although members are typically oligomeric ring assemblies with inter-subunit communication. The central AAA+ motor has been adapted in evolution through structural changes to the core module and through domains added either N- or C-terminal to the AAA+ core. Activities are facilitated by recognition of protein partners functioning in these diverse events. Thus, AAA+ proteins display a variety of macromolecular remodelling events that are energy-driven by nucleotide hydrolysis thought to be occurring throughout what is typically a hexameric complex assembly [28, 29]. The conserved Walker A and Walker B motifs within the central module mediate ATP-binding and hydrolysis [7, 30-32]. MCM proteins have two active site motifs, the P-loop domain and the lid. Motifs in the P-loop include Walker A, Walker B and Sensor 1. The lid domain contains the arginine finger and Sensor 2. A catalytic site is created by a dimer interface that employs a *cis* P-loop from one subunit and a *trans* lid from the adjacent subunit [4, 33, 34]. A similar catalytic site created at the interfaces between adjacent monomers is also characteristic of the RecQ ATPase core [21].

Walker A		Walker B	
hWRN	DNVAVMATGYGKSLCFQYPPVY	hWRN	ITLIAVDEAHCISEWGHDFRDSFRKL
hBLM	DCFILMPTGGGKSLCYQLPACV	hBLM	LARFVIDEAHCVSQWGHDFRQDYKRM
hRECQL4	STLLVLPTGAGKSLCYQLPALL	hRECQL4	VAFACIDEAHCLSQWSHNFRPCYLRV
hMCM8	HILVVGDPGLGKSQMLQAACNV	hMCM8	QGICGIDEFDKMGNQHQALLEAMEQQ

Figure 1. Comparison of conserved ATPase motifs in RecQ proteins WRN, BLM and RECQL4 to the Walker A and Walker B boxes of MCM8. MCM8 was chosen as the MCM for comparison because it has a canonical GKS Walker A [12] and the signature MCM IDEFDKM Walker B ATP-binding domains. In the Walker B motif, note the conserved structural features and the conserved DE motif (containing aspartate, D, and the catalytic glutamate, E).

3. RecQ and MCM helicases: association with disease and aging

3.1. BLM and Bloom syndrome, WRN and Werner syndrome, RECQL4 and Rothmund-Thomson syndrome

As a group, mutations in the RecQ helicases lead to adult segmental progeria with abnormalities in development, predisposition to cancer and acceleration of aging processes. Three of the RecQ family members are associated with rare autosomal recessive diseases [19]. These disorders Werner syndrome (WS) [35], Bloom syndrome (BS) [36] and Rothmund-Thomson syndrome (RTS) [37, 38] are caused by mutations in the genes coding for WRN, BLM and RECQL4, respectively. RTS is a heterogeneous disorder with about 60% of cases resulting from mutations in the *RECQL4* gene [37]. Mutations in *RECQL4* can also lead to two other disease phenotypes [39], but only RTS will be discussed here. The RecQ deficiency diseases are associated with cancer predisposition and several characteristics of aging [8, 20, 26, 40]. In BS cells, there is a 10-fold elevation in frequency of homologous recombination (HR), and reciprocal exchanges occur between homologous chromosomes and sister chromatids [41, 42]. WS cells, on the other hand, display large

chromosome deletions and an increase in illegitimate recombination [43]. A higher frequency of chromosomal aberrations is reported for cells from RTS [44]. These deficiencies thus provide hints as to the cellular activities of these three helicases. Clinical
features of these diseases, as referenced above, are as follows.

BS manifests in pleiotropic phenotypes such as growth retardation leading to proportional
dwarfism, erythema with light sensitivity, skin lesions with hypo- and hyperpigmentation,
immunodeficiency, susceptibility to type II diabetes, male infertility, female sub-fertility, reports of mental retardation, cancer predisposition (all types but at an earlier age of diagnosis
than in the normal population).

WS leads to short stature and early onset age-related diseases, including greying hair, alopecia, bilateral cataracts, osteoporosis, arteriosclerosis, atherosclerosis, skin atrophy, hypogonadism, type II diabetes mellitus and susceptibility to tumors, especially those of
mesenchymal origin (sarcomas).

RTS manifests as early growth deficiency, congenital bone defects, poikiloderma, cataracts,
greying hair, alopecia, hypogonadism, and some increased susceptibility to cancer, especially osteogenic sarcomas.

3.2. MCMs and genomic stability

The MCM proteins 2-7 are necessary for DNA replication in yeast [45, 46], and this basic life
function extends in evolution to a single MCM protein in archaea [47-49]. MCM2-7 are also
essential for replication in *Xenopus* [50, 51] and have been proposed as a licensing factor for
initiation of eukaryotic replication [52, 53]. The MCM proteins 2-8 have an identical Walker
B-box motif of IDEFDKM. The MCM2-7 complex is enigmatic in that MCMs 4, 6 and 7 function alone as a heterohexameric helicase [54, 55]. For a discussion of individual MCM subunit arrangements and activities, see the references [33, 34]. MCM2-7 have now been shown
to have helicase activity *in vitro* [56], and to be components of a holo-helicase Cdc45/
MCM2-7/GINS (CMG) complex [57, 58]. The MCMs require a clamp-loading factor to assemble as a multimeric ring on DNA, and this function is fulfilled in known cases by the
protein Cdc6 [59-64] although the regulation of this step in the formation of the CMG complex proceeds through multiple pathways [57, 58]. Various papers have dealt with the function of MCM proteins in DNA replication [10, 46, 65-68], and regarding their processive
mechanism of DNA unwinding [27, 56, 58, 69, 70] and only certain lingering, disease-related
questions will be considered here. A summary statement can be made regarding known relationships between MCM and RecQ helicases. Members of the MCM protein family are essential for the life-creating process of DNA replication, whereas members of the RecQ
family are essential for the life-prolonging maintenance of the genome.

Due to their essential roles, it is not surprising that there are few diseases directly ascribed to
defects in MCMs 2-7. This does, however, bring up an unresolved MCM enigma: there are
more MCM proteins than are required to form initiation complexes at cellular origins active
within a given round of DNA replication [71]. In addition, MCM proteins in human cells remain at peak levels in G2 phase of the cell cycle, after DNA replication is complete [59, 72, 73]

This high copy number could be an aspect of securing sufficient protein quantities for basic function. A role for MCM proteins in transcription has also been proposed, and this remains to be defined [74]; for further discussion see reference [46]. Redundancy of functions among MCM members is also a factor to be considered. Intriguingly, knockdown of MCM8 [75], not even present in yeast [12], retards S-phase approximately 25% in human cells in culture [75], suggesting a specialized function in higher eukaryotes critical for basic replication. It has been suggested that the excess MCMs license dormant origins, which are used under conditions of stress upon normally functional origins [71]. An overview provides a discussion of recent work connecting dormant origins of replication and tumor suppression based on the role of dormant origins during fork restart after repair of DNA damage [76].

Mutations in MCM family members have been studied in yeast and mouse models, and they confirm an essential role for MCM2-7 in DNA replication [34, 45, 77, 78]. A human MCM4 mutation (destabilizing the MCM2-7 complex) was recently reported concurrently by two groups who studied consanguineous families, and the resulting phenotype was found to be associated with immune deficiency (NK cells), adrenal insufficiency and short stature [79, 80]. Patient fibroblasts showed chromosome fragility [79]. The susceptibility of these patients to cancer is not currently known [80]. An *MCM8* disruption and alternative splice form have been noted in hepatic carcinoma [11] and choriocarcinoma [12], respectively. Although not necessarily a cause of disease, the MCM proteins may be useful tools in diagnoses [81-85]. Elevated levels of MCM proteins 2-7 have been observed in several cancers [81, 84, 85]. In contrast, reduced levels of MCM8 mRNA have been reported in colon carcinoma [12]. Nuclear MCM7 is a good marker for proliferating cells [86], and MCMs 2, 5 or 7 may be an alternative to the Ki-67 marker to distinguish certain hyperproliferative disorders [83].

3.2.1. A structural domain deleted from MCM8 and present in WRN and BLM

A brief discussion of MCM8 is included in this section because it contains a motif that may be structurally similar to one found in BLM and WRN and is linked to neoplasia. Human MCM8 has a splice variant that results in a 16 amino acid (aa) deletion in a location between the Zn finger of MCM8 and its Walker A box [12], Fig. 2. Thus far this deletion has been detected by various different groups only in cases of choriocarcinoma. MCM8 with this same sequence deleted (Fig. 2) has, however, been detected in several higher eukaryotes other than humans, suggesting that the variant does have, or perhaps lacks, a function. That function is as yet unknown, but clues to it may be gleaned from a comparison of the MCM8 deletion with sequences from WRN and BLM. These RecQ proteins have a counterpart 16 aa domain with partial sequence homology and notable structural homology, as denoted in Fig. 2. This sequence in the RecQ proteins is located in a different orientation to that of MCM8 with regard to the helicase Walker A and B boxes (Fig. 2). In each case the first 8 aa of this sequence are highly charged and, in WRN, BLM and MCM8, contain a polar S. The 9th aa in this deletion is a conserved C. The remaining 7 aa contain a preponderance of aromatic and hydrophobic residues. This configuration of charged and aromatic residues is characteristic of known single-stranded nucleic acid binding proteins [87]. Among MCMs, this 16 aa domain is identifiable in the single MCM of *Sulfolobus solfataricus*, in which it has been implicated as a single-stranded DNA-binding "finger" [88]. Mutations of the positively charged amino acids strongly reduce single-stranded DNA binding of this MCM. In contrast, in BLM the polar S residue is thought to be involved in ATP binding [89]. Because of the aromatic nature of a portion of this domain, binding to one or more DNA nucleotide bases may be involved as a common link between functions of these 16 aa in RECQ and MCM helicases.

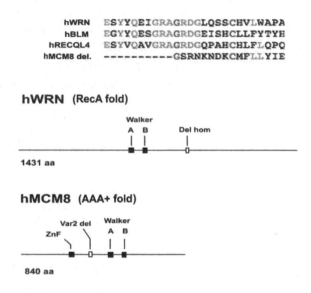

Figure 2. Comparison of placement of selected structural motifs in human RecQ and MCM8 helicases. MCM8 has a splice variant that results in a 16 aa deletion in choriocarcinoma [12]. The Walker A and B boxes and the MCM Zn finger (ZnF) domains are indicated by filled rectangles. The MCM8 variant 2 deletion (Var2 del) is indicated by an open rectangle, as is its partially-homologous counterpart (del hom) in WRN and BLM. This conservatively structured domain consists of an N-terminal highly charged sequence followed by a conserved C and an aromatic-hydrophobic sequence. The open box is located in a different orientation relative to Walker A and B boxes in WRN vs. MCM8. The positions of black and white boxes are approximately to scale.

4. Supportive roles for WRN, BLM helicases and RECQL4 during replication elongation

During normal metabolism, such as in mitochondrial respiration, endogenous reactive oxygen species (ROS) are produced that lead to oxidative DNA modifications. In addition to endogenous mutagens, there are also environmental mutagens that damage the DNA [90]. Furthermore, during replication and transcription, duplex DNA is transiently opened, and there is an opportunity for non-B DNA structures to form in the genomic DNA [40]. RecQ helicases act on recombination intermediates, on preferred substrates including those resembling G-quadruplex DNA [91, 92], D-loops [93], HJ [94] and double HJ [95]. RecQ helicases may play a regulatory role in both pro- and anti-HR events. They function at the interface of HR with the stressed replication fork and may affect repair pathway choice. There is little evidence for a specific clear cut role, but for a discussion of proposed mechanistic models of RecQ protein function, see the references [19, 20].

4.1. BLM and RECQL4

In BS cells, there are S-phase defects in DNA replication involving abnormal replication intermediates, and replication elongation is slower [96-98]. These cells don't recover well from induced fork stalling and accumulate DSB [99, 100]. In human cells, RECQL4 interacts with the MCM2-7 complex, Cdc45 and GINS. The interaction is facilitated by MCM10 [25]. BLM and RECQL4 are at their highest levels during S phase. At stalled replication forks, BLM physically associates with Rad51 and p53, and BLM and p53 function synergistically in HR [101]. BLM colocalizes in foci with PCNA and with the BASC (BRCA1-associated genome surveillance complex) [102]. BLM is phosphorylated by ATR [103]. When replication forks are stalled by use of hydroxyurea (HU), BLM colocalizes with Chk1 and p53BP1 foci. Chk1 is required for BLM and 53BP1 foci formation. Thus Chk1 may recruit BLM to stalled forks [104]. This implicates RecQ proteins as DNA damage checkpoint mediators in response to stalled forks. *In vitro* studies with use of substrates similar to Okazaki fragments showed BLM stimulation of flap endonuclease [105] (a protein that functions in lagging strand synthesis [106]). BLM functions to prevent the association of homologous sequences in the displaced flap DNA of the Okazaki fragment and the sister chromatid [107-110]. D-loops are formed when a ssDNA tail invades a homologous duplex, and BLM has a preference for dissociating D-loops with a 5' invaded end suggesting a selection of recombination intermediates that are not extended by polymerase [93, 111, 112]. BLM is able to disrupt the initial Rad51 filament formation step to destabilize recombinase-nucleoprotein filaments. D-loops are susceptible to BLM activity when Rad51 is in an inactive form (ADP-bound) [112, 113]. BLM physically associates with CAF-1 (chromatin assembly factor I) largest subunit, and the colocalization of these two proteins occurs at sites of DNA synthesis. BLM inhibits CAF-1 function in chromatin assembly during DNA repair *in vitro*, and inhibits its mobilization after damage induction *in vivo* [114]. Mammalian WRN and BLM interact [115], and they both interact with RPA [116-118] and p53 [119-121]. Based on coimmunoprecipitation there is limited BLM and WRN interaction, but they may function in the same pathway during HR [20]. BLM helicase activity is stimulated by its binding to the RPA70 kDa subunit [116]. In mouse spermatocytes during meiotic prophase, BLM and RPA are nuclear colocalized [122]. This suggests a potential role for these proteins in resolution of recombination intermediates during meiosis [8]. BLM has a preference for unwinding G-quadruplex structures versus HJ [92]. Both mammalian WRN and BLM bind to G-quadruplex structures, which are roadblocks to polymerases [123, 124].

Aberrant replication intermediates arise in cells lacking WRN and BLM [97, 125]. Such unresolved replication or recombination structures lead to incomplete chromosome segregation. BLM, topoisomerase 3 alpha (Topo3α) and BLAP75/RMI1 (for BLM-associated polypeptide/RecQ-mediated genome instability) or a BLM-Topo3α-BLAP75/RMI1 complex localizes to resulting anaphase bridges [126]. A helicase known as PICH arrives first, followed by the resolution activity of BLM [127]. As helicases unwind duplex DNA, torsional stress produced in the DNA may require relief through topoisomerase activity, and such activity finally decatenates interlocked DNA molecules [128, 129]. *In vitro* studies show that BLM can partner with Topo3α to resolve dHJ and prevent sequence exchange through resolution of this recombination intermediate [95]. The double-junction dissolution reaction requires the HRDC domain of BLM [24]. An additional protein, BLAP75/DMI1 mediates formation of the "dissolvasome" (BLM, Topo3α and BLAP75/RMI1) [130-132]. Mammalian BLAP18/RMI2 has also been found to be part of this complex [133].

4.2. WRN

Over 50 distinct *WRN* mutations have been reported, most of which lead to premature termination of translation [19]. Recent missense mutations in the exonuclease domain in one patient compromised protein stability [134]. Most mutations in WS patients occur in the WRN C-terminal domain, which could disrupt the WRN/p53 interaction [20, 134]. Such premature termination could also disrupt the Del hom sequence shown in Fig. 2. No WS mutations have been reported that eliminate only helicase or only exonuclease activity. Both activities are compromised in the development of WS [134]. WS fibroblasts undergo replicative senescence prematurely [135-139]. Telomere defects in WS cells suggest WRN activity in human telomere homeostasis [19]. Telomeres are needed to avoid loss of genetic material. They are important for chromosome end replication and for protection of the ends from enzymatic attack [140]. Human telomeres contain 5 to 20 kb of the repetitive sequence TTAGGG [141]. At the terminal there are 100-200 bp of 3' ssDNA overhang. This overhang can anneal with telomere DNA to form a stable D-loop leading to a structure referred to as a 't-loop' [142, 143]. Alternatively, this free unannealed end may form G-quadruplex DNA [144]. Human WRN functions in lagging-strand synthesis, and in the replication of telomeric G-rich DNA ends [145]. *C. elegans* WRN-1 can disrupt D-loops [146] and human WRN can prevent aberrant recombination [147]. WRN 3' to 5' exonuclease is stimulated by the interacting Ku70/Ku86 complex supporting a role for WRN in DNA repair [148]. Evidence suggests that in the absence of telomerase, WRN and BLM have a role in the ALT (Alternate Lengthening of Telomeres) pathway for telomere maintenance [149, 150]. In biochemical experiments, WRN releases a 3' invading tail from a telomeric type D-loop by coordinated WRN helicase and exonuclease activities [149].

WRN and BLM catalytic activities are comparable except for 3' to 5' exonuclease activity of WRN [151-153], which BLM lacks. On dsDNA and on RNA-DNA hybrids, the WRN exonuclease activity degrades a 3' recessed end. This activity can remove only one mismatched NT at the end of the recessed 3' DNA and can initiate exonuclease activity from a gap or nick [154, 155]. The exonuclease activity of WRN can degrade abnormal DNA structures suggesting that WRN helicase and exonuclease activities are involved in resolution of aberrant DNA structures at stalled forks [156]. Human WRN interacts with proteins involved in DNA replication. WRN coimmunoprecipitates with PCNA and topoisomerase 1 [157]. WRN functionally and physically interacts with RPA [117], and it functionally interacts with DNA polymerase delta [152]. WS cells accumulate recombination intermediates that impede cell growth [158]. In cells treated with HU, WRN colocalizes with RPA foci and is thought to dissociate recombination intermediates at the stalled forks [94, 147]. WRN stimulates polymerase delta activity in the absence of its processivity factor PCNA. This suggests a role for WRN in recruiting polymerase delta for replication restart at blocked or collapsed forks [152]. RPA can stimulate the processivity of WRN. The stimulation of WRN by RPA is due to protein-protein interactions as opposed to enhanced ATPase activity [117, 118].

5. Unification of BLM, WRN, RECQL4 and MCM2-7 activities in DNA replication and recombination/repair

5.1. BLM: Role in DNA damage response with a complex role in inhibiting or promoting HR

BLM is found mostly in fine granules throughout the nucleoplasm at highest levels during S and G2 phases of the cell cycle. Its focal localization is in PML nuclear bodies (PML-NB). The name PML derives from the promyelocytic leukemia protein, PML [159-162]. This protein forms the structural groundwork of the PML bodies, which store various nuclear proteins [163]. These PML-NB store repair proteins (e.g., Topo3α, MRN and p53) and may be involved in sensing DNA damage [163]. By regulating the availability of repair proteins, response can be directed to DNA damage sites. Trafficking of proteins to the PML-NB is regulated by sumoylation [164]. The sumoylation pathway involves E1, E2 and E3 enzymes, which regulate respectively, SUMO activation, SUMO conjugation and targeting of specific substrates for sumoylation through ligation [165]. BLM contains a motif for SUMO binding that would facilitate its integration into this repair protein storage network [166]. In addition, BLM is SUMO-1 and SUMO-2 modified [167]. When mutants are prepared in which the SUMO-binding sequences are deleted, BLM cannot localize to PML-NB [168]. When mutants are prepared that do not allow BLM localization to PML-NB, there is about a two-fold increase in sister chromatid exchange. These findings indicate that there is a need for BLM-SUMO interaction in order for BLM to localize to PML-NB, and that BLM activity, such as its accumulation at stalled replication forks, may be regulated by this specific localization [168].

At sites of DSB, repair foci form. A central player, H2AX, is phosphorylated when DSBs are induced, and this phosphorylation involves ATM, ATR and DNA-PK. Over one million bp are then marked by phosphorylated H2AX (γH2AX) on each side of the break [112, 168-171]. γH2AX recruits additional repair proteins to the damage site [172]. In studies where normal S-phase cells are treated with DNA damaging agents (HU, UV and cross-linking agents), BLM responds by leaving the PML-NBs to relocate to repair foci and colocalize with the marker γH2AX [101, 173]. BLM interacts physically and functionally with γH2AX as well as with ATM and ATR [103, 173-175]. In damage that is S-phase specific, BLM associates with the complex ATR/CHK1/53BP1, which gathers at repair foci as an early response to the damage. Based on kinetic studies, BLM may facilitate BRCA1 and MRN complex localization at repair foci in S phase [173], which may involve BLM regulation of p53 in these foci [101]. This early function of BLM at repair foci may allow for BML to influence the choice of repair pathways and facilitate a BLM function in anti-recombination at stalled forks that perhaps involves SU-MO regulation [19].

As discussed in the previous section, BLM interacts with the recombinase, Rad51. Rad51 catalyzes the pairing of a ssDNA tail and a homologous stretch of dsDNA to promote strand exchange early in the HR pathway [111]. Following ionizing radiation, Rad51 foci contained BLM [159]. Rad51 functions in HR, and localizes to ssDNA when DSB are induced [176]. BML can displace the recombinase from the ssDNA filament [112], which can be viewed as an anti-recombinagenic function [19]. In Rad51-associated D-loops, BLM can interact with Rad51 and unwind DNA in front of the polymerase [177]. This could favor SDSA leading to pro-recombinagenic function [19, 177]. BLM, however, also functions in resolution of G-quadruplex DNA structure and has higher binding affinity for it compared to HJ. BLM helicase activity is required for resolution of this structure [92]. Whereas the BLM/Rad51 interaction would represent an early event in a DNA damage/repair process, the formation of HJ, on the other hand, is a late event in HR. As discussed in section 5.1, the BLM-Topo3α-BLAP75/RMI1 complex functions to "dissolve" dHJ by convergent fork migration to generate non-cross-over products [178]. This is facilitated by Topo3α relief of superhelicity and by its ability to cleave and rejoin one strand of a DNA duplex. In the absence of BLM, Topo3α activity would involve break and rejoining activities instead of dissolution, which could lead to crossover events and an increase in sister chromatid exchange in BS cells [19]. BLM that is mutated to be unable to interact with Topo3α can only partially rescue the frequency of sister chromatid exchange [168]. Thus, together in a complex, BLM and Topo3α achieve dissolution of a recombinogenic intermediate. BLM has been proposed to regulate ploidy based on a role along with other members of this complex in resolution of anaphase bridges [126]. BLM has been found in complexes with mismatch repair protein MLH1 [179, 180]. It has been found to be present in large complexes containing not only BLM-Topo3a-BLAF75, but also additional factors. These additional factors could include several BLAF factors, proteins from the FA (Fanconi anemia) pathway, RPA, and mismatch repair protein MLH1. [181, 182]. The interactive role of these pathway components remains to be determined.

5.2. WRN: Helicase and exonuclease activities in concert

When WRN biochemical activities were compared using nonhydrolyzable ATPγS to inhibit only WRN helicase activity or aa substitutions to eliminate only the WRN exonuclease activity, the exonuclease activity was shown to function in degradation of the leading strand on replication fork-type substrates and in degradation of the annealed telomere overhangs on substrates resembling D-loop structures. WRN binding proteins were inhibitory [149, 183]. In addition, WRN was found to degrade ssDNA substrates longer than 40 nt with dependence upon the helicase activity [183]. WRN activities were also explored using WS fibroblasts. In WS cells, WRN and the enzyme telomerase are able to reverse the phenotype of excess chromosome fusion. In these cells, the anaphase bridges were missing telomere DNA. Dominant negative telomerase was not able to rescue the phenotype indicating that a stable telomere length was needed for rescue [184]. Telomeres are stabilized by a complex of proteins that bind DNA, known as the shelterin complex [185]. This complex consists of TRF1, TRF2 (double-strand DNA binding proteins), and POT1 (single-strand DNA binding protein) as well as adaptor proteins. The coordinate action of these proteins in the presence of telomerase is needed to regulate telomere length. Without telomerase, telomeres shorten in length each cell division [185]. In S phase, WRN colocalizes with TRF2 [145, 186]. TRF1 and TRF2 limit WRN exonuclease activity on synthetic telomere D-loops [149]. WRN helicase is stimulated by single-stranded DNA binding proteins, RPA or POT1. These proteins modulate WRN exonuclease degradation of the 3' overhang [149, 187]. The hypothesis is that WRN could unwind D-loops to facilitate leading strand synthesis through telomeric DNA, and there is also the possibility that WRN prevents interchromosomal interactions between telomeres [19].

5.2.1. WRN activities in response to G-quadruplex structures in the lagging strand

At the telomere, the G-rich strand is duplicated by lagging-strand synthesis. It may assume a G-quadruplex structure, which would interrupt the replication fork. Experimentally, when inhibiting WRN by use of overexpression of a dominant-negative WRN, the helicase deficiency leads to loss almost entirely affecting the sister telomere on the lagging strand [145]. WRN interaction with FEN-1 (the flap endonuclease involved in the processing of Okazaki fragments, [188]) could assist in the maturation of these fragments [105, 189]. Data suggest that WRN may function in concert with lagging strand synthesis perhaps in unwinding complex structure at the telomere. WRN interacts physically with DNA polymerase delta to stimulate its activity [190]. WRN can prevent stalling of polymerases delta at telomere sequence in vitro [191]. WRN stimulation of polymerase delta happens only in the absence of PCNA, which suggests that WRN has a role at stalled forks rather than in the regulation of processive DNA synthesis [19]. WRN, as opposed to BLM, is specific for G-quadruplex structure in the trinucleotide repeat of Fragile X syndrome [192].

5.2.2. WRN and Ku in suppression of aberrant recombination at telomeres

Proteins involved in the NHEJ path for repair of DSBs that are found at telomeres include Ku heterodimer, DNA-PK, MRN complex and ATM [193]. Through interaction with NBS1, WRN colocalizes with two of these components, Ku and the MRN complex [194]. Under unstable conditions, Ku can suppress sister chromatid exchange at telomeres [195]. Mouse knockout studies show that WRN normally suppresses aberrant recombination at telomeres [196].

5.2.3. WRN activities in base excision repair and interstrand cross link repair by HR

WS cells show increased sensitivity to alkylating agents and to agents that increase ROS. Human fibroblasts in which WRN has been knocked down respond to oxidative stress with an increase in DNA damage [197]. These observations support a role for WRN in repair of such damage [198]. DNA damage resulting from oxidation, alkylation, methylation and deamination is repaired by the base excision repair (BER) pathway. WRN interacts with proteins in this pathway. In addition, WRN helicase activity stimulates and is stimulated by polymerase beta activity [199]. There is evidence that WRN activity in BER is regulated by PARP-1, but the reader is referred to references [200, 201]. BRCA1 and BRCA2 play an important role in DSB repair in the HR subpathway [202, 203]. HR is a pathway for repairing interstrand crosslinks (ICL). Cells deficient in either WRN or BRCA1 are hypersensitive to induction of ICLs [204, 205]. A physical interaction of WRN with BRCA1 enhances the activities of WRN [206]. In addition, data from WRN and/or BRCA1 knockdown studies indicate that BRCA1 may act cooperatively with WRN in HR during ICL repair [206].

5.3. RECQL4

Data from *Recql4-/-* mice indicate a role for mouse RECQL4 in genomic stability and in the promoting cohesion between sister chromatids [207]. Depleting the *X. laevis* homologue of RTS, xRTS, in *Xenopus* egg extracts led to reduced DNA synthesis and an inhibition of RPA stabilization of ssDNA prior to polymerase loading at unwound origins. The addition of purified human RECQL4 could reverse this effect [208]. A nonhelical region in the N terminus of xRTS could be important in initiation of DNA replication based on its interaction with the Cut5 protein and the importance of xRTS in loading DNA polymerase alpha onto chromatin [209]. The N terminus of xRTS is not homologous to that of RECQL4 in mammals, however, and this role may not be conserved [19]. RECQL4, along with Ctf4 and MCM10, has been shown to be required for stable association of the CMG complex in human cells. In this study, Cut5/TopBP1 was not required for CMG stabilization [210]. In the *Xenopus* replication model, RECQL4 binds to chromatin that has been processed to resemble DSBs with a dependence on RPA and ATM activity [211]. Chromatin immunoprecipitation experiments show that RECQL4 functions on DNA in close association with Ku [212]. In HeLa cells human RECQL4 forms a complex with Rad51 and colocalizes with Rad51 foci formed after treatment with etoposide [212]. PML-NB contain a portion of RECQL4 [212]. It is also found in the nucleolus [213]. When using a T7 phage display screen, RECQL4 was found to interact with PARP-1, and the association influenced the nuclear localization of RECQL4. PARP-1 has a role in RECQL4 movement to the nucleolus from the nucleoplasm. When oxidative stress is induced, as opposed to other types of damage induction, RECQL4 increasingly localizes in the nucleolus. This trafficking is inhibited by inhibition of PARP [213]. RTS cells decrease proliferation and synthesis of DNA when exposed to hydrogen peroxide [214]. Lack of proper response to ROS could lead to premature aging as see in RTS. PARP is also involved in an end-joining pathway of DNA repair [215, 216]. The role of RECQL4 in its interaction with PARP-1 is not known (for further discussion see reference [19]).

5.4. MCM2-7

Replication fork stalling, can lead to DSB and chromosomal rearrangements. An S phase checkpoint is triggered by these events and there is a block to elongation. When this occurs, the proteins Mrc1, Tof1, and Csm3 (M/T/C complex) interact with the MCM2-7 complex to stabilize the replication fork. When the M/T/C complex is missing, the replisome continues, but synthesis stops. This may be partially due to loss of DNA polymerase epsilon from the fork [217-219]. Studies in yeast provide insight into these activities. The M/T/C complex associates with MCM2-7 [218, 220-222] and also with polymerase epsilon [221, 223]. These physical interactions permit communication between polymerase epsilon and the MCM complex [223]. The M/T/C complex may be part of the normal replication fork protein entourage [218, 220, 221, 224]. Mrc1 and Cdc45 coimmunoprecipitate, indicating an interaction of Mrc1 with the The Replication Progression Complex core, which includes Cdc45, the MCMs and GINS [223]. In each cell cycle Mrc loads onto replication origins along with the polymerases. This occurs after the Replication Progression Complex forms. Mrc migrates with the replication forks. [218, 219, 224-226]. Tof1, like Mrc, also coimmunoprecipitates with Cdc45 [227]. The exact mechanism of action of the M/T/C complex is not known. In a yeast study, the Tof1 homologue could switch regulation between pro- and anti-recombination activities in a site-specific manner [228]. Data indicate that a Mrc1 and MCM6-C terminal interaction senses alkylated DNA damage [221]. The other two subunits Tof1 and Csm3, may function to sense other types of damage. Although the helicase domains of MCM2-7 are conserved, the N and C terminals are divergent. Other negative regulators could differentially bind to these regions to regulate powering of elongation by the MCM2-7 helicase during times of stress [10]. A future question relates to the extent to which leading or lagging strand polymerase arrest is associated with formation of ssDNA, fork regression and formation of abnormal DNA structures [66]. These data indicate a functional connection between the MCM proteins, which act at stalled forks, and the RecQ proteins, which facilitate repair of the resulting damage.

6. Summary

BLM, WRN and RECQL4 act during events that stress the advancing replication fork providing relief through DNA damage repair and through resolution of aberrant replication/recombination intermediates, including those present at the telomere. At checkpoint, the replication proteins at a stalled fork are held stable through communication that occurs due to proteins that bind and signal to both the MCM complex and polymerase(s). This would allow repair proteins such as WRN and BLM helicases and RECQL4 to resolve the stress and thus aid in fork restart. Advancing our knowledge of the RecQ and MCM family activities and the mechanisms and signalling behind these activities will increase our understanding of cancer and aging and perhaps enlighten us regarding how to accommodate these challenges to human health.

Acknowledgments

This. work was supported by NIH funding to EMJ and resources derived from grant funding from Virginia's Commonwealth Health Research Board to DCD.

Author details

Dianne C. Daniel*, Ayuna V. Dagdanova and Edward M. Johnson

*Address all correspondence to: danieldc@evms.edu

Department of Microbiology and Molecular Cell Biology, Eastern Virginia Medical School, Norfolk, Virginia, USA

References

[1] Berger JM. SnapShot: nucleic acid helicases and translocases. Cell. 2008 Sep 5;134(5): 888- e1. 18775318

[2] Bernstein KA, Gangloff S, Rothstein R. The RecQ DNA helicases in DNA repair. Annu Rev Genet. 2010;44:393-417. 21047263

[3] Kelman Z, Hurwitz J. Structural lessons in DNA replication from the third domain of life. Nat Struct Biol. 2003 Mar;10(3):148-50. 12605215

[4] Erzberger JP, Berger JM. Evolutionary relationships and structural mechanisms of AAA+ proteins. Annu Rev Biophys Biomol Struct. 2006;35:93-114. 16689629

[5] Koonin EV. A common set of conserved motifs in a vast variety of putative nucleic acid-dependent ATPases including MCM proteins involved in the initiation of eukaryotic DNA replication. Nucleic Acids Res. 1993 Jun 11;21(11):2541-7. 8332451

[6] Iyer LM, Leipe DD, Koonin EV, Aravind L. Evolutionary history and higher order classification of AAA+ ATPases. J Struct Biol. 2004 Apr-May;146(1-2):11-31. 15037234

[7] Leipe DD, Koonin EV, Aravind L. Evolution and classification of P-loop kinases and related proteins. J Mol Biol. 2003 Oct 31;333(4):781-815. 14568537

[8] Mohaghegh P, Hickson ID. DNA helicase deficiencies associated with cancer predisposition and premature ageing disorders. Hum Mol Genet. 2001 Apr;10(7):741-6. 11257107

[9] Brosh RM, Jr., Bohr VA. Human premature aging, DNA repair and RecQ helicases. Nucleic Acids Res. 2007;35(22):7527-44. 18006573

[10] Bochman ML, Schwacha A. The Mcm complex: unwinding the mechanism of a replicative helicase. Microbiol Mol Biol Rev. 2009 Dec;73(4):652-83. 19946136

[11] Gozuacik D, Chami M, Lagorce D, Faivre J, Murakami Y, Poch O, Biermann E, Knippers R, Brechot C, Paterlini-Brechot P. Identification and functional characterization of a new member of the human Mcm protein family: hMcm8. Nucleic Acids Res. 2003 Jan 15;31(2):570-9. 12527764

[12] Johnson EM, Kinoshita Y, Daniel DC. A new member of the MCM protein family encoded by the human MCM8 gene, located contrapodal to GCD10 at chromosome band 20p12.3-13. Nucleic Acids Res. 2003 Jun 1;31(11):2915-25. 12771218

[13] Lutzmann M, Maiorano D, Mechali M. Identification of full genes and proteins of MCM9, a novel, vertebrate-specific member of the MCM2-8 protein family. Gene. 2005 Dec 5;362:51-6. 16226853

[14] Yoshida K. Identification of a novel cell-cycle-induced MCM family protein MCM9. Biochem Biophys Res Commun. 2005 Jun 3;331(2):669-74. 15850810

[15] Fien K, Cho YS, Lee JK, Raychaudhuri S, Tappin I, Hurwitz J. Primer utilization by DNA polymerase alpha-primase is influenced by its interaction with Mcm10p. J Biol Chem. 2004 Apr 16;279(16):16144-53. 14766746

[16] Ricke RM, Bielinsky AK. Mcm10 regulates the stability and chromatin association of DNA polymerase-alpha. Mol Cell. 2004 Oct 22;16(2):173-85. 15494305

[17] Sawyer SL, Cheng IH, Chai W, Tye BK. Mcm10 and Cdc45 cooperate in origin activation in Saccharomyces cerevisiae. J Mol Biol. 2004 Jul 2;340(2):195-202. 15201046

[18] Yang X, Gregan J, Lindner K, Young H, Kearsey SE. Nuclear distribution and chromatin association of DNA polymerase alpha-primase is affected by TEV protease cleavage of Cdc23 (Mcm10) in fission yeast. BMC Mol Biol. 2005;6:13. 15941470

[19] Ouyang KJ, Woo LL, Ellis NA. Homologous recombination and maintenance of genome integrity: cancer and aging through the prism of human RecQ helicases. Mech Ageing Dev. 2008 Jul-Aug;129(7-8):425-40. 18430459

[20] Bachrati CZ, Hickson ID. RecQ helicases: suppressors of tumorigenesis and premature aging. Biochem J. 2003 Sep 15;374(Pt 3):577-606. 12803543

[21] Singleton MR, Dillingham MS, Wigley DB. Structure and mechanism of helicases and nucleic acid translocases. Annu Rev Biochem. 2007;76:23-50. 17506634

[22] Killoran MP, Keck JL. Sit down, relax and unwind: structural insights into RecQ helicase mechanisms. Nucleic Acids Res. 2006;34(15):4098-105. 16935877

[23] Macris MA, Krejci L, Bussen W, Shimamoto A, Sung P. Biochemical characterization of the RECQ4 protein, mutated in Rothmund-Thomson syndrome. DNA Repair (Amst). 2006 Feb 3;5(2):172-80. 16214424

[24] Wu L, Chan KL, Ralf C, Bernstein DA, Garcia PL, Bohr VA, Vindigni A, Janscak P, Keck JL, Hickson ID. The HRDC domain of BLM is required for the dissolution of double Holliday junctions. Embo J. 2005 Jul 20;24(14):2679-87. 15990871

[25] Xu X, Rochette PJ, Feyissa EA, Su TV, Liu Y. MCM10 mediates RECQ4 association with MCM2-7 helicase complex during DNA replication. Embo J. 2009 Oct 7;28(19): 3005-14. 19696745

[26] Chu WK, Hickson ID. RecQ helicases: multifunctional genome caretakers. Nat Rev Cancer. 2009 Sep;9(9):644-54. 19657341

[27] Mogni ME, Costa A, Ioannou C, Bell SD. The glutamate switch is present in all seven clades of AAA+ protein. Biochemistry. 2009 Sep 22;48(37):8774-5. 19702328

[28] Snider J, Thibault G, Houry WA. The AAA+ superfamily of functionally diverse proteins. Genome Biol. 2008;9(4):216. 18466635

[29] Davey MJ, Jeruzalmi D, Kuriyan J, O'Donnell M. Motors and switches: AAA+ machines within the replisome. Nat Rev Mol Cell Biol. 2002 Nov;3(11):826-35. 12415300

[30] Ogura T, Wilkinson AJ. AAA+ superfamily ATPases: common structure--diverse function. Genes Cells. 2001 Jul;6(7):575-97. 11473577

[31] Saraste M, Sibbald PR, Wittinghofer A. The P-loop--a common motif in ATP- and GTP-binding proteins. Trends Biochem Sci. 1990 Nov;15(11):430-4. 2126155

[32] Walker JE, Saraste M, Runswick MJ, Gay NJ. Distantly related sequences in the alpha- and beta-subunits of ATP synthase, myosin, kinases and other ATP-requiring enzymes and a common nucleotide binding fold. Embo J. 1982;1(8):945-51. 6329717

[33] Davey MJ, Indiani C, O'Donnell M. Reconstitution of the Mcm2-7p heterohexamer, subunit arrangement, and ATP site architecture. J Biol Chem. 2003 Feb 14;278(7): 4491-9. 12480933

[34] Schwacha A, Bell SP. Interactions between two catalytically distinct MCM subgroups are essential for coordinated ATP hydrolysis and DNA replication. Mol Cell. 2001 Nov;8(5):1093-104. 11741544

[35] Yu CE, Oshima J, Fu YH, Wijsman EM, Hisama F, Alisch R, Matthews S, Nakura J, Miki T, Ouais S, Martin GM, Mulligan J, Schellenberg GD. Positional cloning of the Werner's syndrome gene. Science. 1996 Apr 12;272(5259):258-62. 8602509

[36] Bloom D. Congenital telangiectatic erythema resembling lupus erythematosus in dwarfs; probably a syndrome entity. AMA Am J Dis Child. 1954 Dec;88(6):754-8. 13206391

[37] Kitao S, Lindor NM, Shiratori M, Furuichi Y, Shimamoto A. Rothmund-thomson syndrome responsible gene, RECQL4: genomic structure and products. Genomics. 1999 Nov 1;61(3):268-76. 10552928

[38] Siitonen HA, Sotkasiira J, Biervliet M, Benmansour A, Capri Y, Cormier-Daire V, Crandall B, Hannula-Jouppi K, Hennekam R, Herzog D, Keymolen K, Lipsanen-Nyman M, Miny P, Plon SE, Riedl S, Sarkar A, Vargas FR, Verloes A, Wang LL, Kaariainen H, Kestila M. The mutation spectrum in RECQL4 diseases. Eur J Hum Genet. 2009 Feb;17(2):151-8. 18716613

[39] Hanada K, Hickson ID. Molecular genetics of RecQ helicase disorders. Cell Mol Life Sci. 2007 Sep;64(17):2306-22. 17571213

[40] Karow JK, Wu L, Hickson ID. RecQ family helicases: roles in cancer and aging. Curr Opin Genet Dev. 2000 Feb;10(1):32-8. 10679384

[41] German J. Bloom syndrome: a mendelian prototype of somatic mutational disease. Medicine (Baltimore). 1993 Nov;72(6):393-406. 8231788

[42] German J. Bloom's syndrome. Dermatol Clin. 1995 Jan;13(1):7-18. 7712653

[43] Shen JC, Loeb LA. The Werner syndrome gene: the molecular basis of RecQ helicase-deficiency diseases. Trends Genet. 2000 May;16(5):213-20. 10782115

[44] Vasseur F, Delaporte E, Zabot MT, Sturque MN, Barrut D, Savary JB, Thomas L, Tho-
      mas P. Excision repair defect in Rothmund Thomson syndrome. Acta Derm Venere-
      ol. 1999 Mar;79(2):150-2. 10228638

[45] Tye BK. MCM proteins in DNA replication. Annu Rev Biochem. 1999;68:649-86.
      10872463

[46] Forsburg SL. Eukaryotic MCM proteins: beyond replication initiation. Microbiol Mol
      Biol Rev. 2004 Mar;68(1):109-31. 15007098

[47] Chong JP, Hayashi MK, Simon MN, Xu RM, Stillman B. A double-hexamer archaeal
      minichromosome maintenance protein is an ATP-dependent DNA helicase. Proc
      Natl Acad Sci U S A. 2000 Feb 15;97(4):1530-5. 10677495

[48] Kelman Z, Lee JK, Hurwitz J. The single minichromosome maintenance protein of
      Methanobacterium thermoautotrophicum DeltaH contains DNA helicase activity.
      Proc Natl Acad Sci U S A. 1999 Dec 21;96(26):14783-8. 10611290

[49] Tye BK. Insights into DNA replication from the third domain of life. Proc Natl Acad
      Sci U S A. 2000 Mar 14;97(6):2399-401. 10716976

[50] Madine MA, Swietlik M, Pelizon C, Romanowski P, Mills AD, Laskey RA. The roles
      of the MCM, ORC, and Cdc6 proteins in determining the replication competence of
      chromatin in quiescent cells. J Struct Biol. 2000 Apr;129(2-3):198-210. 10806069

[51] Romanowski P, Madine MA, Rowles A, Blow JJ, Laskey RA. The Xenopus origin rec-
      ognition complex is essential for DNA replication and MCM binding to chromatin.
      Curr Biol. 1996 Nov 1;6(11):1416-25. 8939603

[52] Blow JJ, Hodgson B. Replication licensing--defining the proliferative state? Trends
      Cell Biol. 2002 Feb;12(2):72-8. 11849970

[53] Madine MA, Khoo CY, Mills AD, Musahl C, Laskey RA. The nuclear envelope pre-
      vents reinitiation of replication by regulating the binding of MCM3 to chromatin in
      Xenopus egg extracts. Curr Biol. 1995 Nov 1;5(11):1270-9. 8574584

[54] Ishimi Y. A DNA helicase activity is associated with an MCM4, -6, and -7 protein
      complex. J Biol Chem. 1997 Sep 26;272(39):24508-13. 9305914

[55] Lee JK, Hurwitz J. Processive DNA helicase activity of the minichromosome mainte-
      nance proteins 4, 6, and 7 complex requires forked DNA structures. Proc Natl Acad
      Sci U S A. 2001 Jan 2;98(1):54-9. 11136247

[56] Bochman ML, Schwacha A. The Mcm2-7 complex has in vitro helicase activity. Mol
      Cell. 2008 Jul 25;31(2):287-93. 18657510

[57] Boos D, Frigola J, Diffley JF. Activation of the replicative DNA helicase: breaking up
      is hard to do. Curr Opin Cell Biol. 2012 Jun;24(3):423-30. 22424671

[58] Kang YH, Galal WC, Farina A, Tappin I, Hurwitz J. Properties of the human Cdc45/
      Mcm2-7/GINS helicase complex and its action with DNA polymerase epsilon in roll-

ing circle DNA synthesis. Proc Natl Acad Sci U S A. 2012 Apr 17;109(16):6042-7. 22474384

[59] Kinoshita Y, Johnson EM. Site-specific loading of an MCM protein complex in a DNA replication initiation zone upstream of the c-MYC gene in the HeLa cell cycle. J Biol Chem. 2004 Aug 20;279(34):35879-89. 15190069

[60] Schepers A, Diffley JF. Mutational analysis of conserved sequence motifs in the budding yeast Cdc6 protein. J Mol Biol. 2001 May 11;308(4):597-608. 11350163

[61] Nevis KR, Cordeiro-Stone M, Cook JG. Origin licensing and p53 status regulate Cdk2 activity during G(1). Cell Cycle. 2009 Jun 15;8(12):1952-63. 19440053

[62] Coverley D, Pelizon C, Trewick S, Laskey RA. Chromatin-bound Cdc6 persists in S and G2 phases in human cells, while soluble Cdc6 is destroyed in a cyclin A-cdk2 dependent process. J Cell Sci. 2000 Jun;113 (Pt 11):1929-38. 10806104

[63] Alexandrow MG, Hamlin JL. Cdc6 chromatin affinity is unaffected by serine-54 phosphorylation, S-phase progression, and overexpression of cyclin A. Mol Cell Biol. 2004 Feb;24(4):1614-27. 14749377

[64] Shin JH, Grabowski B, Kasiviswanathan R, Bell SD, Kelman Z. Regulation of minichromosome maintenance helicase activity by Cdc6. J Biol Chem. 2003 Sep 26;278(39):38059-67. 12837750

[65] Daniel DC, Johnson EM. Addressing the enigma of MCM8 in DNA replication. In: Kušić-Tišma J, ed. Fundamental Aspects of DNA Replication: INTECH 2011:37-52. DOI: 10.5772/21177. http://www.intechopen.com/books/fundamental-aspects-of-dna-replication/addressing-the-enigma-of-mcm8-in-dna-replication

[66] Forsburg SL. The MCM helicase: linking checkpoints to the replication fork. Biochem Soc Trans. 2008 Feb;36(Pt 1):114-9. 18208397

[67] Masai H, You Z, Arai K. Control of DNA replication: regulation and activation of eukaryotic replicative helicase, MCM. IUBMB Life. 2005 Apr-May;57(4-5):323-35. 16036617

[68] Costa A, Onesti S. The MCM complex: (just) a replicative helicase? Biochem Soc Trans. 2008 Feb;36(Pt 1):136-40. 18208401

[69] Brewster AS, Wang G, Yu X, Greenleaf WB, Carazo JM, Tjajadi M, Klein MG, Chen XS. Crystal structure of a near-full-length archaeal MCM: functional insights for an AAA+ hexameric helicase. Proc Natl Acad Sci U S A. 2008 Dec 23;105(51):20191-6. 19073923

[70] Makarova KS, Koonin EV, Kelman Z. The CMG (CDC45/RecJ, MCM, GINS) complex is a conserved component of the DNA replication system in all archaea and eukaryotes. Biol Direct. 2012;7:7. 22329974

[71] Woodward AM, Gohler T, Luciani MG, Oehlmann M, Ge X, Gartner A, Jackson DA, Blow JJ. Excess Mcm2-7 license dormant origins of replication that can be used under conditions of replicative stress. J Cell Biol. 2006 Jun 5;173(5):673-83. 16754955

[72] Claycomb JM, MacAlpine DM, Evans JG, Bell SP, Orr-Weaver TL. Visualization of replication initiation and elongation in Drosophila. J Cell Biol. 2002 Oct 28;159(2): 225-36. 12403810

[73] Hirai K, Shirakata M. Replication licensing of the EBV oriP minichromosome. Curr Top Microbiol Immunol. 2001;258:13-33. 11443858

[74] Yankulov K, Todorov I, Romanowski P, Licatalosi D, Cilli K, McCracken S, Laskey R, Bentley DL. MCM proteins are associated with RNA polymerase II holoenzyme. Mol Cell Biol. 1999 Sep;19(9):6154-63. 10454562

[75] Volkening M, Hoffmann I. Involvement of human MCM8 in prereplication complex assembly by recruiting hcdc6 to chromatin. Mol Cell Biol. 2005 Feb;25(4):1560-8. 15684404

[76] Klotz-Noack K, Blow JJ. A role for dormant origins in tumor suppression. Mol Cell. 2011 Mar 4;41(5):495-6. 21362544

[77] Gomez EB, Catlett MG, Forsburg SL. Different phenotypes in vivo are associated with ATPase motif mutations in Schizosaccharomyces pombe minichromosome maintenance proteins. Genetics. 2002 Apr;160(4):1305-18. 11973289

[78] You Z, Komamura Y, Ishimi Y. Biochemical analysis of the intrinsic Mcm4-Mcm6-mcm7 DNA helicase activity. Mol Cell Biol. 1999 Dec;19(12):8003-15. 10567526

[79] Gineau L, Cognet C, Kara N, Lach FP, Dunne J, Veturi U, Picard C, Trouillet C, Ei-denschenk C, Aoufouchi S, Alcais A, Smith O, Geissmann F, Feighery C, Abel L, Smogorzewska A, Stillman B, Vivier E, Casanova JL, Jouanguy E. Partial MCM4 defi-ciency in patients with growth retardation, adrenal insufficiency, and natural killer cell deficiency. J Clin Invest. 2012 Mar 1;122(3):821-32. 22354167

[80] Hughes CR, Guasti L, Meimaridou E, Chuang CH, Schimenti JC, King PJ, Costigan C, Clark AJ, Metherell LA. MCM4 mutation causes adrenal failure, short stature, and natural killer cell deficiency in humans. J Clin Invest. 2012 Mar 1;122(3):814-20. 22354170

[81] Chatrath P, Scott IS, Morris LS, Davies RJ, Rushbrook SM, Bird K, Vowler SL, Grant JW, Saeed IT, Howard D, Laskey RA, Coleman N. Aberrant expression of minichro-mosome maintenance protein-2 and Ki67 in laryngeal squamous epithelial lesions. Br J Cancer. 2003 Sep 15;89(6):1048-54. 12966424

[82] Davidson EJ, Morris LS, Scott IS, Rushbrook SM, Bird K, Laskey RA, Wilson GE, Kitchener HC, Coleman N, Stern PL. Minichromosome maintenance (Mcm) proteins, cyclin B1 and D1, phosphohistone H3 and in situ DNA replication for functional analysis of vulval intraepithelial neoplasia. Br J Cancer. 2003 Jan 27;88(2):257-62. 12610511

112   DNA Replication

[83] Freeman A, Morris LS, Mills AD, Stoeber K, Laskey RA, Williams GH, Coleman N. Minichromosome maintenance proteins as biological markers of dysplasia and malignancy. Clin Cancer Res. 1999;5(8):2121-32.

[84] Gonzalez MA, Pinder SE, Callagy G, Vowler SL, Morris LS, Bird K, Bell JA, Laskey RA, Coleman N. Minichromosome maintenance protein 2 is a strong independent prognostic marker in breast cancer. J Clin Oncol. 2003 Dec 1;21(23):4306-13. 14645419

[85] Hunt DP, Freeman A, Morris LS, Burnet NG, Bird K, Davies TW, Laskey RA, Coleman N. Early recurrence of benign meningioma correlates with expression of minichromosome maintenance-2 protein. Br J Neurosurg. 2002 Feb;16(1):10-5. 11928726

[86] Khalili K, Del Valle L, Muralidharan V, Gault WJ, Darbinian N, Otte J, Meier E, Johnson EM, Daniel DC, Kinoshita Y, Amini S, Gordon J. Puralpha Is Essential for Postnatal Brain Development and Developmentally Coupled Cellular Proliferation As Revealed by Genetic Inactivation in the Mouse. Mol Cell Biol. 2003 Oct 1;23(19): 6857-75. 12972605

[87] Bergemann AD, Johnson EM. The HeLa Pur factor binds single-stranded DNA at a specific element conserved in gene flanking regions and origins of DNA replication. Mol Cell Biol. 1992;12:1257-65.

[88] Liu W, Pucci B, Rossi M, Pisani FM, Ladenstein R. Structural analysis of the Sulfolobus solfataricus MCM protein N-terminal domain. Nucleic Acids Res. 2008 Jun; 36(10):3235-43. 18417534

[89] Rong SB, Valiaho J, Vihinen M. Structural basis of Bloom syndrome (BS) causing mutations in the BLM helicase domain. Mol Med. 2000 Mar;6(3):155-64. 10965492

[90] Lindahl T. Instability and decay of the primary structure of DNA. Nature. 1993 Apr 22;362(6422):709-15. 8469282

[91] Sun H, Karow JK, Hickson ID, Maizels N. The Bloom's syndrome helicase unwinds G4 DNA. J Biol Chem. 1998 Oct 16;273(42):27587-92. 9765292

[92] Huber MD, Lee DC, Maizels N. G4 DNA unwinding by BLM and Sgs1p: substrate specificity and substrate-specific inhibition. Nucleic Acids Res. 2002 Sep 15;30(18): 3954-61. 12235379

[93] van Brabant AJ, Ye T, Sanz M, German IJ, Ellis NA, Holloman WK. Binding and melting of D-loops by the Bloom syndrome helicase. Biochemistry. 2000 Nov 28;39(47):14617-25. 11087418

[94] Karow JK, Constantinou A, Li JL, West SC, Hickson ID. The Bloom's syndrome gene product promotes branch migration of holliday junctions. Proc Natl Acad Sci U S A. 2000 Jun 6;97(12):6504-8. 10823897

[95] Wu L, Hickson ID. The Bloom's syndrome helicase suppresses crossing over during homologous recombination. Nature. 2003 Dec 18;426(6968):870-4. 14685245

[96] Hand R, German J. A retarded rate of DNA chain growth in Bloom's syndrome. Proc Natl Acad Sci U S A. 1975 Feb;72(2):758-62. 1054854

[97] Lonn U, Lonn S, Nylen U, Winblad G, German J. An abnormal profile of DNA replication intermediates in Bloom's syndrome. Cancer Res. 1990 Jun 1;50(11):3141-5. 2110504

[98] Han H, Hurley LH. G-quadruplex DNA: a potential target for anti-cancer drug design. Trends Pharmacol Sci. 2000 Apr;21(4):136-42. 10740289

[99] Davies SL, North PS, Hickson ID. Role for BLM in replication-fork restart and suppression of origin firing after replicative stress. Nat Struct Mol Biol. 2007 Jul;14(7): 677-9. 17603497

[100] Rao VA, Conti C, Guirouilh-Barbat J, Nakamura A, Miao ZH, Davies SL, Sacca B, Hickson ID, Bensimon A, Pommier Y. Endogenous gamma-H2AX-ATM-Chk2 checkpoint activation in Bloom's syndrome helicase deficient cells is related to DNA replication arrested forks. Mol Cancer Res. 2007 Jul;5(7):713-24. 17634426

[101] Sengupta S, Linke SP, Pedeux R, Yang Q, Farnsworth J, Garfield SH, Valerie K, Shay JW, Ellis NA, Wasylyk B, Harris CC. BLM helicase-dependent transport of p53 to sites of stalled DNA replication forks modulates homologous recombination. Embo J. 2003 Mar 3;22(5):1210-22. 12606585

[102] Wang Y, Cortez D, Yazdi P, Neff N, Elledge SJ, Qin J. BASC, a super complex of BRCA1-associated proteins involved in the recognition and repair of aberrant DNA structures. Genes Dev. 2000 Apr 15;14(8):927-39. 10783165

[103] Davies SL, North PS, Dart A, Lakin ND, Hickson ID. Phosphorylation of the Bloom's syndrome helicase and its role in recovery from S-phase arrest. Mol Cell Biol. 2004 Feb;24(3):1279-91. 14729972

[104] Sengupta S, Robles AI, Linke SP, Sinogeeva NI, Zhang R, Pedeux R, Ward IM, Celeste A, Nussenzweig A, Chen J, Halazonetis TD, Harris CC. Functional interaction between BLM helicase and 53BP1 in a Chk1-mediated pathway during S-phase arrest. J Cell Biol. 2004 Sep 13;166(6):801-13. 15364958

[105] Brosh RM, Jr., Driscoll HC, Dianov GL, Sommers JA. Biochemical characterization of the WRN-FEN-1 functional interaction. Biochemistry. 2002 Oct 8;41(40):12204-16. 12356323

[106] Kao HI, Veeraraghavan J, Polaczek P, Campbell JL, Bambara RA. On the roles of Saccharomyces cerevisiae Dna2p and Flap endonuclease 1 in Okazaki fragment processing. J Biol Chem. 2004 Apr 9;279(15):15014-24. 14747468

[107] Bartos JD, Wang W, Pike JE, Bambara RA. Mechanisms by which Bloom protein can disrupt recombination intermediates of Okazaki fragment maturation. J Biol Chem. 2006 Oct 27;281(43):32227-39. 16950766

[108] Sharma S, Otterlei M, Sommers JA, Driscoll HC, Dianov GL, Kao HI, Bambara RA, Brosh RM, Jr. WRN helicase and FEN-1 form a complex upon replication arrest and

together process branchmigrating DNA structures associated with the replication fork. Mol Biol Cell. 2004 Feb;15(2):734-50. 14657243

[109] Sharma S, Sommers JA, Gary RK, Friedrich-Heineken E, Hubscher U, Brosh RM, Jr. The interaction site of Flap Endonuclease-1 with WRN helicase suggests a coordination of WRN and PCNA. Nucleic Acids Res. 2005;33(21):6769-81. 16326861

[110] Sharma S, Sommers JA, Wu L, Bohr VA, Hickson ID, Brosh RM, Jr. Stimulation of flap endonuclease-1 by the Bloom's syndrome protein. J Biol Chem. 2004 Mar 12;279(11):9847-56. 14688284

[111] Bachrati CZ, Borts RH, Hickson ID. Mobile D-loops are a preferred substrate for the Bloom's syndrome helicase. Nucleic Acids Res. 2006;34(8):2269-79. 16670433

[112] Bugreev DV, Yu X, Egelman EH, Mazin AV. Novel pro- and anti-recombination activities of the Bloom's syndrome helicase. Genes Dev. 2007 Dec 1;21(23):3085-94. 18003860

[113] Bugreev DV, Brosh RM, Jr., Mazin AV. RECQ1 possesses DNA branch migration activity. J Biol Chem. 2008 Jul 18;283(29):20231-42. 18495662

[114] Jiao R, Bachrati CZ, Pedrazzi G, Kuster P, Petkovic M, Li JL, Egli D, Hickson ID, Stagljar I. Physical and functional interaction between the Bloom's syndrome gene product and the largest subunit of chromatin assembly factor 1. Mol Cell Biol. 2004 Jun;24(11):4710-9. 15143166

[115] von Kobbe C, Karmakar P, Dawut L, Opresko P, Zeng X, Brosh RM, Jr., Hickson ID, Bohr VA. Colocalization, physical, and functional interaction between Werner and Bloom syndrome proteins. J Biol Chem. 2002 Jun 14;277(24):22035-44. 11919194

[116] Brosh RM, Jr., Li JL, Kenny MK, Karow JK, Cooper MP, Kureekattil RP, Hickson ID, Bohr VA. Replication protein A physically interacts with the Bloom's syndrome protein and stimulates its helicase activity. J Biol Chem. 2000 Aug 4;275(31):23500-8. 10825162

[117] Brosh RM, Jr., Orren DK, Nehlin JO, Ravn PH, Kenny MK, Machwe A, Bohr VA. Functional and physical interaction between WRN helicase and human replication protein A. J Biol Chem. 1999 Jun 25;274(26):18341-50. 10373438

[118] Shen JC, Gray MD, Oshima J, Loeb LA. Characterization of Werner syndrome protein DNA helicase activity: directionality, substrate dependence and stimulation by replication protein A. Nucleic Acids Res. 1998 Jun 15;26(12):2879-85. 9611231

[119] Blander G, Kipnis J, Leal JF, Yu CE, Schellenberg GD, Oren M. Physical and functional interaction between p53 and the Werner's syndrome protein. J Biol Chem. 1999 Oct 8;274(41):29463-9. 10506209

[120] Spillare EA, Robles AI, Wang XW, Shen JC, Yu CE, Schellenberg GD, Harris CC. p53-mediated apoptosis is attenuated in Werner syndrome cells. Genes Dev. 1999 Jun 1;13(11):1355-60. 10364153

[121] Wang XW, Tseng A, Ellis NA, Spillare EA, Linke SP, Robles AI, Seker H, Yang Q, Hu P, Beresten S, Bemmels NA, Garfield S, Harris CC. Functional interaction of p53 and BLM DNA helicase in apoptosis. J Biol Chem. 2001 Aug 31;276(35):32948-55. 11399766

[122] Walpita D, Plug AW, Neff NF, German J, Ashley T. Bloom's syndrome protein, BLM, colocalizes with replication protein A in meiotic prophase nuclei of mammalian spermatocytes. Proc Natl Acad Sci U S A. 1999 May 11;96(10):5622-7. 10318934

[123] Li JL, Harrison RJ, Reszka AP, Brosh RM, Jr., Bohr VA, Neidle S, Hickson ID. Inhibition of the Bloom's and Werner's syndrome helicases by G-quadruplex interacting ligands. Biochemistry. 2001 Dec 18;40(50):15194-202. 11735402

[124] Popuri V, Bachrati CZ, Muzzolini L, Mosedale G, Costantini S, Giacomini E, Hickson ID, Vindigni A. The Human RecQ helicases, BLM and RECQ1, display distinct DNA substrate specificities. J Biol Chem. 2008 Jun 27;283(26):17766-76. 18448429

[125] Poot M, Hoehn H, Runger TM, Martin GM. Impaired S-phase transit of Werner syndrome cells expressed in lymphoblastoid cell lines. Exp Cell Res. 1992 Oct;202(2): 267-73. 1327851

[126] Chan KL, North PS, Hickson ID. BLM is required for faithful chromosome segregation and its localization defines a class of ultrafine anaphase bridges. Embo J. 2007 Jul 25;26(14):3397-409. 17599064

[127] Baumann C, Korner R, Hofmann K, Nigg EA. PICH, a centromere-associated SNF2 family ATPase, is regulated by Plk1 and required for the spindle checkpoint. Cell. 2007 Jan 12;128(1):101-14. 17218258

[128] Johnson FB, Lombard DB, Neff NF, Mastrangelo MA, Dewolf W, Ellis NA, Marciniak RA, Yin Y, Jaenisch R, Guarente L. Association of the Bloom syndrome protein with topoisomerase IIIalpha in somatic and meiotic cells. Cancer Res. 2000 Mar 1;60(5): 1162-7. 10728666

[129] Wu L, Davies SL, North PS, Goulaouic H, Riou JF, Turley H, Gatter KC, Hickson ID. The Bloom's syndrome gene product interacts with topoisomerase III. J Biol Chem. 2000 Mar 31;275(13):9636-44. 10734115

[130] Bussen W, Raynard S, Busygina V, Singh AK, Sung P. Holliday junction processing activity of the BLM-Topo IIIalpha-BLAP75 complex. J Biol Chem. 2007 Oct 26;282(43):31484-92. 17728255

[131] Raynard S, Zhao W, Bussen W, Lu L, Ding YY, Busygina V, Meetei AR, Sung P. Functional role of BLAP75 in BLM-topoisomerase IIIalpha-dependent holliday junction processing. J Biol Chem. 2008 Jun 6;283(23):15701-8. 18390547

[132] Wu L, Bachrati CZ, Ou J, Xu C, Yin J, Chang M, Wang W, Li L, Brown GW, Hickson ID. BLAP75/RMI1 promotes the BLM-dependent dissolution of homologous recombination intermediates. Proc Natl Acad Sci U S A. 2006 Mar 14;103(11):4068-73. 16537486

[133] Singh TR, Ali AM, Busygina V, Raynard S, Fan Q, Du CH, Andreassen PR, Sung P, Meetei AR. BLAP18/RMI2, a novel OB-fold-containing protein, is an essential compo-nent of the Bloom helicase-double Holliday junction dissolvasome. Genes Dev. 2008 Oct 15;22(20):2856-68. 18923083

[134] Huang S, Lee L, Hanson NB, Lenaerts C, Hoehn H, Poot M, Rubin CD, Chen DF, Yang CC, Juch H, Dorn T, Spiegel R, Oral EA, Abid M, Battisti C, Lucci-Cordisco E, Neri G, Steed EH, Kidd A, Isley W, Showalter D, Vittone JL, Konstantinow A, Ring J, Meyer P, Wenger SL, von Herbay A, Wollina U, Schuelke M, Huizenga CR, Leistritz DF, Martin GM, Mian IS, Oshima J. The spectrum of WRN mutations in Werner syn-drome patients. Hum Mutat. 2006 Jun;27(6):558-67. 16673358

[135] Davis T, Singhrao SK, Wyllie FS, Haughton MF, Smith PJ, Wiltshire M, Wynford-Thomas D, Jones CJ, Faragher RG, Kipling D. Telomere-based proliferative lifespan barriers in Werner-syndrome fibroblasts involve both p53-dependent and p53-inde-pendent mechanisms. J Cell Sci. 2003 Apr 1;116(Pt 7):1349-57. 12615976

[136] Epstein CJ, Martin GM, Schultz AL, Motulsky AG. Werner's syndrome a review of its symptomatology, natural history, pathologic features, genetics and relationship to the natural aging process. Medicine (Baltimore). 1966 May;45(3):177-221. 5327241

[137] Faragher RG, Kill IR, Hunter JA, Pope FM, Tannock C, Shall S. The gene responsible for Werner syndrome may be a cell division "counting" gene. Proc Natl Acad Sci U S A. 1993 Dec 15;90(24):12030-4. 8265666

[138] Schulz VP, Zakian VA, Ogburn CE, McKay J, Jarzebowicz AA, Edland SD, Martin GM. Accelerated loss of telomeric repeats may not explain accelerated replicative de-cline of Werner syndrome cells. Hum Genet. 1996 Jun;97(6):750-4. 8641691

[139] Tahara H, Tokutake Y, Maeda S, Kataoka H, Watanabe T, Satoh M, Matsumoto T, Su-gawara M, Ide T, Goto M, Furuichi Y, Sugimoto M. Abnormal telomere dynamics of B-lymphoblastoid cell strains from Werner's syndrome patients transformed by Ep-stein-Barr virus. Oncogene. 1997 Oct 16;15(16):1911-20. 9365237

[140] Sandell LL, Zakian VA. Loss of a yeast telomere: arrest, recovery, and chromosome loss. Cell. 1993 Nov 19;75(4):729-39. 8242745

[141] Nurnberg P, Thiel G, Weber F, Epplen JT. Changes of telomere lengths in human in-tracranial tumours. Hum Genet. 1993 Mar;91(2):190-2. 8462979

[142] Griffith JD, Comeau L, Rosenfield S, Stansel RM, Bianchi A, Moss H, de Lange T. Mammalian telomeres end in a large duplex loop. Cell. 1999 May 14;97(4):503-14. 10338214

[143] Murti KG, Prescott DM. Telomeres of polytene chromosomes in a ciliated protozoan terminate in duplex DNA loops. Proc Natl Acad Sci U S A. 1999 Dec 7;96(25):14436-9. 10588723

[144] Williamson JR, Raghuraman MK, Cech TR. Monovalent cation-induced structure of telomeric DNA: the G-quartet model. Cell. 1989 Dec 1;59(5):871-80. 2590943

[145] Crabbe L, Verdun RE, Haggblom CI, Karlseder J. Defective telomere lagging strand synthesis in cells lacking WRN helicase activity. Science. 2004 Dec 10;306(5703): 1951-3. 15591207

[146] Hyun M, Bohr VA, Ahn B. Biochemical characterization of the WRN-1 RecQ helicase of Caenorhabditis elegans. Biochemistry. 2008 Jul 15;47(28):7583-93. 18558712

[147] Constantinou A, Tarsounas M, Karow JK, Brosh RM, Bohr VA, Hickson ID, West SC. Werner's syndrome protein (WRN) migrates Holliday junctions and co-localizes with RPA upon replication arrest. EMBO Rep. 2000 Jul;1(1):80-4. 11256630

[148] Cooper MP, Machwe A, Orren DK, Brosh RM, Ramsden D, Bohr VA. Ku complex interacts with and stimulates the Werner protein. Genes Dev. 2000 Apr 15;14(8):907-12. 10783163

[149] Opresko PL, Otterlei M, Graakjaer J, Bruheim P, Dawut L, Kolvraa S, May A, Seidman MM, Bohr VA. The Werner syndrome helicase and exonuclease cooperate to resolve telomeric D loops in a manner regulated by TRF1 and TRF2. Mol Cell. 2004 Jun 18;14(6):763-74. 15200954

[150] Bhattacharyya S, Sandy A, Groden J. Unwinding protein complexes in ALTernative telomere maintenance. J Cell Biochem. 2010 2010;109(1):7-15. 19911388

[151] Huang S, Li B, Gray MD, Oshima J, Mian IS, Campisi J. The premature ageing syndrome protein, WRN, is a 3'-->5' exonuclease. Nat Genet. 1998 Oct;20(2):114-6. 9771700

[152] Kamath-Loeb AS, Johansson E, Burgers PM, Loeb LA. Functional interaction between the Werner Syndrome protein and DNA polymerase delta. Proc Natl Acad Sci U S A. 2000 Apr 25;97(9):4603-8. 10781066

[153] Shen JC, Gray MD, Oshima J, Kamath-Loeb AS, Fry M, Loeb LA. Werner syndrome protein. I. DNA helicase and dna exonuclease reside on the same polypeptide. J Biol Chem. 1998 Dec 18;273(51):34139-44. 9852073

[154] Huang S, Beresten S, Li B, Oshima J, Ellis NA, Campisi J. Characterization of the human and mouse WRN 3'-->5' exonuclease. Nucleic Acids Res. 2000 Jun 15;28(12): 2396-405. 10871373

[155] Kamath-Loeb AS, Shen JC, Loeb LA, Fry M. Werner syndrome protein. II. Characterization of the integral 3' --> 5' DNA exonuclease. J Biol Chem. 1998 Dec 18;273(51): 34145-50. 9852074

[156] Shen JC, Loeb LA. Werner syndrome exonuclease catalyzes structure-dependent degradation of DNA. Nucleic Acids Res. 2000 Sep 1;28(17):3260-8. 10954593

[157] Lebel M, Spillare EA, Harris CC, Leder P. The Werner syndrome gene product copurifies with the DNA replication complex and interacts with PCNA and topoisomerase I. J Biol Chem. 1999 Dec 31;274(53):37795-9. 10608841

118

DNA Replication

[158] Prince PR, Emond MJ, Monnat RJ, Jr. Loss of Werner syndrome protein function promotes aberrant mitotic recombination. Genes Dev. 2001 Apr 15;15(8):933-8. 11316787

[159] Bischof O, Kim SH, Irving J, Beresten S, Ellis NA, Campisi J. Regulation and localization of the Bloom syndrome protein in response to DNA damage. J Cell Biol. 2001 Apr 16;153(2):367-80. 11309417

[160] Dutertre S, Ababou M, Onclercq R, Delic J, Chatton B, Jaulin C, Amor-Gueret M. Cell cycle regulation of the endogenous wild type Bloom's syndrome DNA helicase. Oncogene. 2000 May 25;19(23):2731-8. 10851073

[161] Sanz MM, Proytcheva M, Ellis NA, Holloman WK, German J. BLM, the Bloom's syndrome protein, varies during the cell cycle in its amount, distribution, and co-localization with other nuclear proteins. Cytogenet Cell Genet. 2000;91(1-4):217-23. 11173860

[162] Zhong S, Hu P, Ye TZ, Stan R, Ellis NA, Pandolfi PP. A role for PML and the nuclear body in genomic stability. Oncogene. 1999 Dec 23;18(56):7941-7. 10637504

[163] Dellaire G, Ching RW, Ahmed K, Jalali F, Tse KC, Bristow RG, Bazett-Jones DP. Promyelocytic leukemia nuclear bodies behave as DNA damage sensors whose response to DNA double-strand breaks is regulated by NBS1 and the kinases ATM, Chk2, and ATR. J Cell Biol. 2006 Oct 9;175(1):55-66. 17030982

[164] Matunis MJ, Zhang XD, Ellis NA. SUMO: the glue that binds. Dev Cell. 2006 Nov; 11(5):596-7. 17084352

[165] Johnson ES. Protein modification by SUMO. Annu Rev Biochem. 2004;73:355-82. 15189146

[166] Shen TH, Lin HK, Scaglioni PP, Yung TM, Pandolfi PP. The mechanisms of PML-nuclear body formation. Mol Cell. 2006 Nov 3;24(3):331-9. 17081985

[167] Eladad S, Ye TZ, Hu P, Leversha M, Beresten S, Matunis MJ, Ellis NA. Intra-nuclear trafficking of the BLM helicase to DNA damage-induced foci is regulated by SUMO modification. Hum Mol Genet. 2005 May 15;14(10):1351-65. 15829507

[168] Hu P, Beresten SF, van Brabant AJ, Ye TZ, Pandolfi PP, Johnson FB, Guarente L, Ellis NA. Evidence for BLM and Topoisomerase IIIalpha interaction in genomic stability. Hum Mol Genet. 2001 Jun 1;10(12):1287-98. 11406610

[169] Paull TT, Rogakou EP, Yamazaki V, Kirchgessner CU, Gellert M, Bonner WM. A critical role for histone H2AX in recruitment of repair factors to nuclear foci after DNA damage. Curr Biol. 2000 Jul 27-Aug 10;10(15):886-95. 10959836

[170] Pilch DR, Sedelnikova OA, Redon C, Celeste A, Nussenzweig A, Bonner WM. Characteristics of gamma-H2AX foci at DNA double-strand breaks sites. Biochem Cell Biol. 2003 Jun;81(3):123-9. 12897845

[171] Rogakou EP, Boon C, Redon C, Bonner WM. Megabase chromatin domains involved in DNA double-strand breaks in vivo. J Cell Biol. 1999 Sep 6;146(5):905-16. 10477747

[172] Bassing CH, Alt FW. H2AX may function as an anchor to hold broken chromosomal DNA ends in close proximity. Cell Cycle. 2004 Feb;3(2):149-53. 14712078

[173] Davalos AR, Campisi J. Bloom syndrome cells undergo p53-dependent apoptosis and delayed assembly of BRCA1 and NBS1 repair complexes at stalled replication forks. J Cell Biol. 2003 Sep 29;162(7):1197-209. 14517203

[174] Beamish H, Kedar P, Kaneko H, Chen P, Fukao T, Peng C, Beresten S, Gueven N, Purdie D, Lees-Miller S, Ellis N, Kondo N, Lavin MF. Functional link between BLM defective in Bloom's syndrome and the ataxia-telangiectasia-mutated protein, ATM. J Biol Chem. 2002 Aug 23;277(34):30515-23. 12034743

[175] Franchitto A, Pichierri P. Bloom's syndrome protein is required for correct relocalization of RAD50/MRE11/NBS1 complex after replication fork arrest. J Cell Biol. 2002 Apr 1;157(1):19-30. 11916980

[176] Raderschall E, Golub EI, Haaf T. Nuclear foci of mammalian recombination proteins are located at single-stranded DNA regions formed after DNA damage. Proc Natl Acad Sci U S A. 1999 Mar 2;96(5):1921-6. 10051570

[177] Adams MD, McVey M, Sekelsky JJ. Drosophila BLM in double-strand break repair by synthesis-dependent strand annealing. Science. 2003 Jan 10;299(5604):265-7. 12522255

[178] Plank JL, Wu J, Hsieh TS. Topoisomerase IIIalpha and Bloom's helicase can resolve a mobile double Holliday junction substrate through convergent branch migration. Proc Natl Acad Sci U S A. 2006 Jul 25;103(30):11118-23. 16849422

[179] Langland G, Kordich J, Creaney J, Goss KH, Lillard-Wetherell K, Bebenek K, Kunkel TA, Groden J. The Bloom's syndrome protein (BLM) interacts with MLH1 but is not required for DNA mismatch repair. J Biol Chem. 2001 Aug 10;276(32):30031-5. 11325959

[180] Pedrazzi G, Perrera C, Blaser H, Kuster P, Marra G, Davies SL, Ryu GH, Freire R, Hickson ID, Jiricny J, Stagljar I. Direct association of Bloom's syndrome gene product with the human mismatch repair protein MLH1. Nucleic Acids Res. 2001 Nov 1;29(21):4378-86. 11691925

[181] Meetei AR, Sechi S, Wallisch M, Yang D, Young MK, Joenje H, Hoatlin ME, Wang W. A multiprotein nuclear complex connects Fanconi anemia and Bloom syndrome. Mol Cell Biol. 2003 May;23(10):3417-26. 12724401

[182] Yin J, Sobeck A, Xu C, Meetei AR, Hoatlin M, Li L, Wang W. BLAP75, an essential component of Bloom's syndrome protein complexes that maintain genome integrity. Embo J. 2005 Apr 6;24(7):1465-76. 15775963

[183] Machwe A, Xiao L, Lloyd RG, Bolt E, Orren DK. Replication fork regression in vitro by the Werner syndrome protein (WRN): holliday junction formation, the effect of leading arm structure and a potential role for WRN exonuclease activity. Nucleic Acids Res. 2007;35(17):5729-47. 17717003

[184] Crabbe L, Jauch A, Naeger CM, Holtgreve-Grez H, Karlseder J. Telomere dysfunc-
       tion as a cause of genomic instability in Werner syndrome. Proc Natl Acad Sci U S A.
       2007 Feb 13;104(7):2205-10. 17284601

[185] de Lange T. Shelterin: the protein complex that shapes and safeguards human telo-
       meres. Genes Dev. 2005 Sep 15;19(18):2100-10. 16166375

[186] Opresko PL, von Kobbe C, Laine JP, Harrigan J, Hickson ID, Bohr VA. Telomere-
       binding protein TRF2 binds to and stimulates the Werner and Bloom syndrome heli-
       cases. J Biol Chem. 2002 Oct 25;277(43):41110-9. 12181313

[187] Opresko PL, Mason PA, Podell ER, Lei M, Hickson ID, Cech TR, Bohr VA. POT1
       stimulates RecQ helicases WRN and BLM to unwind telomeric DNA substrates. J Bi-
       ol Chem. 2005 Sep 16;280(37):32069-80. 16030011

[188] Kao HI, Campbell JL, Bambara RA. Dna2p helicase/nuclease is a tracking protein,
       like FEN1, for flap cleavage during Okazaki fragment maturation. J Biol Chem. 2004
       Dec 3;279(49):50840-9. 15448135

[189] Sharma S, Sommers JA, Brosh RM, Jr. In vivo function of the conserved non-catalytic
       domain of Werner syndrome helicase in DNA replication. Hum Mol Genet. 2004 Oct
       1;13(19):2247-61. 15282207

[190] Szekely AM, Chen YH, Zhang C, Oshima J, Weissman SM. Werner protein recruits
       DNA polymerase delta to the nucleolus. Proc Natl Acad Sci U S A. 2000 Oct
       10;97(21):11365-70. 11027336

[191] Kamath-Loeb AS, Loeb LA, Johansson E, Burgers PM, Fry M. Interactions between
       the Werner syndrome helicase and DNA polymerase delta specifically facilitate
       copying of tetraplex and hairpin structures of the d(CGG)n trinucleotide repeat se-
       quence. J Biol Chem. 2001 May 11;276(19):16439-46. 11279038

[192] Fry M, Loeb LA. Human werner syndrome DNA helicase unwinds tetrahelical struc-
       tures of the fragile X syndrome repeat sequence d(CGG)n. J Biol Chem. 1999 Apr
       30;274(18):12797-802. 10212265

[193] Riha K, Heacock ML, Shippen DE. The role of the nonhomologous end-joining DNA
       double-strand break repair pathway in telomere biology. Annu Rev Genet.
       2006;40:237-77. 16822175

[194] Cheng WH, von Kobbe C, Opresko PL, Arthur LM, Komatsu K, Seidman MM, Car-
       ney JP, Bohr VA. Linkage between Werner syndrome protein and the Mre11 complex
       via Nbs1. J Biol Chem. 2004 May 14;279(20):21169-76. 15026416

[195] Celli GB, Denchi EL, de Lange T. Ku70 stimulates fusion of dysfunctional telomeres
       yet protects chromosome ends from homologous recombination. Nat Cell Biol. 2006
       Aug;8(8):885-90. 16845382

[196] Laud PR, Multani AS, Bailey SM, Wu L, Ma J, Kingsley C, Lebel M, Pathak S, DePin-
       ho RA, Chang S. Elevated telomere-telomere recombination in WRN-deficient, telo-

mere dysfunctional cells promotes escape from senescence and engagement of the ALT pathway. Genes Dev. 2005 Nov 1;19(21):2560-70. 16264192

[197] Harrigan JA, Wilson DM, 3rd, Prasad R, Opresko PL, Beck G, May A, Wilson SH, Bohr VA. The Werner syndrome protein operates in base excision repair and cooperates with DNA polymerase beta. Nucleic Acids Res. 2006;34(2):745-54. 16449207

[198] Blank A, Bobola MS, Gold B, Varadarajan S, D DK, Meade EH, Rabinovitch PS, Loeb LA, Silber JR. The Werner syndrome protein confers resistance to the DNA lesions N3-methyladenine and O6-methylguanine: implications for WRN function. DNA Repair (Amst). 2004 Jun 3;3(6):629-38. 15135730

[199] Harrigan JA, Opresko PL, von Kobbe C, Kedar PS, Prasad R, Wilson SH, Bohr VA. The Werner syndrome protein stimulates DNA polymerase beta strand displacement synthesis via its helicase activity. J Biol Chem. 2003 Jun 20;278(25):22686-95. 12665521

[200] Li B, Navarro S, Kasahara N, Comai L. Identification and biochemical characterization of a Werner's syndrome protein complex with Ku70/80 and poly(ADP-ribose) polymerase-1. J Biol Chem. 2004 Apr 2;279(14):13659-67. 14734561

[201] von Kobbe C, Harrigan JA, Schreiber V, Stiegler P, Piotrowski J, Dawut L, Bohr VA. Poly(ADP-ribose) polymerase 1 regulates both the exonuclease and helicase activities of the Werner syndrome protein. Nucleic Acids Res. 2004;32(13):4003-14. 15292449

[202] Bryant HE, Schultz N, Thomas HD, Parker KM, Flower D, Lopez E, Kyle S, Meuth M, Curtin NJ, Helleday T. Specific killing of BRCA2-deficient tumours with inhibitors of poly(ADP-ribose) polymerase. Nature. 2005 Apr 14;434(7035):913-7. 15829966

[203] Farmer H, McCabe N, Lord CJ, Tutt AN, Johnson DA, Richardson TB, Santarosa M, Dillon KJ, Hickson I, Knights C, Martin NM, Jackson SP, Smith GC, Ashworth A. Targeting the DNA repair defect in BRCA mutant cells as a therapeutic strategy. Nature. 2005 Apr 14;434(7035):917-21. 15829967

[204] Bohr VA, Souza Pinto N, Nyaga SG, Dianov G, Kraemer K, Seidman MM, Brosh RM, Jr. DNA repair and mutagenesis in Werner syndrome. Environ Mol Mutagen. 2001;38(2-3):227-34. 11746759

[205] Yun J, Zhong Q, Kwak JY, Lee WH. Hypersensitivity of Brca1-deficient MEF to the DNA interstrand crosslinking agent mitomycin C is associated with defect in homologous recombination repair and aberrant S-phase arrest. Oncogene. 2005 Jun 9;24(25):4009-16. 15782115

[206] Cheng WH, Kusumoto R, Opresko PL, Sui X, Huang S, Nicolette ML, Paull TT, Campisi J, Seidman M, Bohr VA. Collaboration of Werner syndrome protein and BRCA1 in cellular responses to DNA interstrand cross-links. Nucleic Acids Res. 2006;34(9): 2751-60. 16714450

[207] Mann MB, Hodges CA, Barnes E, Vogel H, Hassold TJ, Luo G. Defective sister-chromatid cohesion, aneuploidy and cancer predisposition in a mouse model of type II Rothmund-Thomson syndrome. Hum Mol Genet. 2005 Mar 15;14(6):813-25. 15703196

[208] Sangrithi MN, Bernal JA, Madine M, Philpott A, Lee J, Dunphy WG, Venkitaraman AR. Initiation of DNA replication requires the RECQL4 protein mutated in Rothmund-Thomson syndrome. Cell. 2005 Jun 17;121(6):887-98. 15960976

[209] Matsuno K, Kumano M, Kubota Y, Hashimoto Y, Takisawa H. The N-terminal noncatalytic region of Xenopus RecQ4 is required for chromatin binding of DNA polymerase alpha in the initiation of DNA replication. Mol Cell Biol. 2006 Jul;26(13):4843-52. 16782873

[210] Im JS, Ki SH, Farina A, Jung DS, Hurwitz J, Lee JK. Assembly of the Cdc45-Mcm2-7-GINS complex in human cells requires the Ctf4/And-1, RecQL4, and Mcm10 proteins. Proc Natl Acad Sci U S A. 2009 Sep 15;106(37):15628-32. 19805216

[211] Kumata Y, Tada S, Yamanada Y, Tsuyama T, Kobayashi T, Dong YP, Ikegami K, Murofushi H, Seki M, Enomoto T. Possible involvement of RecQL4 in the repair of double-strand DNA breaks in Xenopus egg extracts. Biochim Biophys Acta. 2007 Apr;1773(4):556-64. 17320201

[212] Petkovic M, Dietschy T, Freire R, Jiao R, Stagljar I. The human Rothmund-Thomson syndrome gene product, RECQL4, localizes to distinct nuclear foci that coincide with proteins involved in the maintenance of genome stability. J Cell Sci. 2005 Sep 15;118(Pt 18):4261-9. 16141230

[213] Woo LL, Futami K, Shimamoto A, Furuichi Y, Frank KM. The Rothmund-Thomson gene product RECQL4 localizes to the nucleolus in response to oxidative stress. Exp Cell Res. 2006 Oct 15;312(17):3443-57. 16949575

[214] Werner SR, Prahalad AK, Yang J, Hock JM. RECQL4-deficient cells are hypersensitive to oxidative stress/damage: Insights for osteosarcoma prevalence and heterogeneity in Rothmund-Thomson syndrome. Biochem Biophys Res Commun. 2006 Jun 23;345(1):403-9. 16678792

[215] Wang M, Wu W, Wu W, Rosidi B, Zhang L, Wang H, Iliakis G. PARP-1 and Ku compete for repair of DNA double strand breaks by distinct NHEJ pathways. Nucleic Acids Res. 2006;34(21):6170-82. 17088286

[216] Audebert M, Salles B, Calsou P. Involvement of poly(ADP-ribose) polymerase-1 and XRCC1/DNA ligase III in an alternative route for DNA double-strand breaks rejoining. J Biol Chem. 2004 Dec 31;279(53):55117-26. 15498778

[217] Bailis JM, Luche DD, Hunter T, Forsburg SL. Minichromosome maintenance proteins interact with checkpoint and recombination proteins to promote s-phase genome stability. Mol Cell Biol. 2008 Mar;28(5):1724-38. 18180284

[218] Gambus A, Jones RC, Sanchez-Diaz A, Kanemaki M, van Deursen F, Edmondson RD, Labib K. GINS maintains association of Cdc45 with MCM in replisome progression complexes at eukaryotic DNA replication forks. Nat Cell Biol. 2006 Apr;8(4):358-66. 16531994

[219] Katou Y, Kanoh Y, Bando M, Noguchi H, Tanaka H, Ashikari T, Sugimoto K, Shira-hige K. S-phase checkpoint proteins Tof1 and Mrc1 form a stable replication-pausing complex. Nature. 2003 Aug 28;424(6952):1078-83. 12944972

[220] Calzada A, Hodgson B, Kanemaki M, Bueno A, Labib K. Molecular anatomy and reg-ulation of a stable replisome at a paused eukaryotic DNA replication fork. Genes Dev. 2005 Aug 15;19(16):1905-19. 16103218

[221] Komata M, Bando M, Araki H, Shirahige K. The direct binding of Mrc1, a checkpoint mediator, to Mcm6, a replication helicase, is essential for the replication checkpoint against methyl methanesulfonate-induced stress. Mol Cell Biol. 2009 Sep;29(18): 5008-19. 19620285

[222] Nedelcheva MN, Roguev A, Dolapchiev LB, Shevchenko A, Taskov HB, Shevchenko A, Stewart AF, Stoynov SS. Uncoupling of unwinding from DNA synthesis implies regulation of MCM helicase by Tof1/Mrc1/Csm3 checkpoint complex. J Mol Biol. 2005 Apr 1;347(3):509-21. 15755447

[223] Lou H, Komata M, Katou Y, Guan Z, Reis CC, Budd M, Shirahige K, Campbell JL. Mrc1 and DNA polymerase epsilon function together in linking DNA replication and the S phase checkpoint. Mol Cell. 2008 Oct 10;32(1):106-17. 18851837

[224] Szyjka SJ, Viggiani CJ, Aparicio OM. Mrc1 is required for normal progression of rep-lication forks throughout chromatin in S. cerevisiae. Mol Cell. 2005 Sep 2;19(5):691-7. 16137624

[225] Alcasabas AA, Osborn AJ, Bachant J, Hu F, Werler PJ, Bousset K, Furuya K, Diffley JF, Carr AM, Elledge SJ. Mrc1 transduces signals of DNA replication stress to activate Rad53. Nat Cell Biol. 2001 Nov;3(11):958-65. 11715016

[226] Bjergbaek L, Cobb JA, Tsai-Pflugfelder M, Gasser SM. Mechanistically distinct roles for Sgs1p in checkpoint activation and replication fork maintenance. Embo J. 2005 Jan 26;24(2):405-17. 15616582

[227] Bando M, Katou Y, Komata M, Tanaka H, Itoh T, Sutani T, Shirahige K. Csm3, Tof1, and Mrc1 form a heterotrimeric mediator complex that associates with DNA replica-tion forks. J Biol Chem. 2009 Dec 4;284(49):34355-65. 19819872

[228] Pryce DW, Ramayah S, Jaendling A, McFarlane RJ. Recombination at DNA replica-tion fork barriers is not universal and is differentially regulated by Swi1. Proc Natl Acad Sci U S A. 2009 Mar 24;106(12):4770-5. 19273851

Chapter 5

Proposal for a Minimal DNA Auto-Replicative System

Agustino Martinez-Antonio,
Laura Espindola-Serna and Cesar Quiñones-Valles

Additional information is available at the end of the chapter

1. Introduction

DNA replication allows cell division and population growth of living organisms. Here we will focus on DNA replication in prokaryotic single celled microorganisms. Several excellent reviews of the molecular processes that carry out DNA replication in bacteria already exist, *E. coli* being the model described in most detail (Langston LD et al., 2009; Quiñones-Valles et al., 2011). Briefly, the process begins when DnaA (DNA initiator replication protein) in its activated form (DnaA-ATP) recognizes and binds the *oriC* (origin of replication on the bacterial chromosome). In the following step, the replisome is assembled and binds to the complex of DnaA-ATP at the *oriC*. Next, the DNA strands are separated and synthesis of the complementary strands initiates followed by elongation steps. The molecular mechanisms of elongation differ depending on the strand used as a template; the leading strand is replicated continuously starting from a unique RNA primer, whereas on the lagging strand DNA polymerase III must recognize several RNA primers, previously synthesized by DnaG, and then replicate each DNA fragment (Okazaki fragments). This is followed by the replacement of RNA primers by DNA polymerase I, and removal of nicks by a DNA ligase. The whole process concludes when replisomes reach the *ter* site, almost opposite to *oriC* on the circular DNA molecule. Tus proteins are attached to the *ter* sites and when replisomes reach these complexes, they collide and finally are disassembled (see Figure 1 for an overview of the whole process).

From another aspect, one of the more challenging areas of Synthetic Biology is the design and construction of minimal cells. The accomplishment of this aim might contribute to answering basic questions about the minimal components necessary to sustain life systems, in addition to cell auto-organization, function and evolution. In a practical application, minimal cells can be used as a background chassis for the generation of dedicated biological systems designed for the synthesis or degradation of diverse compounds of interest.

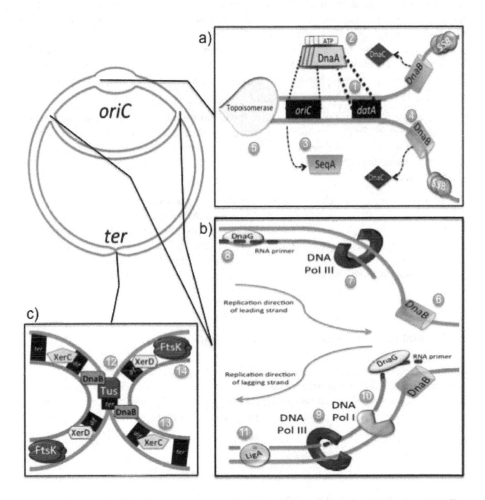

Figure 1. Main steps of DNA replication in bacteria. a) Initiation of DNA replication; the *datA* locus has a high affinity for binding DnaA (1). DnaA binds to ATP, homo-multimers of DnaA-ATP are formed (2). These homo-multimers bind to *oriC* and once replication is initiated SeqA binds this region and prevents initiation of a new replication event (3). The SSB (single strand binding) protein and DnaB assist the complex to open the DNA strands and release DnaC (4). A DNA topoisomerase helps to further unfold the DNA strands (5). b) The elongation phase; the replication fork is formed and the replisome is assembled (6). DNA polymerase III replicates the leading strand (7). DnaG incorporates RNA primers as primers for replication of the lagging strand (8). Polymerase III can now replicate Okazaki fragments on the lagging strand (9). DNA polymerase I replaces RNA nucleotides for DNA nucleotides (10). A DNA ligase (LigA) seals the nicks on contiguous DNA fragments (11). c) Termination of DNA replication; The protein Tus binds to the *ter* sites, when replisomes reach Tus, replication ceases (12). The recombinases XerC and XerD resolve the replicated DNA strands (13). Finally, FtsK translocates the DNA strands and each double-stranded DNA molecule can be liberated (14).

In recent years, the essential properties and capabilities necessary to develop minimum cells have been broadly speculated (MacDonald et al., 2011). Among these characteristics it is evident that DNA replication should be a fundamental property of these biosystems. Many genes for DNA replication are found to be conserved when comparative analysis of bacterial genomes is carried out. These types of genes are considered as informational genes, in charge of maintaining the genetic code, and are among the genes less frequently found be horizontally transferable (Jain et al, 1999). Therefore by genomic comparisons and functional analyses it is possible to propose a minimum core of genes capable of supporting the process of DNA replication.

From a genetic point of view, and for the purpose of this study it is important to state our definition of a minimal DNA auto-replicative system (MiDARS) as: *a genetic system comprising the minimum number of DNA components, including regulatory elements and gene products necessary for the auto-replication of the DNA molecule on which they are encoded, functioning in an in vitro condition.*

In this chapter we will develop a proposal for the construction of such an auto-replicative DNA system. This system is designed to serve as a scaffold for the incorporation of additional biological functions such as transcription and translation, etc. For the scaffold design we exploit information of genes necessary for replication in *E. coli* that are highly conserved in bacteria with extremely reduced genomes and analyze their functional role in DNA replication in order to finally propose a minimal genetic system with a DNA auto-replicatory function.

2. Minimal cells and minimal genetic systems

A minimal cell can be defined as a biological system that has the minimal number of genetic parts and molecular components for supporting life functions under defined growth conditions. In other words, it includes only the necessary number of genes and derived biomolecular machinery that are considered basic to support life functions (Jewett and Forster, 2010).

The concept of life is intrinsically complex; in biochemical terms it could be defined by three basic characteristics (Luisi et al., 2006):

1.   auto-regulation of metabolism,

2.   auto-replication of the genetic material and,

3.   controlled evolution of their components and functions.

The design and synthesis of minimal cells depends on the environmental conditions the systems will be exposed to. Initially, we might consider that a minimal cell should be exposed to the most favorable conditions in order to facilitate its conception and function. These favorable conditions will require an environment where the cell is not suffering any kind of environmental stress. Nonetheless, even this ideal scenario is a challenging condition to direct the rational design of components of a minimal cellular system since the genes for many cellular functions are not yet totally defined. What we could do is to start to reconstruct minimal biological functions that are more or less well defined. These might be the processes relating to the central dogma of molecular biology: DNA replication, DNA transcription and mRNA translation (Figure 2). Some of these functions have been the object of different studies; e. g. transcription and translation were successfully recreated in the experiment of Asahara (2010) by separately expressing the components of the *E. coli* RNA polymerase, including the sigma70 factor and reconstituting the function of the complete enzyme *in vitro*.

Since one of the fundamental characteristics of life systems is the replication of their own genetic material, we can consider the design of minimal genetic systems that sustain DNA auto-replication as an important to starting point.

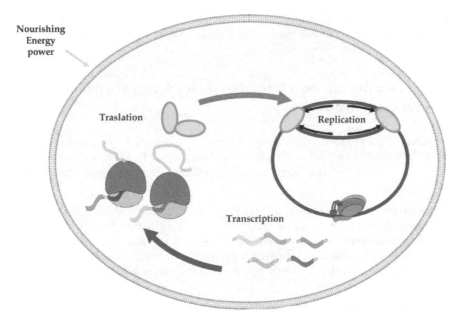

Figure 2. Representation of a hypothetical minimal auto-replicative system. One of the key features of the minimal cell is that it should perform basic functions such as transcription, translation and replication of the genetic information contained in its genome.

3. Approaches for the development of minimal genetic systems

Currently there are two approaches for the study of minimal biological systems. These are the *top down* and *bottom up* strategies (Delaye L & Moya A, 2009; Murtas, 2009). The *top down* approach considers the analysis of existing biological systems and, by following a reductionist approach, looks to minimize the number of components either by searching for conserved genetic elements or by experimentally reducing the genome without losing functionality. This strategy was used to reduce the *E. coli* genome by 15% by deleting non-essential genes, recombinogenic and mobile DNA elements, and cryptic genes. The resulting cells had good growth profiles and showed improved performance for protein production (Pósfai et al., 2006). Another focus of this approach is to carry out comparative genomics and define a set of conserved genes such as those in charge of specific functions (Gil et al., 2004; Forster & Church, 2006).

On the other hand, the *bottom up* approach involves the construction of complex systems starting from relatively simple molecular precursors. A classical example is the experiment of Miller, who obtained amino acids from a mixture of simple organic and inorganic molecules (Miller, 1953).

Considering the design and construction of minimal genetic systems, benefits should be obtained by employing both complementary *top down* and *bottom up* approaches.

4. *Escherichia coli* as a model organism for the design of a minimal DNA auto-replicative system

Escherichia coli is a bacillary Gram-negative, aerobic, facultative and non-sporulating organism. It was discovered in 1885 by the physician Theodore von Escherich and is now classified as part of the Enterobacteriaceae family of the Gamma-proteobacterias (Blattner et al., 1997).

This bacterium lives in the intestine of mammals, and assists its hosts with assimilation of nutrients, providing some vitamins and preventing the establishment of bacterial pathogens. Since its discovery, *E. coli* has been widely used as a working model in the laboratory to study biochemistry and diverse molecular processes. In addition, it has been widely used in biotechnology as a vehicle for the expression of multiple recombinant proteins and whole metabolic pathways.

Arthur Kornberg was one of the most prominent investigators in molecular biology and a pioneer in the description of the replicative process using *E. coli* as a model. For his accomplishments in the field he was awarded the Nobel Prize in Physiology and Medicine in 1959. He discovered DNA polymerase I (Bessman et al., 1958; Lehman et al., 1958a), and describes the synthesis of DNA as a process based on the use of a single strand of DNA as a template (Lehman et al., 1958b). Later, Kornberg and his collaborators discovered additional enzymes involved in DNA replication: DNA primase, DNA helicase, DnaA, PriA among others. Nowadays, the replication process and the replicative enzymes of *E. coli* are the best understood and characterized of any organism.

From a biotechnological standpoint, *E. coli* shows three important characteristics that make it an ideal organism to serve as the platform for the design of a synthetic cellular program (Foley & Shuler, 2010):

1. functionally it is the organism best characterized at the molecular and biochemical levels in terms of components of metabolism,

2. it has proven to be a robust vehicle for the expression of multiple biotechnological processes,

3. it has a short growth cycle and is easy to manipulate genetically.

Additionally, the genome of *E. coli* serves as the principal source of standardized genetic parts for the construction of genetic circuits, the "BioBricks", in a project whose aim is to standardize genetic parts to facilitate biological engineering (http://partsregistry.org), (Smolke, 2009). Most biobricks are designed to function in *E. coli*, therefore, we think *E. coli* is the best organism of choice for the design of a DNA auto-replicative system.

5. Comparison of the DNA replicative machinery of *E. coli* with that of bacteria with reduced genomes

Comparative genomics is a powerful approach that allows the identification of genetic sequences sharing identity/similarity among different organisms. Through these comparisons it is possible to identify conserved genes and predict the components of the replicative machinery in several different organisms.

For our purpose, among the organisms of interest to consider in our design are those with extremely reduced genomes. A characteristic of these organisms is that they are incapable of growth in a free-living manner. The genomes of organisms with these characteristics correspond to those having the minimum number of genes possible in nature. From these we chose the 25 organisms with the most reduced genomes known to date (Table 1). All of these genomes contain less than 1,200 kbp of DNA and all are endosymbiotic bacteria, most of which are thought to survive at the expense of the host.

In these organisms, we searched for genes encoding enzymes involved in DNA replicative functions with orthology to the replicative machinery from *E. coli* (Table 2). To find orthologous genes we followed two complementary strategies: we looked for Clusters of Orthologous Groups (COGs, Tatusov et al., 2003) and also used bidirectional best blast hits (Moreno-Hagelsieb & Latimer, 2008). In Table 2 the blue cells indicate where genes orthologous to *E. coli* are present in the target organism. In the table we show orthologous genes to be present in at least fifteen of these bacteria. Remarkably, bacteria with the most reduced genomes; *Carsonella ruddii* PV (Nakabachi et al., 2006; Tamames et al., 2007), *Hodgkinia cicadicola Dsem* (McCutcheon et al., 2009) and *Tremblaya princeps* PCIT (López-Madrigal et al., 2011; McCutcheon & Moran, 2011) had only 5, 3 and 5 genes related to replication respectively which were orthologous to *E. coli*. These three organisms are strict endosymbionts of insects, with the smallest genomes known to date (Table 2). The fact that these bacteria showed fewer genes related to DNA replication in comparison to bacteria with larger genomes (Figure 3), indicates that the minimal replicative machinery in these organisms might be composed by a small number of constituents. This observation raises many open questions, for instance:

Organism	Abrev	Total length	Total genes	Replic. genes	Classification	Symbiotic Relation/interaction	Host
Escherichia coli K-12 MG1655	eco	4,639,675	4,146	228	Proteobacteria-gamma	Free living	*Homo sapiens*
Anaplasma marginale St. Maries	ama	1,197,687	948	55	Proteobacteria-alpha	Parasitic, Endosymbiotic	*Homo sapiens*
Midichloria mitochondrii IricVA	mmn	1,183,732	1,211	65	Proteobacteria-alpha	Pathogen	*Ixodes ricinus*
Chlamydophila caviae GPIC	cca	1,181,356	1,005	85	Chlamydiae	Parasitic, Endosymbiotic	*Homo sapiens*, swine
Ruthia magnifica Cm	rma	1,160,782	976	61	Proteobacteria-gamma	Endosymbiotic	*Calyptogena magnifica*
Treponema pallidum Nichols	tpa	1,138,011	1,036	74	Spirochaetes	Host-associated Free living	*Homo sapiens*
Rickettsia prowazekii Madrid E	rpr	1,111,523	835	85	Proteobacteria-alpha	Parasitic-endosymbiotic	*Homo sapiens*
Chlamydia trachomatis D/UW-	ctr	1,042,519	895	83	Chlamydiae	Parasitic, Endosymbiotic	*Homo sapiens*
Vesicomyosocius okutanii HA	vok	1,022,154	937	49	Proteobacteria-gamma	Endosymbiotic	*Calyptogena okutanii*
Tropheryma whipplei TW08/27	tws	925,938	783	68	Actinobacteria	Pathogen	*Homo sapiens*
Neorickettsia sennetsu Miyayama	nse	859,006	932	46	Proteobacteria-alpha	Pathogen	*Homo sapiens*
Mesoplasma florum L1; ATCC 33453	mfl	793,224	682	85	Firmicutes	Parasite, non-pathogen	*Citrus limon*
Ureaplasma parvum ATCC 700970	uur	751,719	614	52	Tenericutes	Pathogen	*Homo sapiens*
Blochmannia floridanus	bfl	705,557	583	37	Proteobacteria-gamma	Endosymbiont	*Camponotus floridanus*
Wigglesworthia glossinidia	wbr	703,004	617	41	Proteobacteria-gamma	Endosymbiotic	*Glossina brevipalpis*
Baumannia cicadellinicola Hc	bci	686,194	595	46	Proteobacteria-gamma	Endosymbiotic	Insect endosymbiont
Buchnera aphidicola APS	buc	655,725	574	48	Proteobacteria-gamma	Mutualistic, Endosymbiotic	*Acyrthosiphon pisum*
Blattabacterium sp. BPLAN	bpi	640,442	581	39	Bacteroidetes	Endosymbiotic	*Periplaneta americana*
Riesia pediculicola USDA	rip	582,127	556	31	Proteobacteria-gamma	Endosymbiotic	*Pediculus humanus*
Mycoplasma genitalium G37	mge	580,076	475	40	Tenericutes	Host-associated	*Homo sapiens*
Moranella endobia PCIT	men	538,294	406	41	Proteobacteria-gamma	Host-associated	*Planococcus citri*
Sulcia muelleri GWSS	smg	245,530	227	11	Bacteroidetes	Endosymbiotic	*Glassy-winged sharpshooter*
Zinderia insecticola CARI	zin	208,564	202	15	Proteobacteria beta	Endosymbiotic	Spittlebugs
Carsonella ruddii PV	cpr	159,662	182	5	Proteobacteria-gamma	Endosymbiotic intracellular	*Pachypsylla venusta*
Hodgkinia cicadicola Dsem	hci	143,795	169	3	Proteobacteria-alpha	Endosymbiotic	*Diceroprocta semicincta*
Tremblaya princeps PCIT	tpn	138,927	121	5	Proteobacteria-beta	Endosymbiotic	Mealybug *Planococcus citri* (Risso)

Table 1. Bacteria with genome sizes less than 1200 kbp

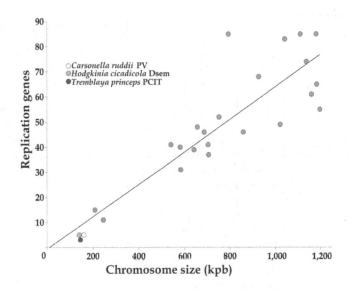

Figure 3. Conservation of DNA replicative machinery in bacteria with reduced genomes. Graph showing the relationship between number of genes annotated with DNA replicative functions versus genome sizes.

i.   Are these genes sufficient to sustain the process of replication of an entire chromosome?;

ii.   Does the host supply the missing elements for replication of the endosymbionts DNA? and,

iii.   Do these organisms use additional proteins in comparison to those currently described for the process of DNA replication?

The apparent requirement of only a handful of genes for DNA replication in extremely reduced genomes, compared with the 228 annotated in *E. coli*, might suggest a parsimonious mechanism of DNA replication in endosymbiont bacteria since they are always living in stable environments. The genes which are more highly conserved in both reduced genomes and *E. coli* are those whose products form the replisome (*dnaE*, *dnaB*, *dnaN*, *dnaG*, *dnaX*, *dnaQ*, *ssb*, *holA* and *holB*), the genes encoding for DNA topoisomerase type II (*gyrA* and *gyrB*), and the gene for the NAD(+)-dependent DNA-ligase, *ligA* (Figure 4).

6. Components of a Minimal DNA Auto-Replicative System (MiDARS)

Of the three organisms with the most reduced genomes in nature, *Carsonella ruddii* is the more closely related phylogenetically to *E. coli* (Nakabachi et al., 2006). For this reason in our design we used the information of the replicative machinery in *Carsonella ruddii* and the functions known in *E. coli*. For the physical construction of the systems, however, we will use genes from *E. coli* for two main reasons:

Organisms/ Gene	eco	rma	tpa	cca	ama	rpr	tws	ctr	vok	mmn	mfl	uur	wbr	bfl	bci	nse	bpi	rip	buc	mge	men	smg	zin	crp	hci	tpn
dnaA																										
priA																										
dnaB																										
dnaG																										
ssb																										
dnaE																										
dnaN																										
dnaQ																										
dnaX																										
holA																										
holB																										
polA																										
ligA																										
gyrA																										
gyrB																										
topA																										
recA																										
rnhA																										
ruvA																										
ruvB																										
tatD																										
ung																										
uvrD																										
yqgF																										

Table 2. Conservation of replicative genetic machinery in bacteria with less than 1200 kbp

1.   it is difficult to obtain genomic DNA from *Carsonella ruddii* since it cannot be cultured *in vitro*.

2.   *E. coli* is the best chassis for applications in synthetic biology as mentioned above and therefore adequate for the incorporation of additional functions.

For the design of the minimal DNA auto-replicative system we will attempt to include the minimal elements present in *Carsonella* and -in a conservative manner- those we presume as necessary to perform the process of DNA replication in *E. coli*. In addition to the coding sequences, it is also necessary to define the regulatory regions of the genes and we propose to conserve the operative regions as defined for *E. coli* with the future aim of expanding the minimal functions of *E. coli* including the regulatory functions. Other important regions to include in the design are: the DNA replication origin (*oriC*) and the signals for termination of replication. Below we propose the genetic components that would constitute a MiDARS.

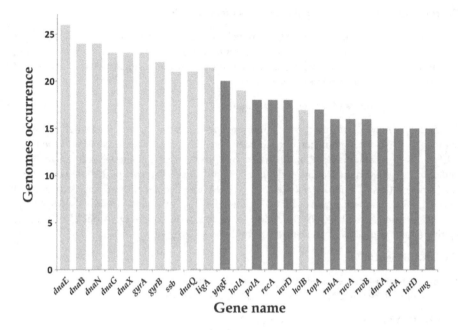

Figure 4. Conservation of genes for DNA replication in 25 reduced genomes and in E. *coli*. Grey bars represent the genes proposed to be essential for DNA auto-replication in a minimal genetic system.

The DNA initiator protein (*dnaA*)

At the beginning of the replication process, check-point proteins have to recognize and un- fold the initiation site for replication at *oriC*. In *E. coli* DnaA is the principal protein em- ployed for this purpose and is highly conserved among bacteria with reduced genomes. Therefore, *dnaA* should be present in the MiDARS.

The DNA helicase (*dnaB*)

The next candidate gene is *dnaB*, which encodes a DNA helicase. The role of the product of this gene is to unwind the DNA strands, a very important process during the elongation stage of replication.

The DNA primase (*dnaG*)

The gene that encodes the primase (*dnaG*) should also be considered. It is important for the synthesis of the RNA primers that permit the elongation of new DNA strands.

The single strand stabilization protein (*ssb*)

Another important function is the stabilization of single strands, carried out by the SSB pro- tein, encoded by the *ssb* gene.

The core components of DNA polymerase III (*dnaE* and *dnaQ*)

The gene for the α subunit (*dnaE*) of DNA polymerase III is present in all 25 organisms with reduced genomes and the gene for the ε subunit (*dnaQ*) in twenty-one. These proteins form part of the core of DNA polymerase III, which carries out the essential polymerization and proofreading activities during DNA synthesis.

The clamp components (*dnaX, holA, holB, dnaN*)

During the elongation stage, two very important structures are formed; the leader and slider clamps. The first has the function of anchorage between DNA polymerase III and the DNA helicase; (Reyes-Lamothe R. et al., 2010) allowing the synthesis of the DNA in a synchronized manner between the leading and lagging strand. It is composed of the following subunits (genes): τ (dnaX), γ (*dnaX*), δ (*holA*) and δ'(*holB*). The circular slider clamp is constituted by two β-subunits (both products of *dnaN* gene), that recognize and bind to DNA-RNA hybrids (Georgescu R. et al., 2010). The slider clamp assists the core of DNA pol III to bind the lagging strand and allows the extension of the Okazaki fragments.

The DNA ligase (*ligA*)

The function of a ligase is needed for sealing nicks formed when the RNA primers are removed and replaced by DNA in the Okazaki fragments on the lagging strand.

Type II DNA topoisomerase (*gyrA* and *gyrB*)

We consider that a relaxing system produced by a DNA helicase may be necessary. This could be provided by the DNA gyrase complex (Type II Topoisomerase) composed of the A (*gyrA*) and B (*gyrB*) subunits.

Protein for termination of replication (*tus*)

Although there are several proteins that could contribute to termination of DNA replication we think that in a minimal system, the action of Tus could be enough to ensure this.

Origin of DNA replication (*oriC*)

This DNA sequence of around 245 bp in *E. coli* (Tabata et al., 1983) is needed to enable the DnaA protein to initiate the process of DNA replication

Termination of DNA replication (*terB* and *terC*)

These sequences are used by the Tus proteins to form the trap which terminates DNA replication.

The proposed elements that constitute the auto-replicative system are also listed in Table 3. This proposal is somewhat similar to previous reports, where genes that could constitute a minimal cell based on a comparative genomics study among various endosymbionts are described (Gil et al., 2004). In the present study however we also considered the inclusion of the DNA regions for initiation and termination of replication, as well as the *dnaA, ssb* and *tus* genes.

Gene/DNA element	Product	Size (bp. *E. coli*)
dnaA	Chromosomal replication initiator protein DnaA	1404
dnaB	Replicative DNA helicase	1416
dnaG	DNA primase	1746
ssb	Single-stranded DNA-binding protein	537
dnaE	DNA polymerase III α subunit	3483
dnaN	DNA polymerase III, β subunit	1101
dnaQ	DNA polymerase III ε subunit	732
dnaX	DNA polymerase III, τ and γsubunits	1932
holA	DNA polymerase III,δ subunit	1032
holB	DNA polymerase III, δ' subunit	1005
polA	DNA pol I 5'-3' and 3'-5' exonuclease ; 3'-5' polymerase	2787
ligA	DNA ligase, NAD(+)-dependent	2016
gyrA	DNA gyrase (type II topoisomerase), subunit A	2628
gyrB	DNA gyrase, subunit B	2415
tus	Termination DNA replication protein	930
oriC	DNA region for initiation, origin of replication	245
ter	DNA region for termination of replication	23

Table 3. Components of a minimal DNA auto-replicative system.

7. Expression of the replicative proteins of the MiDARS

A primary condition for the operation of an auto replicative system is that the protein-machinery encoded in it should be expressed. For transcription of the assembled group of genes, we propose use the *E. coli* RNA polymerase and its transcription factor sigma70 since all the genes of the system have a sigma70 factor promoter. The essential components of the RNA polymerase and their sigma70 factors have previously been successfully expressed separately and their activity reconstituted as mentioned previously (Asahara & Chong, 2010). We propose these components can be assembled as an additional functional module whose activity can be assayed separately and subsequently integrated into the system. The resulting mRNA (16) could be translated in an *in vitro* system such as the Pure System™ (Ueda et al., 1992; Shimizu & Ueda, 2010); containing ribosomes, aminoacyl-tRNAs, chaperones and initiation, elongation and termination factors among other elements essential for translation. Once protein synthesis is completed, the products could initiate replication of the DNA molecule for which the addition of deoxynucleotide triphosphates (dNTPs) and the appropriate buffers will be necessary. The source of energy for the system will be creatine phosphate with the creatine kinase enzyme as the regenerator (Shimizu et al., 2006). An outline for the operation of the DNA auto-replicative system is shown in Figure 5.

8. Perspectives

Previous efforts have been made to propose the design of minimal cells however this objective is still far from being accomplished. From the standpoint of Synthetic Biology, biological systems that are robust, predictable in performance and highly efficient are desired (Jewett & Forster, 2010). In this work, we present a proposal to build an auto-replicative DNA system as the first step toward the development of synthetic biosystems. Additional cellular processes will need to be designed and constructed in a modular way including: transcriptional and translational functions and a minimal metabolism in order to maintain cell growth and produce energy.

Once this first prototype has been constructed and tested for performance, some further reduced combinations of the proposed number of genes could be tested to determine the absolute minimum set of genes sufficient to sustain DNA auto-replication; e.g. the few genes present in *Carsonella ruddii* PV.

The system proposed in this work can be assembled using methodologies such as that used when working with Biobricks (Smolke, 2009). Once the mini-chromosome is assembled it could function in cell-free systems, in anucleated mini-cells (Adler et al., 1967), in spores that lack DNA (Siccardi et al., 1975), in micelles or lipidic vesicles, and in some commercial systems. An important achievement in this sense has previously been reported by another research group, namely DNA replication achieved by using the Phi29 DNA polymerase, inside a lipidic vesicle. In this report only one strand was linearly replicated and circularized (Kurihara, 2011).

The successful development of a DNA auto-replicative system as proposed here could be a very important platform for the development of synthetic biology and the potential for such a system is great:

1.   in the refinement of biotechnological processes since cellular energy could be directed to the desired biosynthetic pathways;

2.   in the study of synthetic or natural circuits at a higher resolution and sharpness due to the minimalization of cellular noise; and

3.   to test some evolutionary hypotheses, such as the proposed components of last common ancestors and components of rudimentary first cells, among others.

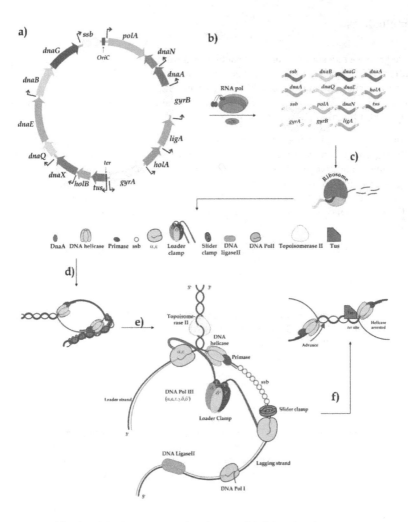

Figure 5. Proposal for the minimal components of a MiDARS and their function. a) The genetic system is a simplified version of a prokaryotic DNA mini-chromosome (25432 bp). The system contains the initiation region (*oriC*) and termination (*ter*) sites for DNA replication as well as a set of genes from *E. coli* (Table 4). The genes can be organized in the same order as in the native chromosome and contain their native operator regions to control expression. b) Transcription and, c) Translation can be carried out in solution using commercial kits (e.g. Pure System), RNA Polymerase and the *E. coli* sigma70 factor. d) The initiation of replication is regulated by DnaA-ATP and the helicase will join to the lagging strand in order to form the replication forks. The primase will bind to the helicase to carry out the synthesis of RNA primers that permit the activity of DNA pol III. The SSB protein stabilizes single strands of DNA. e) Two core subunits (α and ε) of the DNA Pol III, perform the elongation and proofreading of DNA. The DNA ligase and DNA Pol I replace the RNA primers, sealing the nicks between contiguous DNA fragments on the lagging strand. Topoisomerase II will relax the DNA template as the replication fork progresses. f) The Tus protein bound to the *ter* sites serves as a trap for the replicative machinery headed by the DNA helicase, stopping its movement and promoting the separation of the new MiDARS.

9. Conclusions

Here we propose a design for the construction of a minimal genetic system for DNA auto-replication. This proposal is based on the consideration of the latest knowledge of the details of the mechanisms and controls of DNA replication in *E. coli* and by taking into account the conservation of the replicative machinery in bacteria with extremely reduced genomes particularly those present in *Carsonella ruddii* PV.

The proposed auto-replicative device consists of 17 DNA elements (27822 bp including their operator regions) taken from the *E. coli* genome and incorporating the most conserved elements of the replicative machinery found in bacteria with extremely reduced genomes. These genetic elements will maintain their native operator and termination regions. Their products encode proteins encompassing the minimal number of predicted activities involved in DNA replication. Finally we propose some conditions in which the system might function.

Acknowledgments

Authors thank June Simpson for critical comments to the ms. This work was supported by CONACYT grants (102854 and 103686) given to AM-A. LE-S and CQ-V thank to CONACYT for PhD scholarships (208153 and 206011)

Author details

Agustino Martinez-Antonio, Laura Espindola-Serna and Cesar Quiñones-Valles

Departamento de Ingeniería Genética, Cinvestav, Irapuato-León, Irapuato Gto, México

References

[1] Adler, H., Fisher, W. D., Cohen, A., & Hardigree, A. (1967). Miniature Escherichia coli cells deficient in DNA. Proc. Natl. Acad. Sci. USA , 57, 321-326.

[2] Asahara, H., & Chong, S. (2010). In vitro genetic reconstruction of bacterial transcription initiation by coupled synthesis and detection of RNA polymerase holoenzyme. Nucleic Acids Res 3813:e141.

[3] Bessman, M. J., Lehman, I. R., Simms, E. S., & Kornberg, A. (1958). Enzymatic synthesis of deoxyribonucleic acid. II. General properties of the reaction. J Biol Chem ., 233, 171-177.

[4] Blattner, F. R., Plunkett, G., Bloch, C. A., Perna, N. T., Burland, V., Riley, M., Collado-Vides, J., Glasner, J. D., Rode, C. K., Mayhew, G. F., Gregor, J., Davis, N. W., Kirkpatrick, H. A., Goeden, M. A., Rose, D. J., Mau, B., & Shao, Y. (1997). The complete genome sequence of *Escherichia coli* K-12. Science , 277, 1453-1462.

[5] Delaye, L., & Moya, A. (2009). Evolution of reduced prokaryotic genomes and the minimal cell concept: Variations on a theme. *BioEssays*, 32, 281-287.

[6] Foley, P. L., & Shuler, M. L. (2010). Considerations for the design and construction of a synthetic platform cell for biotechnological applications. Biotechnol Bioeng , 105, 26-36.

[7] Forster, A. C., & Church, G. M. (2006). Towards synthesis of a minimal cell. Mol Syst Biol 2:45.

[8] Georgescu, R., Yao, N. Y., & O'Donnell, O. (2010). Single-molecule analysis of the *Escherichia coli* replisome and use of clamps to bypass replication barriers. *FEBS Letters*, 584, 2596-2605.

[9] Gil, R., Silva, F. J., Peretó, J., & Moya, A. (2004). Determination of the core of a minimal bacterial gene set. Microbiol Mol Biol Rev , 68, 518-37.

[10] Jain, R., Rivera, M. C., & Lake, J. (1999). Horizontal gene transfer among genomes: The complexity hypothesis. PNAS. , 96, 3801-3806.

[11] Jewett, M. C., & Forster, A. C. (2010). Update on designing and building minimal cells. Curr Opin Biotechnol , 21, 697-703.

[12] Keyamura, K., & Katayama, T. (2011). DnaA protein DNA-binding domain binds to Hda protein to promote inter-AAA+ domain interaction involved in regulatory inactivation of DnaA. J Biol Chem , 286, 29336-29346.

[13] Kurihara, K., Tamura, M., Shohda, K., Toyota, T., Suzuki, K., & Sugawara, T. (2011). Self-reproduction of supramolecular giant vesicles combined with the amplification of encapsulated DNA. *Nature Chemistry*, 3, 775-781.

[14] Langston, L. D., Indiani, C., & O'Donnell, M. (2009). Whither the replisome: emerging perspectives on the dynamic nature of the DNA replication machinery. *Cell Cycle*, 8, 2686-91.

[15] Lehman, I. R., Bessman, M. J., Simms, E. S. ., & Kornberg, A. (1958a). Enzymatic synthesis of deoxyribonucleic acid. I. Preparation of substrates and partial purification of an enzyme from *Escherichia coli*. J. Biol Chem , 233, 163-170.

[16] Lehman, I. R., Zimmerman, S. B., Adler, J., Bessman, M. J., Simms, E. S., & Kornberg, A. (1958b). Enzymatic synthesis of deoxyribonucleic acid. V. Chemical composition of enzymatically synthesized deoxyribonucleic acid. Proc Natl Acad. Sci U S A , 44, 1191-1196.

[17] López-Madrigal, S., Latorre, A., Porcar, M., Moya, A., & Gil, R. (2011). Complete genome sequence of "Candidatus *Tremblaya princeps*" strain PCVAL, an intriguing translational machine below the living-cell status. J Bacteriol. , 193, 5587-8.

[18] Luisi, P. L., Ferri, F., & Stano, P. (2006). Approaches to semi-synthetic minimal cells: a review. *Naturwissenschaften*, 93, 1-13.

[19] Mac, Donald. J. T., Barnes, C., Kitney, R. I., Freemont, P. S., & Stan, G. B. (2011). Computational design approaches and tools for synthetic biology. Integr Biol (Camb). , 3, 97-108.

[20] Mc Cutcheon, J., & Moran, N. (2011). Extreme genome reduction in symbiotic bacteria. Nat Rev Microbiol , 10, 13-26.

[21] Mc Cutcheon, J. P., Mc Donald, B. R., & Moran, N. A. (2009). Origin of an alternative genetic code in the extremely small and GC-rich genome of a bacterial symbiont. PLoS Genet. 5:, e1000565 EOF.

[22] Miller, S. L. (1953). A Production of Amino Acids Under Possible Primitive Earth Conditions. Science. , 117, 528-529.

[23] Moreno-Hagelsieb, G., & Latimer, K. (2008). Choosing BLAST options for better detection of orthologs as reciprocal best hits. Bioinformatic , 24, 319-234.

[24] Murtas, G. (2009). Artificial assembly of a minimal cell. Mol BioSyst , 5, 1292-1297.

[25] Nakabachi, A., Yamashita, A., Toh, H., Ishikawa, H., Dunbar, H. E., Moran, N. A., & Hattori, M. (2006). The 160 -Kilobase Genome of the Bacterial Endosymbiont Carsonella. Science. 314:267.

[26] Pósfai, G., Plunkett, G., 3rd , , Fehér, T., Frisch, D., Keil, G. M., Umenhoffer, K., Kolisnychenko, V., Stahl, B., Sharma, S. S., de Arruda, M., Burland, V., Harcum, S. W., & Blattner, F. R. (2006). Emergent properties of reduced-genome *Escherichia coli*. Science , 312, 1044-1046.

[27] Quiñones-Valles, C., Espíndola-Serna, L., & Martínez-Antonio, A. (2011). Mechanisms and Controls of DNA Replication in Bacteria. Fundamental Aspects of DNA Replication. InTech , 219-244.

[28] Reyes-Lamothe, R., Sherratt, D. J., & Leake, M. C. (2010). Stoichiometry and Architecture of Active DNA Replication Machinery in *Escherichia coli*. *Science*, 328, 498-501.

[29] Shimizu, Y., Kuruma, Y., Ying, B. W., Umekage, S., & Ueda, T. (2006). Cell-free translation systems for protein engineering. FEBS Journal , 273, 4133-4140.

[30] Shimizu, Y., & Ueda, T. (2010). PURE technology. Methods Mol Biol. , 607, 11-21.

[31] Siccardi, A. G., Galizzi, A., Mazza, G., Clivio, A., & Albertini, A. M. (1975). Synchronous germination and outgrowth of fractionated *Bacillus subtilis* spores: tool for the analysis of differentiation and division of bacterial cells. J Bacteriol. , 121, 13-19.

[32] Smolke, C. D. (2009). Building outside of the box: iGEM and the BioBricks Founda-
      tion. Nat Biotechnol , 27, 1099-1102.

[33] Tabata, S., Oka, A., Sugimoto, K., Takanami, M., Yasuda, S., & Hirota, Y. (1983). The
      245 base-pair oriC sequence of the *E. coli* chromosome directs bidirectional replica-
      tion at an adjacent region. Nucleic Acids Res. , 11, 2617-26.

[34] Tamames, J., Gil, R., Latorre, A., Peretó, J., Silva, F. J., & Moya, A. (2007). The frontier
      between cell and organelle: genome analysis of Candidatus Carsonella ruddii. BMC
      Evolutionary Biology 7:181.

[35] Tatusov, R. L., Fedorova, N. D., Jackson, J. D., Jacobs, A. R., Kiryutin, B., Koonin, E.
      V., Krylov, D. M., Mazumder, R., Mekhedov, S. L., Nikolskaya, A. N., Rao, B. S.,
      Smirnov, S., Sverdlov, A. V., Vasudevan, S., Wolf, Y. I., Yin, J. J., & Natale, D. A.
      (2003). The COG database: an updated version includes eukaryotes. BMC Bioinfor-
      matics 4:41.

[36] Ueda, T., Tohda, H., Chikazumi, N., Eckstein, F., & Watanabe, K. (1991). Cell-free
      translation system using phosphorothioate-containing mRNA. Nucleic Acids Symp
      Ser. , 25, 151-2.

[37] Williams, K. P., Gillespie, J. J., Sobra, B. W., Nordberg, E. K., Snyder, E. E., Shallom, J.
      M., & Dickerman, A. W. (2010). Phylogeny of gammaproteobacteria. J Bacteriol ,
      192(9), 2305-1234.

Chapter 6

Intrinsically Disordered Proteins in Replication Process

Apolonija Bedina Zavec

Additional information is available at the end of the chapter

1. Introduction

Intrinsically disordered proteins (IDPs) are proteins that lack stable tertiary conformation (3D structure) under physiological conditions and are biologically active in their unstructured form. IDPs are disordered either along their entire lengths, but more often they are disordered only in localized regions, intrinsically disordered regions (IDRs).

IDRs often undergo transitions to more ordered states after binding to their targets and adopt a fixed three dimensional structures. Folding transition enables specificity without excessive binding strength. Important characteristic of IDRs is multispecificity. One IDR is able to bind multiple targets (multispecific recognition) because it can adopt different conformations upon interaction with different binding partners [1]. IDPs are able to simultaneously bind their partners, which enable the assembly of large complexes. An additional functional advantage of IDPs is increased speed of the interaction due to greater capture radius and larger interaction surfaces.

The level of IDPs is tightly regulated in a cell and diverse post-translational modifications facilitate regulation of their function [2].

IDRs with multispecific recognition capabilities are especially important for the complex recognition processes. Therefore, IDRs are particularly enriched in proteins implicated in cell signalling. It is known that the majority of transcription factors and proteins involved in signal transduction contain long disordered segments [3]. How about IDPs in replication process? The analysis of the yeast proteome showed that IDPs are often located in the cell nucleus [4]. In addition, IDRs are abundant in DNA-binding proteins and many replication and recombination proteins are DNA-binding proteins. Many IDPs are involved in recognition and regulation pathways, because interactions with multiple partner molecules and high-specificity/low-affinity interactions are extremely important in these pathways. Additional interesting feature of IDRs is that they are very sensitive to the environment (Subchapter 4.2.). Summarizing these findings, a high level of protein disorder is to be expected in processes that take place in the cell nucleus and the highest level of disorder is expected in processes involved in responses to environmental changes. Therefore it is expected that in the nucleus, transcription is a process with the highest level of IDPs. Recombination and repair processes are also expected to have many IDPs; however, these processes are tightly linked to DNA replication and many proteins are used by all three processes. DNA replication is a process that proceeds by a precise program with a defined temporal order. The structural and functional properties of IDPs indicate that a disordered structure is likely present to a lesser extent in DNA replication process. Because of the need for responsiveness to the environment, the initiation of DNA replication should engage more IDPs than the elongation of DNA replication. It is expected that the majority of IDPs in these processes are regulatory proteins.

In this chapter, the binding mechanism of IDRs, the level of IDRs in replication and recombination proteins, and the role of IDPs in replication and recombination processes are discussed.

2. Intrinsically disordered proteins

IDPs contain one or more long intrinsically disordered regions (IDRs) or they are disordered along their entire lengths. Structural disorder can span from short stretches, through long regions, to entire proteins [5]. The majority of IDRs is not fully disordered, but contains some secondary structure and sometimes even partial tertiary structure. IDRs are dynamic fluctuating systems that exist as structural ensembles of rapidly interconverting alternative conformations and perform their biological functions in a highly dynamic disordered state; however, they often have more compact configurations than simply a random coil and contain sites of molecular recognition [6]. The structure of IDPs is similar to a molten globule or pre-molten globule, which preserve the main elements of the native secondary structure and the approximate positions of the folded state, while the loops and ends are flexible. Structural flexibility is a major feature and a major functional advantage of these proteins. IDRs are rich in binding sites for various partners and these binding sites mean that many IDPs with flexible structure are polyfunctional proteins.

The disordered structure gives IDPs specific properties. They need no stable conformation to remain functional; therefore, they are more robust to different changes. Contrary to globular proteins, IDPs are stable at extreme temperatures and extreme pHs [7]. Increases and decreases in temperature or pH can even induce partial folding of IDPs. It has been shown that IDPs partially fold at extreme pH due to minimization of their large net charge present at neutral pH. An increase in temperature can also induce the partial folding of IDPs; in addition, they are resistant to freeze-thaw treatment [8].

2.1. Amino acid (AA) composition of IDPs

IDPs have a specific AA composition that differs from the AA composition of ordered proteins. In particular, IDPs are depleted in hydrophobic (Ile, Leu, Val) and aromatic AA (Trp, Tyr, Phe) that stabilize the structure of folded proteins, while they are enriched in hydrophilic and charged AAs. The charge/hydropathy (C/H) ratio has been suggested to govern the degree of compaction in IDPs [9]. The combination of low hydrophobicity and high net charge represents an important prerequisite for the disordered structure under physiological conditions.

2.2. Evolution

IDPs are more abundant in eukaryotes than in archaea and prokaryotes, while multicellular eukaryotes have much more predicted disorder than unicellular eukaryotes [4,10]. IDPs play an important role in complex organisms by participating in recognition and in various signalling and regulatory pathways.

IDRs show higher robustness against mutations [11], presumably because changes in protein sequence do not affect protein stability and function as severely. IDRs are more tolerant of mutations than structured proteins. It was found that flexible proteins exhibiting functional promiscuity are the foundation stones of protein evolvability [12]. They are able to accumulate a large number of mutations and thereby facilitate adaptation. Structural disorder seems to enable the rapid appearance of novel, 'less-evolved' proteins [13]. It has been shown that in alternative splicing both alternative proteins have high disorders, because the chance is very low that dual coding would result in two sequences that are both capable of folding into well-defined, functional, 3D structures [14].

3. The binding mechanism of IDRs

IDPs bind to their molecular partners and perform their biological functions by regulation of the function of their binding partners or by promotion the assembly of multi-molecular complexes. One IDR is able to bind many different partners because of its flexible structure; on the other hand, some IDRs do not bind to any partner, but they provide flexible linkers between domains that maintain constant motion during functioning or they provide flexible tails that regulate the structured domains [7,15].

IDPs have functionally relevant characteristics:

• They frequently fold up upon binding to their biological targets [16]. The interaction of a disordered protein with a structured partner, very often induces a disorder-to-order transition thereby forming stable structures, enabling high-specificity-low-affinity interactions [17,18].

• They have possibility of overlapping binding sites (binding diversity) due to extended linear conformation [19]. Structural flexibility of IDPs enables their interactions with numerous biological targets.

• IDRs enable a very large accessible surface area [20]. Greater capture radius and larger interaction surfaces enable increased speed of interactions [15].

- IDPs undergo tighter regulation by post-translational modification as compared to structured proteins [2].

3.1. Complexes with IDPs

Molecular complexes with IDPs are diverse: the IDR may bind on the surface of the binding partner (Figure 1), by wrapping around the binding partner, or by penetrating deep inside the binding partner [21].

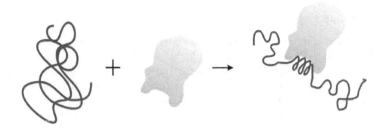

Figure 1. Intrinsically disordered protein forms complex with structured protein.

IDRs in complexes may control the degree of motion between domains, mask binding sites, enable transient binding of different binding partners, and be targets of post-translational modifications. IDPs are often involved in the binding of large partners or they are proteins involved in the binding of large number of small partners. In the latter case, they often function as scaffold proteins that enable the assembly of the relevant proteins into specific multi-molecular complexes and increase the efficiency of the interaction between partner molecules (Figure 2).

Figure 2. Intrinsically disordered proteins often function as scaffold proteins that enable assembling the relevant proteins into multi-molecular complexes.

The majority of intrinsically disorder-based complexes are ordered and relatively static due to disorder-to-order transitions; however, there are also dynamic complexes where IDRs go through an ensemble of rapidly interconverting conformations. Dynamic complexes do not involve significant ordering of the interacting protein segments but rely exclusively on transient contacts.

3.2. Mechanisms of formation of the complexes with IDPs

The primary mechanism by which disorder is utilized in molecular interactions is that the same IDR may fold differently and bind to several structurally diverse partners. On the other hand, different IDRs with different AA sequences may use their flexibility to bind to the same protein partner [6,22]. Their associations are dynamic. The lack of structure of highly flexible IDRs enables more diverse functionality [23]. IDRs are ensembles of conformations and each individual conformation has a dynamic structure. The binding partner selects the most binding-compatible conformation from this ensemble to form a complex [21,15]. The equilibrium is thus shifted towards this interaction-prone member of the conformational ensemble.

Models of IDRs interaction processes:

- The 'binding and folding' mechanism with disorder-to-order transition is the most accepted model for the binding of IDR, where a highly structured conformation is formed by binding to the partner molecule. A structured conformation is formed on binding IDR (the local disorder-to-order transition) or on the entire molecule of IDP (the global disorder-to-order transition) [25,26]. An IDR binds weakly at a relatively large distance followed by folding when the protein comes close to the binding site. One model utilizes a prediction that an IDR with an open structure has a larger binding surface and a greater capture radius for a specific binding site than the ordered protein and therefore the binding rate is significantly enhanced over the binding rate of the ordered proteins [21]. The binding induced disorder-to-order transition is accompanied by a dramatic decrease in accessible surface area and by the release of a large number of water molecules [6]. A large decrease in conformation entropy during this process enables highly specific but easily reversible interactions.

- The 'polyelectrostatic' model describes the interaction of highly charged IDR with several similar binding motifs and a folded partner with one binding site [27,28]. Multiple disordered binding motifs interact with the partner's folded binding site in a dynamic equilibrium. The flexibility of the IDR makes all binding motifs equally accessible. Weak affinities of the individual interactions permit their efficient exchange. In this model, the IDR generates an electrostatic field representing the cumulative electrostatic interaction of all charges in the IDR.

- The 'multi-step interaction' model describes the binding of an IDR that depends on the conformational selection of the structural ensemble via the pre-formed elements that dominate the ensemble [29]. When the IDP comes close to the binding site of the partner molecule, an encounter complex is formed that either proceeds towards the final complex or dissociates again. Electrostatic forces are the most important for encounter complex formation [30]. Interacting partners in the encounter complex affect the conformational landscapes of each other. Consecutive steps depend on the preceding steps and cooperation between protein partners. This process is called an interdependent protein dance [31,32]. The structural variability of complexes with IDPs can be considered a reflection of interdependent protein dynamics, where the structure of the complex is a result of coordinated mutual co-folding [21]. In such encounters 'preorganized' complexes, mainly non-specific electrostatic interactions are involved and multiple conformations and orientations are employed. In the 'multi-step interaction' model, IDPs interacts with their partners by a biphasic process with a fast Phase I leading to the formation of disordered complexes and slower Phase II leading to the formation of ordered complexes. Phase II includes the 'Binding and folding' model that may or may not (binding without folding) follow a Phase I [33]. 'Polyelectrostatic' complexes are probably the stopped stages of encounter complexes [21].

It is the most likely that the IDR contains a conformational preference for the structure it will take upon binding.

3.3. Levels of IDPs in the cell and modulation of their activity

The level of IDPs inside the cell is precisely controlled. IDPs are more tightly regulated as compared to structured proteins. Obviously it is very important that they are available at the appropriate time and in the appropriate amount. The level of IDPs is controlled at the synthesis and clearance levels and their activity is further modulated via interaction with specific binding partners and post-translational modifications. IDRs are more solvent-accessible then folded regions and therefore suitable for diverse post-translational modifications, such as phosphorylation, sumoylation, ubiquitination, acetylation, etc. Such modification can change the electrostatic properties of IDRs and affect their affinity for charged molecules like DNA.

The predicted intrinsic disorder is the strongest determinant of dosage sensitivity - proteins become harmful when they are overexpressed [34]. The likely cause of dosage sensitivity is the binding promiscuity of IDPs [11]. IDRs are prone to make promiscuous interactions when their concentration is increased; it has been demonstrated that this is a likely cause of pathology when genes are overexpressed [34].

4. The role of IDPs in replication processes

This chapter refers to the proteins of budding yeast *Saccharomyces cerevisiae*, because *S.cerevisiae* is the best studied eukaryotic model organism that providing the most integrated view of replication and recombination processes.

4.1. DNA replication process is tightly linked to recombination process

DNA replication and DNA recombination are central characteristics of life that cooperate to maintain biological inheritance and genomic integrity. Replication enables the formation of two identical DNA molecules from a single double-stranded DNA, while recombination enables accurate repair of errors that occur on both strands of DNA, as well as the formation of new combinations of genes. Both processes are tightly intertwined [35]. The recombination system plays a crucial role in DNA replication ensuring that the replication machines can complete their task of genome duplication. DNA replication forks stall or collapse at DNA lesions or problematic genomic regions. When replication forks collapse, recombination is the most important rescue mechanism. The recombination mechanism forms substrates for the assembly of a new replication fork thus allowing continued DNA replication. On the other hand, DNA synthesis is a crucial step during the recombination process. After Rad51-mediated DNA strand invasion, DNA synthesis is the next step in recombination to restore the integrity of the chromosome. Repair DNA synthesis during the recombination process is similar to normal S-phase replication, but has specific properties. Thus recombination is part of DNA replication and, vice versa, DNA synthesis is part of the recombination process.

Clearly then, the replication process requires both, replication and recombination proteins, but then again so does the recombination process. This is why replication and recombination proteins are discussed within the same functional group.

4.2. Predicted level of IDPs in replication and recombination processes

There are some facts to consider when predicting the level of IDPs in replication and recombination processes:

- An analysis of the yeast proteome showed that IDPs are often located in the cell nucleus [4]. IDRs are abundant in DNA-binding proteins, while many replication and recombination proteins are DNA-binding proteins. IDRs play a crucial role in DNA-binding proteins by increasing the affinity and specificity of DNA binding [36]. The ability of IDRs to interact with DNA is tightly linked to the high content of charged residues in IDRs; IDRs that bind to DNA are rich in positively charged residues and their positive charges are highly clustered.

- Many IDPs are involved in recognition and regulation pathways, because interactions with multiple partner molecules and high-specificity/low-affinity interactions are extremely important in these pathways [2].

- Interesting feature of IDRs is that they are very sensitive to the environment. Flexible IDPs more readily undergo conformational change in response to environmental perturbations than rigid proteins [37,38]. Due to flexible structure, their local and global structures can easily be shaped by their environment. High-specificity/low-affinity interactions with their partners enable extremely sensitive functioning of IDPs, which is favourable for responses to the environmental changes. In addition, the level of IDPs inside the cell is precisely controlled (Subchapter 3.4.) allowing rapid and accurate responses of the cell to changing environmental conditions. Higher and more regulated synthesis, higher degradation rates, and tightly regulated activity make the levels of IDPs very sensitive to the environment.

Summarizing these findings, a high level of protein disorder is to be expected in processes that take place in the cell nucleus, especially within regulatory proteins. The highest level of disorder is expected in processes involved in responses to environmental changes. According to those findings, in the nucleus, transcription should be the process with the highest level of IDPs. Recombination and repair processes are also expected to have many IDPs; however, these processes are tightly linked to DNA replication and many proteins are used by all these processes. DNA replication is a process that proceeds by a precise program with a defined temporal order. The structural and functional properties of IDPs indicate that a disordered structure is likely present to a lesser extent in DNA replication process. However, the initiation of DNA replication would be expected to engage more IDPs than the elongation of DNA replication due to the need for responsiveness to the environment (Figure 3). It is expected that the majority of IDPs in these processes are regulatory proteins.

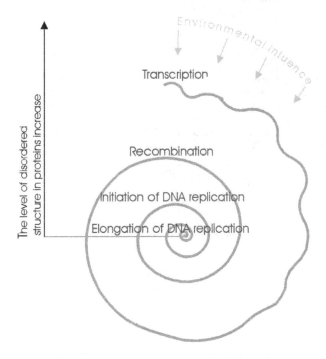

Figure 3. Processes in the nucleus: global prediction of protein disordered structure in the processes linked to DNA, considering responsiveness to changes in the environment.

4.3. IDRs in replication proteins

Analysis of predicted IDRs within proteins that have a role in DNA replication was done (Table 1). The majority of them functions also in recombination and repair processes.

It was found that proteins with the role in initiation of DNA replication have more predicted disordered structure (26%) than proteins with the role in elongation of DNA replication (20%). Difference is significant among proteins with very short IDRs; there is 35% proteins with the role in initiation of DNA replication that contain less than 10% disordered structure, while there is as much as 60% such proteins with the role in elongation of DNA replication. However, difference is tiny among proteins with very large IDRs; 22% proteins with the role in initiation of DNA replication contain more than 50% disordered structure and there is 20% such proteins with the role in elongation of DNA replication.

Protein	Protein Length (AA)	*Location of IDR (AA)	*IDR Length (AA)	% of disorder	Recombination and Repair
		Elongation of DNA replication			
Pol2	2222	0-44, 1186-1263	44, 77	5.4	R
Dpb2	689	92-159	67	9.7	R
Dpb3	201	96-201	105	**52.2**	R
Dpb4	196	0-26, 119-196	26, 77	**52.6**	R
Pol3	1097	0-100	100	9.1	R
Pol31	487	0-32	32	6.6	R
Pol32	350	118-350	232	**66.3**	R
PCNA	258	-	0	0	R
Cdc9	755	0-144	144	19.1	R
Rfa1	621	126-183	57	9.2	R
Rfa2	273	0-39, 177-235	39, 58	35.5	R
Rfa3	122	-	0	0	R
Rfc1	861	0-155, 230-296, 780-861	155, 66, 81	35.1	R
Rfc2	353	0-25	25	7.1	
Rfc3	340	-	0	0	
		Initiation of DNA replication			
Pol1	1468	82-341	259	17.6	R
Pol12	705	70-203	133	18.8	R
Pri1	409	-	0	0	R
Pri2	528	0-43	43	8.1	R
Orc1	914	0-42, 195-428	42, 233	30.1	
Orc2	620	0-267	267	43.1	
Orc3	616	0-40	40	6.5	
Mcm2	868	0-68, 112-189	68, 77	16.7	R
Mcm3	971	742-897	155	16.0	R
Mcm4	933	0-177	177	19.0	R
Mcm10	571	41-154, 338-571	113, 233	**60.6**	
Sld2	453	0-453	453	**100**	
Sld3	668	87-155, 292-335, 417-668	68, 43, 271	**57.2**	
Sld5	294	17-45	28	9.5	
Psf1	208	-	0	0	
Psf2	213	195-213	18	7.8	
Psf3	194	-	0	0	
Dpb11	764	226-321, 567-764	95, 197	38.2	R
Cdt1	604	427-507	80	13.2	R
Cdc6	513	0-61	61	11.9	
Cdc7	507	-	0	0	R
Dbf4	704	0-110, 317-658	110, 341	**64.1**	R
Ecm11	302	0-183	183	**60.6**	R

Data about proteins were obtained from *Saccharomyces genome* database [39] and references therein.
Server Disopred2 [40] was used for protein disorder prediction.
*Predicted disorder regions more than 30 AA long (more than 20 AA long for very small proteins).
R - proteins involved in recombination and repair processes;
Disordered regions at N- or C- terminus are underlined.

Table 1. Predicted disorder regions of replication proteins.

The majority (63%) of IDRs are at N- or C- terminus in proteins that have a role in DNA replication. Most often IDRs are at the N-terminus, in 44% of replication proteins; 19% of replication proteins have IDRs at the C-terminus.

4.4. IDRs in DNA polymerases

4.4.1. DNA polymerase α

Polymerase α is the only enzyme that can synthesize DNA de novo. It is required for initiation of chromosomal DNA replication during mitosis and meiosis, intragenic recombination, and repair of double stranded DNA breaks. Pol1, the catalytic subunit of polymerase α complex, has a lot of disordered structure in comparison to Pol3 and Pol2 (Table 1). This fact is consistent with the hypothesis concerning the expected average level of disordered structures, since Pol1 is required for the initiation of DNA replication and Pol3 and Pol2 are required for the elongation of DNA replication.

It was shown that IDR of Pol1 interacts with Cdc13: Pol1 residues 13-392 [41] or Pol1 residues 47-560 [42]. Actually, a short fragment of Pol1 consisting only of residues 215-250 is necessary and sufficient for binding with Cdc13. This disordered region of Pol1 becomes well ordered, folded into a single amphipathic α-helix, when it is in complex with Cdc13, as evidenced by good electron density in the crystals [43]. The interaction between IDR of Pol1 and Cdc13 is primarily mediated by a highly positively charged groove of Cdc13 and a negatively charged acidic convex surface of Pol1. These two surfaces are not only opposite in charge distribution but also complementary in shape.

4.4.2. DNA polymerase δ

DNA polymerase δ is a major replicative DNA polymerase and is primarily required for the lagging strand synthesis. It is a heterotrimeric complex composed of the catalytic subunit Pol3, the structural subunit Pol31, and an additional auxiliary subunit Pol32. Pol32 is highly disordered protein (Figure 4). While structured Pol3 and Pol31 are essential for viability, the disordered Pol32 is not essential. Pol3 and Pol31 are highly conserved in eukaryotes; on the other hand, the disordered Pol32 shows an extreme divergence in its AA sequence [44]. Hydrodynamic studies of polymerase δ have shown an unusually high Stokes radius [45]. This deviation from globularity may be due to the disordered structure of Pol32.

Pol32 is bound to Pol3 through Pol31. The C-terminus of Pol3 interacts with the conserved region of Pol31 [46]. Deletion of the last four C-terminal AAs of Pol3, which are required for the interaction between the Pol3 and Pol31, does not affect DNA replication but leads to defects in homologous recombination and in break-induced replication repair pathways. Deletion of Pol32 leads to signs of DNA replication defects and DNA repair defects, with increased sensitivity to ultraviolet (UV) radiation and methylation damage [47].

Pol32 binds to Pol31 by the N-terminus (92 AA) and to PCNA by the C-terminus [48]. The structured N-terminus of Pol32, which enables binding to Pol3 through Pol31, is essential for damage-induced mutagenesis. Highly disordered C-terminus of Pol32 interacts with the

C-terminus of PCNA during DNA synthesis. Although the C-terminus of Pol32 is highly disordered, there is one motif that is highly conserved in this region: the consensus PCNA-binding motif 338-QGTLESFFKRKAK-350 (conserved amino acids in bold).

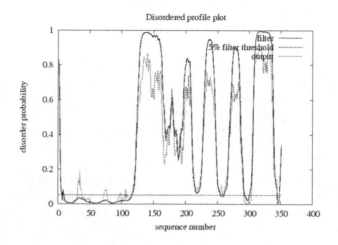

Figure 4. Predicted disordered regions of Pol32 (Disopred2).

It has also been shown that Pol32 interacts also with Pol1 that is a part of polymerase α, suggesting that Pol δ and Pol α interact *via* the Pol32 subunit [48]. These findings show diverse role of Pol32 as typical IDP.

For the replication of the lagging strand where the polymerase must dissociate from the DNA after extension of each Okazaki fragment, Polδ utilizes a collision-release mechanism where the Polδ is released from PCNA. Polδ exhibits a very high processivity in synthesizing DNA with the PCNA sliding clamp. It has been shown that the N-terminal region of Pol3 interacts with PCNA, and that this interaction increases Pol3 processivity [49]. The N-terminal of Pol3 is predicted to be highly disordered (Table 1). Pol31 and Pol32 also have binding sites for PCNA and all three subunits contribute to PCNA-stimulated DNA synthesis by Polδ [50].

4.4.3. DNA polymerase ε

DNA polymerase ε is primarily required for the leading strand synthesizes. Pol2 is the catalytic subunit of the polymerase ε complex. It has been shown that the highly structured C-terminus of Pol2 is essential for DNA replication [51].

Dpb3 and Dpb4 are nonessential small subunits of the DNA polymerase ε complex, which have a histone fold. Both Dpb3 and Dpb4 are highly disordered. They form a sub-assembly that interacts with histones and functions in transcriptional silencing caused by chromatin structures [52].

4.5. Dbf4, IDP with the role in replication and recombination

The complex Cdc7–Dbf4, also known as Dbf4-dependent kinase (DDK), has a role at eukaryotic origins of replication. DDK is required for origin firing and replication fork progression in mitotic S phase, for pre-meiotic DNA replication, meiotic double strand break formation, recruitment of the monopolin complex to kinetochores during meiosis I and as a gene-specific regulator of the meiosis-specific transcription factor Ndt80p. DDK is a Ser/Thr kinase whose activity depends on the association of the Cdc7 catalytic subunit with a regulatory subunit Dbf4. The level of Dbf4 is changes during the cell cycle and is the highest during metaphase I [53]. Both subunits Cdc7 and Dbf4 are essential for growth.

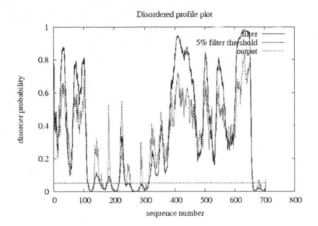

Figure 5. Predicted disordered regions of Dbf4 (Disopred2)

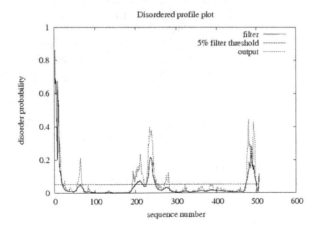

Figure 6. Predicted disordered regions of Cdc7 (Disopred2).

As expected, the regulatory subunit Dbf4 is highly disordered (Figure 5), while catalytic subunit Cdc7 is a highly structured protein (Figure 6).

Dbf4 is a highly disordered protein with a disordered N-terminus (110 AA) and an IDR that is half the length of the protein at the C-terminus; it has only a 200 AA long structural region between both IDRs (Figure 5). It was shown that highly disordered C-terminus of Dbf4 has a role in response to mutation by HU and that it is required in meiosis [54]. Superfamily assignments [55] show no confident structure prediction for Dbf4, while Cdc7 has a predicted cyclin-dependent protein kinase function at 1-304 AA with protein, ATP, and DNA binding activity. Cdc7 has well conserved subdomains (30-195, 275-348, 438-465) found in the eukaryotic protein kinase superfamily, while Dbf4 contains only three short conserved regions, termed N (135-179), M (260-309), and C (659-696) [56]. Two of the three conserved regions (N and M) are found in the the structural region of Dbf4.

4.5.1. Dbf4 in initiation of replication during mitosis

DDK phosphorylates the Mcm2-7 helicase, and is probably required for helicase activation or for recruitment of pre-IC factors. DDK preferentially phosphorylates the MCM complexes that are most tightly linked to the DNA [57]. Dbf4 associates with origins in an ORC-dependent manner [58]. The pre-RC components Mcm2, Mcm4, Orc2, and Orc3 have each been identified as binding partners for Dbf4 [59,60,61]. The N-terminal half of Dbf4 is critical for recruitment of DDK to the origin. The highly disordered C-terminal half of Dbf4 is required to bind the Cdc7 kinase [58]; more precisely, region 573-695 is required for interaction with Cdc7, while the structured region of Dbf4 (110-296) is required for binding the Mcm2–7 complex [60].

4.5.2. Dbf4 in checkpoint control

During the replication checkpoint response, Dbf4 is phosphorylated by checkpoint kinase Rad53 allowing inhibition of initiation of replication at late origins. Checkpoint control during S-phase slows the rate of DNA replication in response to DNA damage and blocks the replication fork. This regulation is achieved through the Rad53 kinase-dependent block of late origins of replication [62]. Dbf4 has been shown to be phosphorylated in a Rad53-dependent manner in response to replication stress, which correlates with a reduced DDK activity [63]. It was shown that mutations at predicted Rad53 phosphorylation sites (Ser84, Ser235, Ser377, Thr467, Thr506, Ser507, and Thr551) contribute to bypassing such control [64].

The conserved region N of Dbf4 (66-221) is necessary for the interaction of Cdc7-Dbf4 with the checkpoint kinase Rad53. The core of this binding region folds as a BRCT domain; in addition, it includes an additional N-terminal helix unique to Dbf4 that is essential for the interaction with Rad53 [65]. This unique N-terminal part of the conserved region N is predicted to be an IDR (Figure 6) and probably becomes helix-structured after binding with Rad53.

4.5.3. Dbf4 in meiosis

DDK is required for replication, recombination and segregation events during meiosis in yeast. It has been shown that in addition to the initiation of DNA replication, DDK has an important role in the initiation of meiotic recombination [66]. DDK phosphorylates the double strand break protein Mer2 and facilitates meiotic recombination [67]. CDK-S and DDK function sequentially phosphorylate Mer2 on adjacent serines, Ser30 and Ser29, allowing formation of meiotic double strand breaks.

DDK plays a role in meiotic segregation. DDK allows expression of *NDT80*, a global transcription factor in meiosis, required for the induction of genes required for meiotic progression and spore formation. DDK promotes *NDT80* transcription by relieving repression mediated by a complex of Sum1, Rfm1, and histone deacetylase Hst1. Sum1 exhibits meiosis-specific Cdc7-dependent phosphorylation. By this function, DDK links DNA replication to the segregation of homologous chromosomes in meiosis I [68,69].

DDK is also necessary for recruitment of monopolin Mam1 to sister kinetochores, which is required for mono-orientation of sister kinetochores in the reductional segregation occurring during meiosis I.

The use of the same Cdc7-Dbf4 complex to regulate many distinct meiosis-specific processes could be important for the coordination of these processes during meiosis [68]. DDK is a link between DNA replication, recombination and mono-orientation during meiosis I in budding yeast [70]. In addition to the unifying role in meiosis, DDK has a role in initiation of replication during mitosis and in checkpoint control. Highly flexible structure of Dbf4 is very likely crucial for such a complex role of DDK.

4.6. Ecm11, IDP with the role in replication and recombination

Ecm11 is a protein with a strong meiotic phenotype; it affects meiotic DNA synthesis and recombination [71]. Homozygous deletion of the *ECM11* gene causes delay in a process of meiosis, lower efficiency of ascii formation and lower spore viability.

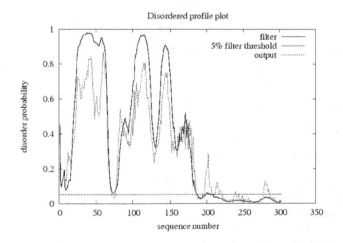

Figure 7. Predicted disordered regions of Ecm11 (Disopred2).

Ecm11 is highly disordered protein; 2/3 of Ecm11 is unstructured (Table 1). Ecm11 has 302 mostly hydrophilic AA as expected for IDP (Subchapter 2.1.). IDR of 183 AA is located at the N-terminal end of Ecm11 (Figure 7). The C-terminal end is predicted to be mostly helical and contain coiled-coil motif at the very C-terminus. Superfamiliy assignments [55] show no confident structure prediction for Ecm11.

4.6.1. The ecm11 mutation affects sporulation efficiency

It was showed that *ecm11* homozygous diploid strains sporulate more slowly and less efficiently than the wild type strains [71]. Wild type strains carrying additional *ECM11* on the centromeric plasmid also showed reduced sporulation efficiency comparing to wild types. Obviously, sporulation efficiency depends on the copy number of Ecm11 protein in the cell during meiosis. As more Ecm11 than usual in the cell make lower sporulation efficiency, Ecm11 is probably a part of heterologous protein complex, demanding exactly correct balance among those proteins.

4.6.2. Ecm11 has a role in meiotic recombination

It was showed that *ecm11* homozygous spores have reduced viability for 50% [71]. The majority of *ecm11* ascii (56%) produced only two viable spores, while only 1% of such ascii were observed in the parental strain. This result shows non-disjunction of homologous chromosomes at the first meiotic division. By recombination tests was demonstrated that *ECM11* is required for crossing over, but not for gene conversion. This result raises the possibility that *ecm11* mutation impairs the crossover process at an early step of recombination, at the differentiation of intermediates into crossovers or non-crossovers.

4.6.3. Ecm11 is required for meiotic DNA replication

Deletion of the *ECM11* gene cause diminished DNA replication in meiosis [71].

In the two-hybrid screen it was found out that Ecm11 strongly interacts with Cdc6 that has a pivotal role in the initiation of DNA replication [72]. Genetic interactions between Cdc6 and Ecm11 were also observed. Moderate supression of *cdc6-1* mutation by overexpression of *ECM11* was detected [72] and deletion of *ECM11* in *cdc6-1* genetic background enhances thermo-sensitivity of *cdc6-1* mutation (Zavec AB, unpublished result). These data suggest direct involving of Ecm11 in initiation of DNA replication process.

4.6.4. Ecm11 is modified by SUMO during meiosis

IDPs are tightly regulated in a cell and diverse post-translational modifications (such as ubiquitination, sumoylation, and phosphorylation) facilitate regulation of their function (Subchapter 3.4.). It was shown that the majority of Ecm11 protein in the cell is sumoylated during meiosis [73]. Lys5 at the highly disordered N-terminus of Ecm11 is modified by SUMO. It was shown that sumoylation is essential for biological role of Ecm11 in meiosis and that sumoylation directly regulates Ecm11 function in meiosis.

5. Conclusion

Cell nuclei contain high levels of IDPs. In this work, a hypothesis has been made, that in the nucleus, transcription is a process with the highest level of IDPs and that a disordered structure is likely present to a lesser extent in DNA replication process. However, the initiation of DNA replication would be expected to engage more IDPs than the elongation of DNA replication due to the need for responsiveness to the environment. By analysis of predicted disordered structure in replication proteins, it was confirmed that proteins with the role in initiation of DNA replication have more disordered structure than proteins with the role in elongation of DNA replication. The majority of IDRs in these proteins are at N- or C- terminus, most often IDRs are at the N-terminus of IDPs.

Acknowledgements

Drago Cerjan provided help with figures.

Financial support was received from the Slovenian Research Agency, Grant P1-0104.

Author details

Apolonija Bedina Zavec*

Address all correspondence to: polona.bedina@ki.si

National Institute of Chemistry, Department for Molecular biology and Nanobiotechnology, Slovenia

References

[1] Uversky, N., Oldfield, C. J., & Dunker, A. K. (2005). Showing your ID: intrinsic disorder as an ID for recognition, regulation and cell signaling. *J Mol Recognit*, 18, 343-384.

[2] Babu, M. M., Lee, R., Groot, S. N., & Gsponer, J. (2011). Intrinsically disordered proteins: regulation and disease. *Curr Opin Struct Bio*, 21(3), 432-40.

[3] Lakoucheva, L. M., Brown, C. J., Lawson, J. D., Obradović, Z., & Dunker, A. K. (2002). Intrinsic disorder in cell-signaling and cancer-associated proteins. *J Mol Biol.*, 323(3), 573-84.

[4] Ward, J. J., Sodhi, J. S., Mc Guffin, L. J., Buxton, B. F., & Jones, D. T. (2004). Prediction and functional analysis of native disorder in proteins from the three kingdoms of life. *J Mol Biol*, 337(3), 635-45.

[5] Tompa, P., Fuxreiter, M., Oldfield, C. J., Simon, I., Dunker, A. K., & Uversky, V. N. (2009). Close encounters of the third kind: disordered domains and the interactions of proteins. *Bioessays*, 31(3), 328-35.

[6] Uversky, V. N., & Dunker, A. K. (2010). Understanding protein non-folding. *Biochim Biophys Acta*, 1804(6), 1231-64.

[7] Uversky, V. N. (2002). Natively unfolded proteins: a point where biology waits for physics. *Protein Sci*, 11(4), 739-56.

[8] Tantos, P., Friedrich, P., & Tompa, P. (2009). Cold stability of intrinsically disordered proteins. *FEBS Lett*, 583(2), 465-9.

[9] Mao, A. H., Crick, S. L., Vitalis, A., Chicoine, C. L., & Pappu, R. V. (2010). Net charge per residue modulates conformational ensembles of intrinsically disordered proteins. *Proc Natl Acad Sci*, USA, 107(18), 8183-8.

[10] Oldfield, C. J., Cheng, Y., Cortese, M. S., Brown, C. J., Uversky, V. N., & Dunker, A. K. (2005). Comparing and combining predictors of mostly disordered proteins. *Biochemistry*, 44(6), 1989-2000.

[11] Bellay, J., Han, S., Michaut, M., Kim, T., Constanzo, M., Andrews, B. J., Boone, C., Bader, G. D., Myers, C. L., & Kim, P. M. (2011). Bringing order to protein disorder through comparative genomics and genetic interactions. *Genome Biol*, 12(2), R14.

[12] Tokuriki, N., & Tawfik, D. S. (2009). Protein Dynamism and Evolvability. *Science*, 324, 203-207.

[13] Tompa, P. (2011). Unstructural biology coming of age. *Current Opinionin Structural Biology*, 21, 1-7.

[14] Kovacs, E., Tompa, P., Liliom, K., & Kalmar, L. (2010). Dual coding in alternative reading frames correlates with intrinsic protein disorder. *Proc Natl Acad Sci*, USA, 107(12), 5429-5434.

[15] Tompa, P. (2002). Intrinsically unstructured proteins. *Trends Biochem Sci*, 27(10), 527-33.

[16] Dyson, H. J., & Wright, P. E. (2002). Coupling of folding and binding for unstructured proteins. *Curr Opin Struct Biol*, 12, 54-60.

[17] Demchenko, A. P. (2001). Recognition between flexible protein molecules: induced and assisted folding. *J Mol Recognit*, 14, 42-61.

[18] Tompa, P., Szasz, C., & Buday, L. (2005). Structural disorder throws new light on moonlighting. Trends Biochem Sci. , 30, 484-489.

[19] Dyson, H. J., & Wright, P. E. (2005). Intrinsically unstructured proteins and their functions. *Nat Rev Mol Cell Biol.*, 6(3), 197-208.

[20] Marsh, J. A., & Teichmann, S. A. (2011). Relative Solvent Accessible Surface Area Predicts Protein Conformational Changes upon Binding Structure. 19(6), 859-867.

[21] Uversky, V. N. (2011). Multitude of binding modes attainable by intrinsically disordered proteins: a portrait gallery of disorder-based complexes. *Chem Soc Rev.*, 40(3), 1623-34.

[22] Oldfield, C. J., Meng, J., Yang, J. Y., Yang, M. Q., Uversky, V. N., & Dunker, A. K. (2008). Flexible nets: disorder and induced fit in the associations of 53 and 14-3-3 with their partners. *BMC Genomics*, 9(1), S1.

[23] Cortese, M. S., Uversky, V. N., & Dunker, A. K. (2008). Intrinsic disorder in scaffold proteins: getting more from less. *Prog Biophys Mol Biol*, 98(1), 85-106.

[24] Tobi, D., & Bahar, I. (2005). Structural changes involved in protein binding correlate with intrinsic motions of proteins in the unbound state. *Proc Natl Acad Sci*, USA, 102(52), 18908-13.

[25] Shoemaker, B. A., Portman, J. J., & Wolynes, P. G. (2000). Speeding molecular recognition by using the folding funnel: the fly-casting mechanism. *Proc Natl Acad Sci*, USA, 97(16), 8868-73.

[26] Pontius, B. W. (1993). Close encounters: why unstructured, polymeric domains can increase rates of specific macromolecular association. *Trends Biochem Sci*, 18(5), 181-6.

[27] Mittag, T., Orlicky, S., Choy, W. Y., Tang, X., Lin, H., Sicheri, F., Kay, L. E., Tyers, M., & Forman-Kay, J. D. (2008). Dynamic equilibrium engagement of a polyvalent ligand with a single-site receptor. *Proc Natl Acad Sci*, USA, 105(46), 17772-7.

[28] Mittag, T., Marsh, J., Grishaev, A., Orlicky, S., Lin, H., Sicheri, F., Tyers, M., & Forman-Kay, J. D. (2010). Structure/function implications in a dynamic complex of the intrinsically disordered Sic1 with the Cdc4 subunit of an SCF ubiquitin ligase. *Structure*, 18(4), 494-506.

[29] Fuxreiter, M., Simon, I., Friedrich, P., & Tompa, P. (2004). Preformed structural elements feature in partner recognition by intrinsically unstructured proteins. *J Mol Biol*, 338(5), 1015-26.

[30] Ubbink, M. (2009). The courtship of proteins: understanding the encounter complex. *FEBS Lett*, 583(7), 1060-6.

[31] Kovács, I.A. , Szalay, M. S., & Csermely, P. (2005). Water and molecular chaperones act as weak links of protein folding networks: energy landscape and punctuated equilibrium changes point towards a game theory of proteins. *FEBS Lett*, 579(11), 2254-60.

[32] Antal, M. A., Böde, C., & Csermely, P. (2009). Perturbation waves in proteins and protein networks: applications of percolation and game theories in signaling and drug design. *Curr Protein Pept Sci*, 10(2), 161-72.

[33] Sigalov , A. B. (2010). Protein intrinsic disorder and oligomericity in cell signaling. *Mol Biosyst*, 6(3), 451-61.

[34] Vavouri, T., Semple, J. I., Garcia-Verdugo, R., & Lehner, B. (2009). Intrinsic protein disorder and interaction promiscuity are widely associated with dosage sensitivity. *Cell*, 138(1), 198-208.

[35] Kusic-Tisma, J. (2011). *DNA Replication and Related Cellular Processes*, Rijeka, InTech.

[36] Vuzman, D., & Levy, Y. (2012). Intrinsically disordered regions as affinity tuners in protein-DNA interactions. *Mol Biosyst*, 8(1), 47-57.

[37] Wright, P. E., & Dyson, H. J. (1999). Intrinsically unstructured proteins: re-assessing the protein structure-function paradigm. *J of Mol Bio.*, 293(2), 321-331.

[38] Romero, P., Obradovic, Z., & Dunker, A. K. (2004). Natively disordered proteins: functions and predictions. *Appl Bioinformatics*, 3(2-3), 105-13.

[39] Saccharomyces genome database. (2012). http://www.yeastgenome.org/, accessed 20 June.

[40] The DISOPRED2 Prediction of Protein Disorder Server. (2012). http://bioinf.cs.ucl.ac.uk/disopred/, accessed 20 June.

[41] Qi, H., & Zakian, V. A. (2000). The Saccharomyces telomere-binding protein Cdc13p interacts with both the catalytic subunit of DNA polymerase alpha and the telomerase-associated est1 protein. *Genes Dev*, 14(14), 1777-88.

[42] Hsu, C. L., Chen, Y. S., Tsai, S. Y., Tu, P. J., Wang, M. J., & Lin, J. J. (2004). Interaction of Saccharomyces Cdc13p with Pol1Imp4pSir4p and Zds2p is involved in telomere replication, telomere maintenance and cell growth control. *Nucleic Acids Res*, 32(2), 511-21.

[43] Sun, J., Yang, Y., Wan, K., Mao, N., Yu, T. Y., Lin, Y. C., De Zwaan, D. C., Freeman, B. C., Lin, J. J., Lue, N. F., & Lei, M. (2011). Structural bases of dimerization of yeast telomere protein Cdc13 and its interaction with the catalytic subunit of DNA polymerase α. *Cell Res*, 21(2), 258-74.

[44] Cliften, P. F., Hillier, L. W., Fulton, L., Graves, T., Miner, T., Gish, W. R., Waterston, R. H., & Johnston, M. (2001). Surveying Saccharomyces genomes to identify functional elements by comparative DNA sequence analysis. *Genome Res*, 11(7), 1175-86.

[45] Johansson, E., Majka, J., & Burgers, P. M. (2001). Structure of DNA polymerase delta from Saccharomyces cerevisiae. *J Biol Chem*, 276(47), 43824-8.

[46] Brocas, C., Charbonnier, J. B., Dhérin, C., Gangloff, S., & Maloisel, L. (2010). Stable interactions between DNA polymerase δ catalytic and structural subunits are essential for efficient DNA repair. *DNA Repair*, 9(10), 1098-111.

[47] Haracska, L., Prakash, S., & Prakash, L. (2000). Replication past O(6)-methylguanine by yeast and human DNA polymerase eta. *Mol Cell Biol*, 20(21), 8001-7.

[48] Johansson, E., Garg, P., & Burgers, P. M. (2004).. The Pol32 subunit of DNA polymerase delta contains separable domains for processive replication and proliferating cell nuclear antigen (PCNA) binding. J Biol Chem , 279(3), 1907-15.

[49] Brown, W. C., & Campbell, J. L. (1993). Interaction of proliferating cell nuclear antigen with yeast DNA polymerase delta. *J Biol Chem*, 268(29), 21706-10.

[50] Acharya, N., Klassen, R., Johnson, R. E., Prakash, L., & Prakash, S. (2011). PCNA binding domains in all three subunits of yeast DNA polymerase δ modulate its function in DNA replication. *Proc Natl Acad Sci*, USA, 108(44), 17927-32.

[51] Dua, R., Levy, D. L., & Campbell, J. L. (1999). Analysis of the essential functions of the C-terminal protein/protein interaction domain of Saccharomyces cerevisiae pol epsilon and its unexpected ability to support growth in the absence of the DNA polymerase domain. *J Biol Chem*, 274(32), 22283-8.

[52] Tsubota, T., Tajima, R., Ode, K., Kubota, H., Fukuhara, N., Kawabata, T., Maki, S., & Maki, H. (2006). Double-stranded DNA binding, an unusual property of DNA polymerase epsilon, promotes epigenetic silencing in Saccharomyces cerevisiae. *J Biol Chem*, 281(43), 32898-908.

[53] Matos, J., Lipp, J. J., Bogdanova, A., Guillot, S., Okaz, E., Junqueira, M., Shevchenko, A., & Zachariae, W. (2008). Dbf4-dependent CDC7 kinase links DNA replication to the segregation of homologous chromosomes in meiosis I. *Cell*, 135(4), 662-78.

[54] Davey, M. J., Andrighetti, H. J., Ma, C. J., & Brandl, X. (2011). A synthetic human kinase can control cell cycle progression in budding yeast. *G3*, 1(4), 317-25.

[55] Malmström, L., Riffle, M., Strauss, C. E., Chivian, D., Davis, T. N., Bonneau, R., & Baker, D. (2007). Superfamily assignments for the yeast proteome through integration of structure prediction with the gene ontology. *PLoS Biol*, e76.

[56] Masai, H., & Arai, K. (2000). Dbf4 motifs: conserved motifs in activation subunits for Cdc7 kinases essential for S-phase. *Biochem Biophys Res Commun.*, 275(1), 228-32.

[57] Francis, L. I., Randell, J. C., Takara, T. J., Uchima, L., & Bell, S. P. (2009). Incorporation into the prereplicative complex activates the Mcm2-7 helicase for Cdc7-Dbf4 phosphorylation. *Genes Dev.*, 23(5), 643-54.

[58] Dowell, S. J., Romanowski, P., & Diffley, J. F. (1994). Interaction of Dbf4, the Cdc7 protein kinase regulatory subunit, with yeast replication origins in vivo. *Science*, 265(5176), 1243-6.

[59] Duncker, B. P., Shimada, K., Tsai-Pflugfelder, M., Pasero, P., & Gasser, S. M. (2002). An N-terminal domain of Dbf4p mediates interaction with both origin recognition complex (ORC) and Rad53p and can deregulate late origin firing. *Proc Natl Acad Sci*, USA, 99(25), 16087-92.

[60] Varrin, A. E., Prasad, A. A., Scholz, R. P., Ramer, M. D., & Duncker, B. P. (2005). A mutation in Dbf4 motif M impairs interactions with DNA replication factors and confers increased resistance to genotoxic agents. *Mol Cell Biol.*, 25(17), 7494-504.

[61] Sheu, Y.J. , & Stillman, B. (1999). Cdc7-Dbf4 phosphorylates MCM proteins via a docking site-mediated mechanism to promote S phase progression. *Mol Cell 2006*, 24(1), 101-13.

[62] Zegerman, P. , & Diffley, J. F. (2010). Checkpoint-dependent inhibition of DNA replication initiation by Sld3 and Dbf4 phosphorylation. *Nature*, 467(7314), 474-8.

[63] Weinreich, M. , & Stillman, B. Cdc7p-Dbf4p kinase binds to chromatin during S phase and is regulated by both the APC and the RAD53 checkpoint pathway. *EMBO J.*, 18(19), 5334-46.

[64] Duch, A. , Palou, G., Jonsson, Z. O., Palou, R., Calvo, E. , Wohlschlegel, J. , & Quintana, D. G. (2011). A Dbf4 mutant contributes to bypassing the Rad53-mediated block of origins of replication in response to genotoxic stress. *J Biol Chem*, 286(4), 2486-91.

[65] Matthews, L. A., Jones, D. R., Prasad, A. A., Duncker, B. P., & Guarné, A. (2012). Saccharomyces cerevisiae Dbf4 has unique fold necessary for interaction with Rad53 kinase. *J Biol Chem*, 287(4), 2378-87.

[66] Wan, L., Niu, H., Futcher, B., Zhang, C., Shokat, K. M., Boulton, S. J., & Hollingsworth, N. M. (2008). Cdc28-Clb5 (CDK-S) and Cdc7-Dbf4 (DDK) collaborate to initiate meiotic recombination in yeast. *Genes Dev.*, 22(3), 386-97.

[67] Sasanuma, H., Hirota, K., Fukuda, T., Kakusho, N., Kugou, K., Kawasaki, Y., Shibata, T., Masai, H., & Ohta, K. (2008). Cdc7-dependent phosphorylation of Mer2 facilitates initiation of yeast meiotic recombination. *Genes Dev.*, 22(3), 398-410.

[68] Lo, H. C., Wan, L., Rosebrock, A., Futcher, B., & Hollingsworth, N. M. (2008). Cdc7-Dbf4 regulates NDT80 transcription as well as reductional segregation during budding yeast meiosis. *Mol Biol Cell*, 19(11), 4956-67.

[69] Matos, J., Lipp, J. J., Bogdanova, A., Guillot, S., Okaz, E., Junqueira, M., Shevchenko, A., & Zachariae, W. (2008). Dbf4-dependent CDC7 kinase links DNA replication to the segregation of homologous chromosomes in meiosis I. *Cell*, 135(4), 662-78.

[70] Marston, A. L. (2009). Meiosis: DDK is not just for replication. *Curr Biol*, 19(2), 74-6.

[71] Zavec, A. B., Lesnik, U., Komel, R., & Comino, A. (2004). The Saccharomyces cerevisiae gene ECM11 is a positive effector of meiosis. *FEMS Microbiol Lett*, 241(2), 193-9.

[72] Zavec, P. B., Comino, A., Watt, P., & Komel, R. (2000). Interaction trap experiment with CDC6. *Pflugers Arch*, 439(3), R 94-6.

[73] Zavec, A. B., Comino, A., Lenassi, M., & Komel, R. (2008). Ecm11 protein of yeast Saccharomyces cerevisiae is regulated by sumoylation during meiosis. *FEMS Yeast Res*, 8(1), 64-70.

Chapter 7

Extending the Interaction Repertoire of FHA and BRCT Domains

Lindsay A. Matthews and Alba Guarné

Additional information is available at the end of the chapter

1. Introduction

All living organisms are connected by the necessity to replicate their DNA. However, the process of unwinding the parental DNA to serve as a template at replication forks is danger-ous. The ssDNA generated by helicases is inherently cytotoxic; not only because it is more prone to damage, but it can also be an inappropriate target for nucleases and recombination proteins leading to loss of genetic material or gross chromosomal rearrangements [1]. Nor-mally, the replication machinery rapidly restores this single-stranded template DNA to its more stable double-stranded form. However, replication forks are prone to stalling if they encounter obstacles that the DNA polymerase is unable to bypass, such as sites of damage or DNA sequences with complex secondary structures [2], resulting in long stretches of ssDNA remaining exposed [3]. Replication stress, therefore, represents an important mecha-nism that erodes the genetic integrity of organisms. Not surprisingly, replication stress has been linked to aging in budding yeast, which can likely be extrapolated to higher eukaryotes as well [4]. Furthermore, inducing replication stress in normal human fibroblasts results in pathogenic changes in copy number due to duplication or deletion events [5]. Therefore, the ability of eukaryotes to detect, stabilize and resolve stalled replication forks using the repli-cation checkpoint represents an important safeguard for genomic stability.

The replication checkpoint response relies on a cascade of kinases that either remain local-ized to the stalled fork or disseminate the stress signal to distal sites resulting in the sup-pression of late origin firing, pausing of the cell cycle, and increasing the expression of DNA repair enzymes [6]. Overall, this checkpoint involves individual proteins and pro-tein complexes coming together to assemble intricate supramolecular complexes triggered by stalled replication forks. For simplicity's sake, this chapter will focus on the *Saccharo-myces cerevisiae* system and nomenclature. However, regardless of the organism being considered, the general recurring theme is that most stress dependent protein interactions involve either BRCT or FHA domains. In fact, BRCT and FHA domains are rarely found in cytosolic proteins, but they are overrepresented in nuclear proteins involved in DNA replication, as well as the detection and response to DNA damage [7, 8]. Both domains share an ability to specifically recognize phosphorylated epitopes, although with different specificities. BRCT domains primarily recognize phospho-serine (pSer) containing epitopes, while FHA domains exclusively recognize phospho-threonine (pThr). However, phosphorylation-independent interactions have recently been described for both domains.

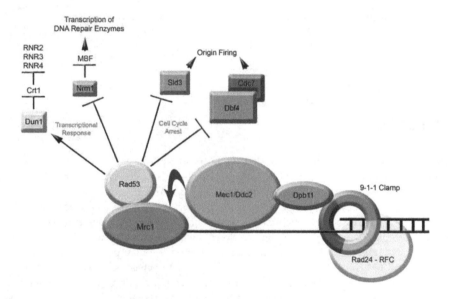

Figure 1. BRCT and FHA domains mediate protein interactions that relay the replication stress signal. A stalled replication fork is represented as a region of ssDNA recognized by a variety of checkpoint proteins colored in shades of blue. Proteins that mediate interactions involved in relaying the stress signal are colored orange if they contain BRCT domains or yellow if they contain FHA domains. The curved arrow indicates a phosphorylation event that takes place on Mrc1 to allow for Rad53 recruitment, while the straight arrows indicate interactions that take place at distal sites to the stalled fork.

Several BRCT-containing proteins are essential for the assembly of the pre-replication complex. For instance, the regulatory subunit of the Cdc7 kinase, Dbf4, contains a single BRCT domain [9, 10] and Dpb11, whose function is essential to activate pre-replication complexes, contains two BRCT pairs [8]. The Rfc1 subunit of the RFC complex, which functions in loading the sliding clamp onto ssDNA/dsDNA junctions during replication [11], also contains a BRCT domain required for DNA binding [12]. Most replication proteins containing BRCT domains are also involved in the replication checkpoint response, suggesting that BRCT domains may have fundamental roles in preserving DNA integrity. During the DNA damage response, BRCT domains are often used to recognize the site of damage, as they can bind to

DNA breaks directly [13, 14] or indirectly by recognizing phosphorylated histone H2A that marks areas of damage [15-20]. In the case of replication stress, at least two first responders are recruited to the ssDNA gap at stalled forks (Figure 1). The first is the protein kinase Mec1 and its targeting subunit Ddc2 that mediates the interaction with RPA-coated ssDNA [21]. The second is the 9-1-1 clamp that encircles the DNA at the ssDNA/dsDNA junction [22]. This requires the action of a clamp loader, which is composed of Rad24 and Rfc2-5. Interestingly, this complex is the alter ego of the clamp loader RFC, which differs only in having Rfc1 instead of Rad24 [23]. In contrast to the sliding clamp, the 9-1-1 complex is held statically by protein-protein interactions at the stalled fork [24].

After the recognition of DNA damage or stalled forks, both FHA- and BRCT-containing proteins feature prominently in bridging protein–protein interactions that disseminate the stress signal (Figure 1). For instance, Dpb11 bridges the interaction between the 9-1-1 clamp and Mec1, leading to the full activation of the Mec1 kinase [24, 25]. Mec1 has many roles at the stalled fork including facilitating the activation of the next downstream kinase in the pathway, Rad53 [26]. This is accomplished after Mec1 phosphorylates Mrc1 — a protein naturally associated with the stalled replisome [27]. This creates phospho-epitopes that act as beacons for the FHA domain of Rad53. Multiple copies of Rad53 are thus recruited, increasing the local concentration of this kinase, which can then be autophosphorylated in trans or phosphorylated by additional kinases present at the stalled fork [28]. Hyperphosphorylation of Rad53 presents phospho-epitopes to other FHA-containing proteins such as Dun1, which leads to increased synthesis of nucleotides [28-30]. Additionally, Rad53 uses its own FHA domains to bind a variety of substrates, including the regulatory subunit of the Dbf4-dependent kinase (DDK) complex [31], consequently suppressing the firing of late origins [32-34]. Rad53 also modulates the activity of Nrm1 leading to a burst of expression of DNA repair enzymes contributing to recovery of the stalled fork [35]. BRCT domains also feature in this recovery process. For example, an important scaffolding protein involved in coordinating the recruitment of repair enzymes to the stalled fork, Rtt107, has six BRCT domains [20].

These themes are echoed in other DNA damage response pathways, where BRCT and FHA domains are known to mediate important interactions. Not surprisingly, mutations in the BRCT and FHA domains of critical damage repair proteins such as Chk2 [36], Nbs1 [37] and BRCA-1 [38, 39], lead to cancer predisposing syndromes. Why does nature rely so heavily on BRCT and FHA domains to respond to stress? The answer may seem to lie in the ability of BRCT and FHA domains to recognize phospho-epitopes, since they are an important cue during replication stress and the DNA damage response, when a number of kinases are awakened. However, their power could also lie in the plasticity of these two domains that can use multiple interaction surfaces to mediate additional interactions beyond phospho-epitope recognition. Such plasticity could, in turn, mediate the interaction network sustaining the formation of the large protein complexes required to promote genome stability in eukaryotes. Along with the well-characterized phosphate recognition ability of FHA and BRCT domains, these varied and unique alternative interaction surfaces will be considered in this chapter. Interactions occurring during the replication checkpoint will be discussed, but examples from other cellular pathways will also be included.

2. Interaction Modes of FHA Domains

An FHA domain consists of an 11-stranded β sandwich connected by loops that often contain short helical regions. The phospho-epitope binding groove is located at the apical surface of the β-sandwich, with the N- and C-termini at the opposite end of the domain. Unlike BRCT domains, that are often present in multiple copies in a single protein, FHA domains are almost always singular. Only two proteins are known to possess two FHA domains in the same polypeptide: Rad53 from *S. cerevisiae* and Rv1747 from *Mycobacterium tuberculosis* (*M. tuberculosis*). In the case of Rad53, these domains (FHA1 and FHA2) are found at opposite ends of the protein and have independent functions [40], reinforcing the idea that FHA domains function as single units. This, however, does not diminish the power of the FHA domain as a scaffold to build large protein complexes in response to stress. FHA domains can bind partners in a phosphorylation dependent or independent manner, the latter of which can utilize either the phospho-epitope binding pocket or alternative surfaces.

2.1. Phospho-epitope dependent interactions

FHA domains recognize phosphorylated proteins with a strict specificity for pThr-containing epitopes. The majority of interactions between FHA domains and their phosphorylated partners have been studied using short peptides including a central phosphorylated threonine [41-46]. These phospho-peptides bind to the apical surface of the FHA domain in an extended conformation using two pockets that determine their binding specificity. The β3–β4 and β4–β5 loops from the FHA domain primarily define the pThr-binding site, where a conserved arginine and serine (Arg70 and Ser85 in Rad53) provide critical contacts with the phosphate group. An extensive hydrogen-bond network mediated by non-conserved residues further stabilizes the interaction with the phosphorylated threonine. The second pocket recognizes the third residue C-terminal to the phospho-threonine (pThr+3), and is usually defined by the β6–β7 and β10–β11 loops of the FHA domain (Figure 2A and B).

A unique aspect of the pThr-binding pocket in FHA domains is its ability to distinguish between phospho-threonine and phospho-serine residues, a talent not shared by other phospho-epitope recognition modules. For instance, MH2 domains share a common ancestor with FHA domains and, yet, MH2 domains can also bind phospho-serine with high affinity [42, 47]. The difference is that the pThr-binding pocket in the FHA domain includes a well-defined hydrophobic nook that provides a docking site for the methyl moiety of the phosphorylated threonine (Figure 2C). While most residues in this nook are not conserved, the hydrophobic nature of this pocket is strictly maintained, thus providing a number of Van der Waals interactions that orient the phospho-threonine such that its phosphate moiety is locked in the most favorable geometry for the interaction with the domain [48]. Thus, the pThr pocket of an FHA domain can be thought of as a glove where only a phospho-threonine can fit (Figure 2C). Despite this rigid mode of binding, certain FHA domains can accommodate deviations. For instance, the FHA2 domain from Rad53 can bind pTyr-containing peptides with low affinity [47] and the FHA domain found in Dun1 has a unique preference for phosphorylated substrates including two pThr residues [49].

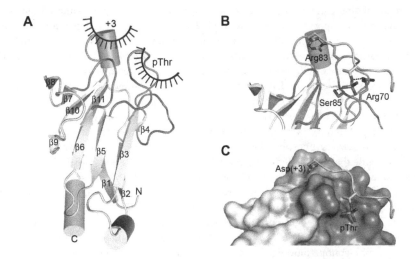

Figure 2. Phospho-peptides interact with FHA domains using at least two pockets. (A) Ribbon diagram of the FHA1 domain from Rad53 (PDB ID: 1G6G) with the loops defining the pThr (β3-β4 and β4-β5) and pThr+3 (β6-β7 and β10-β11) binding pockets highlighted in blue and green, respectively. The N- and C-terminal residues of the domain, which lie on the opposite side of the β-sandwich, are labeled for clarity. (B) Detailed view of a phospho-peptide bound to the pThr and pThr+3 pockets. The side chains of two conserved residues important for the recognition of the phosphate group (Ser85 and Arg70), as well as that of a non-conserved residue important for the specificity of the pThr+3 pocket are shown as sticks. Hydrogen bonds are indicated with dashed lines. (C) Surface representation, shown in the same orientation and color-coding as panel B, indicating the presence of defined pockets for both the phosphate and methyl moieties of the pThr.

In contrast to the pThr pocket, the residues defining the pThr+3 pocket are not conserved allowing for different domains to have different target-sequence specificities. This is called the "pThr+3 rule" wherein different FHA domains have different specificities for the pThr+3 residue. While this provides a convenient way to classify FHA domains, it should be noted that the specificity of the pThr+3 pocket is not fixed. For example, the FHA1 domain from Rad53 prefers aspartic acid as the pThr+3 residue using short peptides *in vitro*, but binds to a bulky, hydrophobic isoleucine in its partner Mdt1 *in vivo*, with the pThr+3 residue (Asp or Ile) occupying physically different pockets in each case [45]. Although FHA domain interactions have been disrupted *in vivo* by mutating the pThr residue, similar experiments with the pThr+3 residue are not available and, hence, the importance of the pThr+3 pocket is unclear. It has been proposed that this ancillary pocket may only have relevance for determining the specificity of the phospho-epitope in small peptides, whereas full-length partners may use different binding mechanisms [50]. Consequently, a detailed understanding of how FHA domains recognize phosphorylated binding partners will necessitate the structural analysis of FHA domains bound to full-length proteins rather than short phosphorylated peptides.

FHA domains often bind only weakly to pThr-containing peptides [43, 51], supporting the idea that additional contact points beyond the phospho-epitope binding site are necessary to form high avidity complexes with their partners. For example, the interaction between the FHA domain from Chk2 and the tandem BRCT repeat from BRCA-1 requires an additional hydrophobic patch on the surface of one of the FHA β-sheets. Mutating either the phosphate binding pocket or this hydrophobic patch destroys the interaction with BRCA-1 even though structurally these two sites are more than 20 Å apart [52]. Reinforcing the importance of this additional interaction surface *in vivo*, mutation of an isoleucine (Ile157Thr) within this hydrophobic patch results in the cancer predisposing Li-Fraumeni syndrome [52]. While this hydrophobic surface is not a common feature of all FHA domains—not even amongst Chk2 homologues [52], it is possible that unique patches exist within the surfaces of the β-sandwich of other FHA domains that provide auxiliary contacts to enhance binding to phosporylated target proteins. Due to the inherent difficulty in obtaining uniformly phosphorylated proteins, solving the structures of FHA domains interacting with their full-length phosphorylated partners is a lofty goal. This is further compounded by the fact that FHA domains cannot be fooled by phospho-mimetic mutations—a trick commonly used to study phosphorylation-dependent interactions—at least when using small peptides [43]. The structure of the Ki67 FHA domain interacting with NIFK1 has recently shed light onto this problem. This interaction was recapitulated with a very long phospho-peptide (consisting of 44 amino acids), which, in addition to occupying the phospho-epitope binding site, also wraps around and extends one of the β-sheets in the Ki67 FHA domain by providing an additional β-strand [50] (Figure 3).

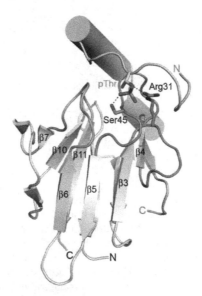

Figure 3. Phosphorylated binding partners of FHA domains can occupy extensive interaction surfaces. Ribbon diagram of the FHA domain of Ki67 bound to a phospho-peptide encompassing residues 226-269 of human NIFK (PDB ID: 2AFF). This long phospho-peptide interacts with three distinct surfaces on the FHA domain, but does not occupy the pThr+3 pocket identified in the structures of FHA domains bound to short phospho-peptides. The phospho-threonine occupies the canonical pThr-binding pocket defined by the β3–β4 and β4–β5 loops (blue), the α helix following the pThr covers a hydrophobic surface partially defined by the β4–β5 and β10–β11 loops (green), and the β-strand at the C-terminus of the peptide extends the β-sheet defined by β7–β10–β11–β1–β2–β4.

This long phospho-peptide does not conform to the "pThr+3" rule. Instead, binding to the FHA domain induces the formation of an α helix that nestles in a hydrophobic pocket formed by the $\beta4$–$\beta5$ and $\beta10$–$\beta11$ loops [50], underscoring the need for additional structural information using full-length phosphorylated binding partners.

2.2. Phospho-epitope independent interactions

Although FHA domains were initially identified as pThr binding domains, it has been predicted that they can also mediate phosphorylation-independent interactions. Members of the kinesin-3 family, a class of motor proteins that transport vesicles to the tips of axons in neural cells [53], contain an N-terminal FHA domain in addition to their motor domain and coiled-coil regions [53]. One member of this family, KIF13B, uses its FHA domain to transport PIP3-rich vesicles in order to facilitate axon development. This involves the formation of a tetrameric complex with CENTA1, which has been studied through X-ray crystallography [54]. This complex has two CENTA1 molecules and two kinesin molecules, with the FHA domain of each kinesin involved in two simultaneous interactions (Figure 4). The first is with the ArfGAP domain of one of the CENTA1 molecules, which contacts the FHA loops that normally recognize a pThr. However, this interaction is phosphorylation-independent because the FHA domain of KIF13B lacks the conserved residues for phospho-threonine recognition [54]. The second CENTA1 molecule in the tetramer uses its Pleckstrin Homology 1 (PH1) domain to contact a surface on the β-sandwich of the same KIF13B FHA domain. This situation is reminiscent of the interaction between Chk2 and BRCA-1, and suggests that auxiliary contacts mediated by the β-sandwich may enhance both phosphorylation dependent and independent interactions. Phosphorylation independent interactions are not exclusive to FHA domains that lack the pThr-recognition residues. Another member of the kinesin-3 family, KIF1A, has a canonical phospho-epitope binding site and, yet, is also suspected of using this pocket for a phosphorylation-independent interaction [55]. Similarly, the FHA domain of S. cerevisiae Rad53 has a canonical phosphoepitope binding site, but is presumed to interact with the BRCT domain of Dbf4 in a phosphorylation-independent manner, though the molecular determinants of this interaction are unclear [10].

Some bacterial proteins can also interact with the phospho-epitope binding site of an FHA domain in a phosphorylation-independent manner. Although bacteria primarily rely on histidine kinases and their associated regulatory responders for phosphorylation-dependent signaling, some also utilize eukaryotic-like Ser/Thr Protein Kinases (STPKs) [56]. Like eukaryotes, bacteria can use STPKs to respond to stress, but they also participate in other processes such as pathogenicity, thereby providing important drug targets [56]. The best characterized bacterium in this regard is M. tuberculosis, the causative agent of tuberculosis, which tops the charts in the prokaryotic kingdom with eleven STPKs [57]. Proteins that work downstream of bacterial STPKs often contain FHA domains, with M. tuberculosis having five such proteins [58]. One of them, Rv1827, is of special interest because it engages a phospho-epitope present in its N-terminal tail intramolecularly [59]. This effectively occludes the phospho-epitope binding site preventing the interaction of other binding partners [59]. Intriguingly, at least three different binding partners can compete with this intramolec-

ular phospho-epitope even though none of them includes a phosphorylation site [59]. Therefore, the interactions mediated by the FHA domain of *M. tuberculosis* Rv1827 reveal two
recurrent features: the ability of an intramolecular interaction to negatively regulate the interactions of an FHA domain, and the use of an FHA phospho-epitope binding site to engage in phosphorylation-independent interactions. These features are also reminiscent of the
interactions mediated by the FHA domains found in KIF1A [55], reinforcing the idea that
FHA domains may use competing interactions to fine-tune cellular processes.

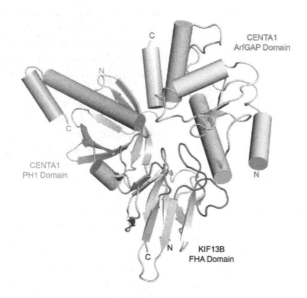

Figure 4. The FHA domain of kinesin KIF13B mediates two phosphorylation independent interactions simultaneously.
Ribbon diagram of the FHA domain of KIF13B bound to CENTA1 (PDB ID: 3MDB). In the crystal structure, the FHA domain of KIF13B (white) contacts the PH1 (tan) and the ArfGAP (cyan) domains of two adjacent CENTA1 molecules simultaneously. The interaction between the FHA and PH1 domains is mediated by one of the faces of the β-sandwich,
while the loops that normally define the pThr-binding pocket are involved on the recognition of the ArfGAP domain.

3. Interaction Modes of BRCT Domains

BRCT domains are named after the breast cancer associated protein 1 (BRCA-1) C-terminus
because they were originally identified at this end of BRCA-1. However, the BRCT is an ancient domain that originates in prokaryotic NAD+ ligases where it is used to bind to DNA.
Eukaryotes obtained the BRCT domain through horizontal gene transfer, and while some
eukaryotic BRCT domains still retain DNA binding function, the vast majority have evolved
to recognize protein partners instead [60]. BRCT domains are defined by a central four-
stranded parallel β-sheet surrounded by three helices: α1 and α3 on one side, and α2 on the
other of the β-sheet [61]. However, additional secondary structure elements have been de-

scribed in the loop regions. There are also BRCT domains that lack elements, notably helix $\alpha 2$ [61]. Rap1 is the most extreme example of this, having only three strands in its central β-sheet and all helices packed against the same side of the sheet, leaving the other side exposed [62]. This unique structure is highly flexible and relatively unstable [62].

BRCT domains can occur both as single or multiple units, which usually consist of two BRCT domains and are referred to as tandem BRCT repeats. The structural diversity of BRCT domains is perhaps best illustrated by the human homologue of Dpb11 (human TopBP1), which contains eight BRCT domains that function as single, double or triple BRCT units [63-65]. Similarly to FHA domains, BRCT domains are overrepresented in DNA damage response proteins where they recognize phosphorylated targets generated during damage recognition and repair [7]. While the molecular intricacies of phospho-epitope recognition by tandem BRCT repeats have been extensively studied, phosphorylation-independent interactions mediated by BRCT tandems or the functions of single BRCT domains remain poorly characterized. Elegant proteolysis studies have shown that mutations in the BRCT tandem repeat of BRCA-1, with a causal link to early onset breast and ovarian cancer, destabilize the BRCT fold [39], suggesting that BRCT domains may work as protein scaffolds. These studies also revealed the hypersensitivity of BRCT domains to mutations, an effect that was attributed to its minimal size (95-100 amino acids) [66] and, consequently, the fact that every residue contributes to either maintaining the domain fold or mediating interactions with BRCT-binding partners [67].

3.1. Tandem BRCT Domains: The Two-Knob Hypothesis

The tandem BRCT repeat was formed through a gene duplication event, in which the binding pocket originally used to bind to the phosphate backbone of DNA evolved to recognize a phospho-epitope in a target protein [60]. Being an α/β fold, the BRCT domain has a topological switch point; a region along the C-terminal edge of the β-sheet whereupon a groove is formed in the connecting loops. Tandem BRCT repeats use the topological switch point—termed the P1 pocket—of their first BRCT domain to bind phospho-serine (pSer) residues in their interaction partners (Figure 5). Similar to FHA domains, this interaction involves the side chain of a conserved serine residue, but in BRCT domains the phosphate moiety of the pSer residue is further stabilized by the interaction with the side chain of a conserved lysine, as well as the backbone atoms from the glycine immediately following the conserved serine [68].

Similar to FHA domains, tandem BRCT repeats also subscribe to a "pSer+3 rule" to enhance phospho-epitope binding specificity [68]. However, the pSer+3-binding pocket—termed the P2 pocket—only forms in tandem BRCT repeats as it is defined by residues at the interface between the first and second BRCT domains. When two BRCT domains coalesce to define a tandem repeat, the central β-sheet of both domains adopt a parallel arrangement that defines an intervening hydrophobic three-helix bundle form by $\alpha 2$ from the N-terminal and $\alpha 1$ and $\alpha 3$ from the C-terminal BRCT domains (Figure 5) [69]. Helical bundles are known to facilitate molecular interactions [70] and, in the case of tandem BRCT domains, it allows for the recognition of a bulky hydrophobic residue at the pSer+3 position [68]. Indeed, the high specificity of tandem BRCT repeats for their phospho-epitopes is primarily due to the presence of the P2 pocket that imposes the need for a second knob in the phospho-epitope, thus precluding non-specific binding.

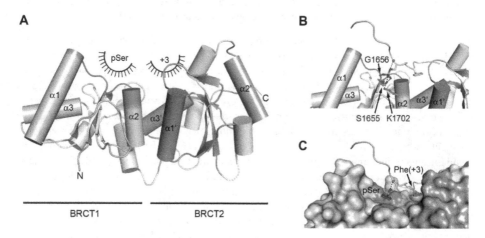

Figure 5. Phospho-epitope recognition by tandem BRCT repeats. (A) Ribbon diagram of the tandem BRCT domain from BRCA-1 (PDB ID: 1T2V) with secondary structure elements defining the BRCT fold shown in tan (BRCT1) and brown (BRCT2), and those not common to the BRCT fold shown in white. The pSer-binding site is located entirely within BRCT1, whereas the pSer+3 pocket is located in a three-helical bundle created at the interface between the two domains. (B) The side chains of Ser1655, Lys1702, as well as the main chain nitrogen of Gly1656 stabilize the phosphate moiety of the pSer. Additionally, the phosphate group engages in hydrogen bonds with the backbone amino groups represented as blue spheres. (C) The surface representation, shown in the same orientation as in panel B, reveals that the pSer binds to a shallow pocket that could not accommodate any other phosphorylated residue, while the conserved aromatic residue at the pSer+3 position is nestled into a well-defined hydrophobic pocket.

In certain tandem BRCT repeats, the P1 pocket is found in the C-terminal rather than the N-terminal BRCT, however due to the absence of the P2 pocket it is presumed that this mode of interaction is weaker than the canonical binding mode [71]. In fact, BRCT repeats containing a P1 pocket on the C-terminal BRCT are known to mediate phosphorylation-independent protein–protein interactions [72], suggesting that the binding specificity for a phospho-epitope may not be as critical. In the structure of the tandem BRCT repeats found in 53BP1 bound to the DNA-binding domain of p53 [73, 74], the inter-domain linker is critical to mediate the interaction between 53BP1 and p53. Similarly, in the structure of the *Schizosaccharo-myces pombe* Crb2 homodimer [75], the linker connecting the two BRCT domains mediates protein dimerization. Collectively, these structures underscore the fact that tandem BRCT repeats define single functional units with multiple interaction surfaces.

3.2. BRCT "Super-domains": Expanding the two-knob model

The individual units in a tandem BRCT repeat are dependent on each other for structural stability due to the hydrophobicity of the α helices that define the intervening helix bundle [69]. Thus, a tandem BRCT repeat can actually be considered one single domain module, distinct from single BRCT domains. Beyond single and tandem BRCT arrangements,

a number of BRCT structures over the past decade have revealed many unexpected tertiary structures formed by the combination with other functional domains. This is not only
in the number of BRCT domains involved, such as the structure of the triple BRCT repeat
from the human TopBP1 [76], but also in the diversity of domains that can be ensnared
by a BRCT neighbor. This includes the Fibronectin Type III domain (FN3) found at the N-
terminus of a tandem BRCT domain in the *S. cerevisiae* protein Chs5 [77], or the FHA domain that does likewise in *S. pombe* Nbs1 [78, 79].

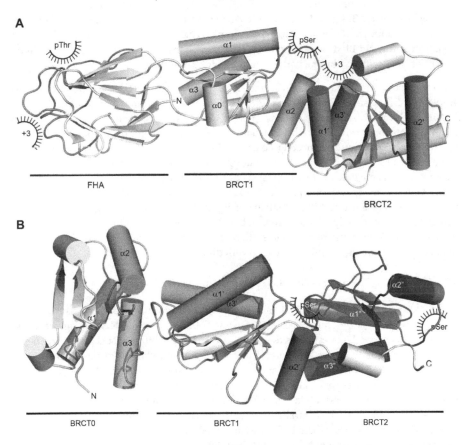

Figure 6. BRCT super-domains. (A) Ribbon diagram of the FHA–BRCT–BRCT super-domain found in Nbs1 (PDB ID:
3ION). Four knobs are present: pThr and pThr+3 pockets in the phospho-epitope binding site in the FHA domain
and pSer and pSer+3 pockets in the phospho-epitope binding site defined by the tandem BRCT repeat. Helix α0 in
the FHA domain interacts with helices α1 and α3 in the BRCT1 domain, however the relative orientation of these
helices does not resemble that of the characteristic three-helix bundle found at the interface of tandem BRCT repeats. (B) Ribbon diagram of the triple BRCT repeat found in human TopBP1 (PDB ID: 2XNH) with the structural elements defining each BRCT domain shown in different shades of brown. The pSer binding sites present in the
second and third BRCT domains are highlighted in blue, while additional structural elements not common to the
BRCT fold are shown in white.

The crystal structure of the FHA–BRCT–BRCT super-domain of Nbs1 emphasizes the ability of BRCT domains to build scaffolds capable of multiple interaction modes. While the two BRCT domains in Nbs1 associate to form a canonical tandem BRCT repeat with a phospho-epitope binding site, the FHA domain interacts with the hydrophobic core of the first BRCT domain leaving helices $\alpha1$ and $\alpha3$ exposed to the solvent (Figure 6) [78, 79]. In contrast to other BRCT domains, these two helices are amphipathic and, hence, break the theme of BRCT domains using hydrophobic three-helix bundles to build super-domains. Surprisingly, the FHA and the first BRCT domain (FHA–BRCT1) form the most stable unit of Nbs1, whereas the second BRCT (BRCT2) is quite flexible despite forming a canonical tandem repeat with the first BRCT domain [78, 79]. Given its tertiary structure, the FHA–BRCT–BRCT super-domain could bind two phospho-epitopes simultaneously, suggesting that the interactions of Nbs1 with its binding partners may be highly regulated.

The recent structure of the triple BRCT repeat in TopBP1 (BRCT0/1/2) provides another example of a BRCT super-domain deviating from the canonical three-helix bundle interface (Figure 6). In this case, neither BRCT0/1 nor BRCT1/2 associate to form canonical tandem BRCT repeats, primarily due to the unusually short inter-domain linkers that connect adjacent BRCTs [76]. Beyond connecting adjacent BRCT domains, the inter-domain linkers in some tandem BRCT repeats actively mediate protein-protein interactions [73, 74] and, in extreme cases, it is the linker rather than the BRCT domains that mediates the interaction. For example, the damage response protein XRCC4 interacts exclusively with the inter-domain linker connecting the two BRCT domains of ligase IV [80]. Collectively, these structures demonstrate that not only the BRCT repeat, but also the length and composition of the inter-domain linker joining the two domains affect the binding plasticity of BRCT repeats and their ability to form higher order structures with diverse binding specificities.

3.3. Single BRCT Domains: Is One the Loneliest Number?

The association of multiple BRCT domains with other functional domains within a single polypeptide chain is becoming a common theme found in many DNA damage response proteins. This poses the question as to whether the increased binding specificity, and hence the underlying ability to fine tune interactions during the checkpoint response, is the driving force for BRCT domains to build such complex structures. Surprisingly, the majority of eukaryotic proteins that possess BRCT domains have at least one that functions solo. The exposed α helices ($\alpha1$, $\alpha2$ and $\alpha3$) in single BRCT domains are amphipathic, with a hydrophobic face interacting with the central β-sheet and a polar face exposed to the solvent, unlike their tandem counterparts where both faces are chiefly hydrophobic. This enhances the stability of single BRCT domains but does not shed light on their functions.

There is mounting evidence indicating that both single and tandem BRCT domains may require additional secondary structural elements to form functional units [10, 12, 81, 82]. Some of these structural elements are required for structural stability, such as the additional C-terminal α-helix in the PARP-1 BRCT [81], while others enhance function (Figure 7). An example of the latter comes from the largest subunit of the budding yeast clamp loader, Rfc1. Rfc1 has a single BRCT domain that is required for DNA-binding [12]. However, as an isolated unit, this domain is unable to recognize DNA despite having a positively charged patch positioned at the conserved P1 pocket [12]. The inclusion of an N-terminal extension recapitulates DNA binding with a K_D in the nM range. The extension encompasses an additional α helix that directly contacts DNA [12]. Rather than being a structural element integrated into the BRCT fold, however, this helix is predicted to act as an auxiliary element to enhance function [12]. Based on this model, both the BRCT domain and its auxiliary helix likely bind synergistically to the DNA leading to a robust interaction (Figure 7). Similarly, the presence of an additional N-terminal helix necessary to bind DNA is also predicted in the single BRCT domain from the translesion polymerase Rev1 [12, 82]. It is conceivable that these additional structural elements play the dual role of stabilizing the BRCT fold and enhancing its function. In fact, this is the case of *S. cerevisiae* Dbf4, the regulatory subunit of the Cdc7 kinase, where an α-helix immediately precedes the $\beta1$ strand and its presence is important for the stability of the domain, as well as the interaction with Rad53 during the checkpoint response [10, 83].

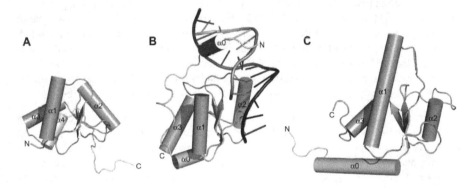

Figure 7. Single BRCT domains including additional structural elements. (A) Ribbon diagram of the BRCT domain of PARP-1 (PDB ID: 2LE0), highlighting an additional C-terminal α helix ($\alpha4$) necessary for the structural integrity of the domain. (B) Ribbon diagram of the molecular model of Rfc1 bound to DNA (PDB ID: 2K7F). Similar to Dbf4, this model predicts two additional helices at the N-terminus of the BRCT domain of Rfc1. Helix $\alpha0$ is not part of the BRCT fold, but it is essential for DNA binding by Rfc1. Conversely, the predicted helix $\alpha0'$ does not appear to be involved in DNA binding but its relative orientation and interaction with the BRCT core resembles that of helix $\alpha0$ in Dbf4. (C) Ribbon diagram of the HBRCT domain of Dbf4 (PDB ID: 3QBZ), highlighting an additional N-terminal α helix ($\alpha0$) necessary to mediate the interaction between this domain of Dbf4 and the FHA1 domain of Rad53. Helix $\alpha0$ is an integral part of the fold as it anchors itself to the BRCT core through hydrophobic interactions, thereby introducing the idea of BRCT domains being building blocks that can be decorated to form super-structures with broader binding specificities.

4. Dbf4/Rad53: A Case Study for phosphorylation-independent BRCT and FHA Interactions

The DDK complex, formed by the association of the Cdc7 kinase and its regulatory subunit Dbf4, is required for initiating DNA replication and, hence, it is essential for the life of all eukaryotes [84]. Like many other replication proteins, DDK is also involved in the replication stress checkpoint [31-34, 85, 86]. The ability of Dbf4 to crossover into the stress response pathway is partly due to a conserved motif at the N-terminus of the protein (motif N). Based on sequence alignments, it was a matter of debate whether motif N was a bona fide BRCT domain [9, 87], however concerns were laid to rest when the crystal structure of the N-terminal region of *S. cerevisiae* Dbf4 was determined [10]. The structure revealed that this region of Dbf4 folds as a modified BRCT domain that requires an additional N-terminal α-helix to form a stable unit (Figure 7). A fragment of Dbf4 consisting of the canonical BRCT domain but missing the additional helix did not support binding to the FHA1 domain of the checkpoint effector kinase, Rad53 [9, 10, 31, 88]. It was proposed that this additional helix (α0) defines, at least in part, the interaction interface [10]. Due to the functional and structural relevance of the α0 helix, this domain of Dbf4 is referred to as H–BRCT to signify the location of the additional helix. In contrast to the additional N-terminal helix identified in the BRCT domain of Rfc1 that is completely independent from the BRCT domain, helix α0 in Dbf4 is physically latched onto the BRCT domain in the crystal structure of Dbf4 [10] (Figure 7). This is through hydrophobic residues from α0 interacting with a hydrophobic pocket between the central β-sheet and α1 of the BRCT domain [10]. Therefore, helix α0 is in a sense decorating the surface of the BRCT and maintained as an integral part of the domain [10, 83]. This serves as a note of caution when studying BRCT domains in general, as the functional and the structurally stable forms of the domain may not necessarily coincide. Thus, functional BRCT units can only be reliably ascertained through empirical assays with different sized protein fragments and not through structure-guided sequence alignments.

The interaction between the H–BRCT domain of Dbf4 and the first FHA domain of Rad53 also poses an interesting paradigm during the checkpoint response because Dbf4 and Rad53 interact in a phosphorylation independent manner using domains notable for recognizing phosphorylated epitopes. This interaction was initially proposed to depend on the recognition of a phospho-epitope in Dbf4 because a point mutation in the conserved arginine (R70A) involved in phospho-threonine recognition by FHA1 effectively abolished the interaction between the two proteins in a yeast two-hybrid experiment [31]. Yeast two-hybrid experiments conducted in the past have demonstrated that phosphorylation-dependent interactions with Rad53 can indeed be detected, likely due to the activity of endogenous kinases [89]. However, all the threonine residues located within or surrounding the H–BRCT domain of Dbf4 can be mutated without abrogating the interaction with Rad53 [10]. Several possible scenarios can explain the critical role of Arg70 in mediating the interaction with Dbf4. First, Arg70 may be part of the interaction interface despite the lack of a phospho-epitope in Dbf4. This scenario would be reminiscent of *M. tuberculosis* Rv1827 that can bind several binding partners using the apical surface of the FHA domain containing the conserved arginine in a phosphorylation-independent manner [59]. Alternatively, the R70A mutation could destabilize the FHA1 fold, in which case the Dbf4-binding defect associated with this mutation would be indirect. This scenario seems unlikely, given that FHA are stable domains and that Arg70 is solvent exposed (Figure 2). Lastly, Dbf4 could have a dual interaction with Rad53, where two independent interactions would need to occur simultaneously to form a high-avidity complex. In this case, the H–BRCT domain of Dbf4 could interact with the FHA1 domain of Rad53 in a phosphorylation-independent manner, while FHA1 recognizes a phospho-epitope located in another region of the DDK complex. While this idea awaits validation, other dual interactions have been previously observed in the structures of other FHA domains, including that of KIF13B and Chk2 [52, 54]. This seems to suggest that simultaneous phosphorylation dependent and independent interactions may be a broader mechanism to regulate interactions mediated by FHA domains than previously anticipated.

5. Conclusion

Through the use of short phospho-peptides, the basic mechanism underlying pThr recognition by FHA domains has been elucidated. However, it is clear that full-length binding partners harboring pThr-epitopes will likely contact the FHA domain at multiple sites in addition to the pThr pocket, many of which will be unique to that particular interaction. Additionally, proteins are capable of interacting with FHA domains in a phosphorylation independent manner, using the phospho-epitope binding site as well as alternative interfaces, most often the surfaces of the β-sandwich. While phosphorylation-dependent interactions have a clear mode for turning the interaction on and off, regulation of phosphorylation-independent interactions remains unexplored. If these interactions are mediated by the phospho-epitope binding site it may simply be a matter of availability of phosphorylated binding partners that can compete with the unmodified protein. However, if the canonical phosphate binding residues are absent or the interaction takes place on an alternative surface, the mechanism for control is less clear.

BRCT domains have previously been divided in tandem repeats, which can interact with phosphorylated partners in a well-defined manner, and the enigmatic single domains. The plethora of interaction mechanisms used by single BRCT domains can seem overwhelming, but like FHA domains a common theme is now emerging. The BRCT fold may serve as a structural core upon which more complex and unique assemblies can be built. In this way, specific interaction surfaces can be created allowing for the BRCT domain to gain function. Therefore, BRCT domains can use extra secondary structural elements, either integrated into the fold as in Dbf4 or as an auxiliary element as in Rfc1, or entire domains such as in the case of Chs5 and Nbs1—or even in tandem BRCT repeats—to modulate their functions.

Functional and structural analyses of FHA and BRCT domains during the last decade have unveiled a complex repertoire of interactions mediated by these two domains. Once regarded as mere phospho-epitope binding units, we now know that they can mediate very sophisticated interactions regulated by multiple binding knobs. Further structural and functional analysis of protein complexes mediated by these two domains will delineate the common mechanisms that regulate the DNA damage response, and will extend the lessons learned from studying the replication stress pathway in yeast to a variety of stress response networks that rely on BRCT and FHA domains across all kingdoms of life.

Author details

Lindsay A. Matthews and Alba Guarné*

*Address all correspondence to: guarnea@mcmaster.ca

Department of Biochemistry and Biomedical Sciences, McMaster University, Hamilton, ON, Canada

References

[1] Longhese, M. P., Clerici, M., & Lucchini, G. (2003). The S-phase checkpoint and its regulation in Saccharomyces cerevisiae. *Mutat Res.*, 532, 41-58.

[2] Branzei, D., & Foiani, M. (2010). Maintaining genome stability at the replication fork. *Nat Rev Mol Cell Biol.*, 11, 208-19.

[3] Sogo, J. M., Lopes, M., & Foiani, M. (2002). Fork reversal and ssDNA accumulation at stalled replication forks owing to checkpoint defects. *Science.*, 297, 599-602.

[4] Burhans, W. C., & Weinberger, M. (2007). DNA replication stress, genome instability and aging. *Nucleic Acids Res.*, 35, 7545-56.

[5] Arlt, M. F., Mulle, J. G., Schaibley, V. M., Ragland, R. L., Durkin, S. G., Warren, S. T., & Glover, T. W. (2009). Replication stress induces genome-wide copy number changes in human cells that resemble polymorphic and pathogenic variants. *Am J Hum Genet.*, 84, 339-50.

[6] Branzei, D., & Foiani, M. (2009). The checkpoint response to replication stress. *DNA Repair Amst*, 8, 1038-46.

[7] Mohammad, D. H., & Yaffe, M. B. (2009). 14-3-3 proteins, FHA domains and BRCT domains in the DNA damage response. *DNA repair*, 8, 1009-17.

[8] Pospiech, H., Grosse, F., & Pisani, F. M. (2010). The initiation step of eukaryotic DNA replication. *Subcell Biochem.*, 50, 79-104.

[9] Gabrielse, C., Miller, C. T., Mc Connell, K. H., De Ward, A., Fox, C. A., & Weinreich, M. (2006). A Dbf4p BRCA1 C-terminal-like domain required for the response to replication fork arrest in budding yeast. *Genetics.*

[10] Matthews, L. A., Jones, D. R., Prasad, A. A., Duncker, B. P., & Guarne, A. (2012). Saccharomyces cerevisiae Dbf4 has unique fold necessary for interaction with Rad53 kinase. *J Biol Chem.*, 287, 2378-87.

[11] Majka, J., & Burgers, P. M. (2004). The PCNA-RFC families of DNA clamps and clamp loaders. *Prog Nucleic Acid Res Mol Biol.*, 78, 227-60.

[12] Kobayashi, M. A. B. E., Bonvin, A. M. J. J., & Siegal, G. (2009). Structure of the DNA-bound BRCA1 C-terminal Region from Human Replication Factor C p140 and Model of the Protein-DNA Complex. *J Biol Chem.*, 285, 10087-97.

[13] Yamane, K., Katayama, E., & Tsuruo, T. (2000). The BRCT regions of tumor suppressor BRCA1 and of XRCC1 show DNA end binding activity with a multimerizing feature. *Biochem Biophys Res Commun.*, 279, 678-84.

[14] Yamane, K., & Tsuruo, T. (1999). Conserved BRCT regions of TopBP1 and of the tumor suppressor BRCA1 bind strand breaks and termini of DNA. *Oncogene*, 18, 5194-203.

[15] Sofueva, S., Du, L. L., Limbo, O., Williams, J. S., & Russell, P. (2010). BRCT domain interactions with phospho-histone H2A target Crb2 to chromatin at double-strand breaks and maintain the DNA damage checkpoint. *Mol Cell Biol.*, 30, 4732-43.

[16] Hammet, A., Magill, C., Heierhorst, J., & Jackson, S. P. (2007). Rad9 BRCT domain interaction with phosphorylated H2AX regulates the G1 checkpoint in budding yeast. *EMBO Rep.*, 8, 851-7.

[17] Kobayashi, J, Tauchi, H, Sakamoto, S, Nakamura, A, Morishima, K, Matsuura, S, Kobayashi, T, Tamai, K, Tanimoto, K, & Komatsu, K. (2002). NBS1 localizes to gamma-H2AX foci through interaction with the FHA/BRCT domain. *Curr Biol.*, 12, 1846-51.

[18] Williams, J. S., Williams, R. S., Dovey, C. L., Guenther, G., Tainer, J. A., & Russell, P. (2010). gammaH2A binds Brc1 to maintain genome integrity during S-phase. *EMBO J.*, 29, 1136-48.

[19] Stucki, M., Clapperton, J. A., Mohammad, D., Yaffe, M. B., Smerdon, S. J., & Jackson, S. P. (2005). MDC1 directly binds phosphorylated histone H2AX to regulate cellular responses to DNA double-strand breaks. *Cell.*, 123, 1213-26.

[20] Li, X., Liu, K., Li, F., Wang, J., Huang, H., Wu, J., & Shi, Y. (2012). Structure of C-terminal tandem BRCT repeats of Rtt107 protein reveals critical role in interaction with phosphorylated histone H2A during DNA damage repair. *J Biol Chem.*, 287, 9137-46.

[21] Zou, L., & Elledge, S. J. (2003). Sensing DNA damage through ATRIP recognition of RPA-ssDNA complexes. *Science.*, 300, 1542-8.

[22] Majka, J., Binz, S. K., Wold, MS, & Burgers, P. M. (2006). Replication protein A directs loading of the DNA damage checkpoint clamp to 5'-DNA junctions. *J Biol Chem.*, 281, 27855-61.

[23] Green, C. M., Erdjument-Bromage, H., Tempst, P., & Lowndes, N. F. (2000). A novel Rad24 checkpoint protein complex closely related to replication factor C. *Curr Biol.*, 10, 39-42.

[24] Navadgi-Patil, V. M., & Burgers, P. M. (2009). A tale of two tails: activation of DNA damage checkpoint kinase Mec1/ATR by the 9-1-1 clamp and by Dpb11/TopBP1. *DNA Repair Amst*, 8, 996-1003.

[25] Mordes, D. A., Nam, E. A., & Cortez, D. (2008). Dpb11 activates the Mec1-Ddc2 complex. *Proc Natl Acad Sci*, U S A, 105, 18730-4.

[26] Chen, S. H., & Zhou, H. (2009). Reconstitution of Rad53 activation by Mec1 through adaptor protein Mrc1. *J Biol Chem.*, 284, 18593-604.

[27] Osborn, A. J., & Elledge, S. J. (2003). Mrc1 is a replication fork component whose phosphorylation in response to DNA replication stress activates Rad53. *Genes Dev.*, 17, 1755-67.

[28] Lee, S. J., Schwartz, M. F., Duong, J. K., & Stern, D. F. (2003). Rad53 phosphorylation site clusters are important for Rad53 regulation and signaling. *Mol Cell Biol.*, 23, 6300-14.

[29] Zhao, X., & Rothstein, R. (2002). The Dun1 checkpoint kinase phosphorylates and regulates the ribonucleotide reductase inhibitor Sml1. *Proc Natl Acad Sci*, U S A, 99, 3746-51.

[30] Huang, M., Zhou, Z., & Elledge, S. J. (1998). The DNA replication and damage checkpoint pathways induce transcription by inhibition of the Crt1 repressor. *Cell*, 94, 595-605.

[31] Duncker, B. P., Shimada, K., Tsai-Pflugfelder, M., Pasero, P., & Gasser, S. M. (2002). An N-terminal domain of Dbf4p mediates interaction with both origin recognition complex (ORC) and Rad53p and can deregulate late origin firing. *Proc Natl Acad Sci.*, U S A, 99, 16087-92.

[32] Duch, A., Palou, G., Jonsson, Z. O., Palou, R., Calvo, E., Wohlschlegel, J., & Quintana, D. G. (2011). A Dbf4 mutant contributes to bypassing the Rad53-mediated block of origins of replication in response to genotoxic stress. *J Biol Chem.*, 286, 2486-91.

[33] Lopez-Mosqueda, J., Maas, N. L., Jonsson, Z. O., Defazio-Eli, L. G., Wohlschlegel, J., & Toczyski, D. P. (2010). Damage-induced phosphorylation of Sld3 is important to block late origin firing. *Nature*, 467, 479-83.

[34] Zegerman, P., & Diffley, J. F. (2010). Checkpoint-dependent inhibition of DNA replication initiation by Sld3 and Dbf4 phosphorylation. *Nature*, 467, 474-8.

[35] Travesa, A., Kuo, D., de Bruin, R. A., Kalashnikova, T. I., Guaderrama, M., Thai, K., Aslanian, A., Smolka, M. B., Yates, J. R., 3rd, Ideker, T., & Wittenberg, C. (2012). DNA replication stress differentially regulates G1/S genes via Rad53-dependent inactivation of Nrm1. *EMBO J.*, 31, 1811-22.

[36] Wu, X., Webster, S. R., & Chen, J. (2001). Characterization of tumor-associated Chk2 mutations. *J Biol Chem.*, 276, 2971-4.

[37] Varon, R., Reis, A., Henze, G., von Einsiedel, H. G., Sperling, K., & Seeger, K. (2001). Mutations in the Nijmegen Breakage Syndrome gene (NBS1) in childhood acute lymphoblastic leukemia (ALL). *Cancer Res.*, 61, 3570-2.

[38] Drikos, I., Nounesis, G., & Vorgias, C. E. (2009). Characterization of cancer-linked BRCA1-BRCT missense variants and their interaction with phosphoprotein targets. *Proteins.*, 77, 464-76.

[39] Williams, R. S., Chasman, D. I., Hau, D. D., Hui, B., Lau, A. Y., & Glover, J. N. (2003). Detection of protein folding defects caused by BRCA1-BRCT truncation and missense mutations. *J Biol Chem.*, 278, 53007-16.

[40] Tam, A. T., Pike, B. L., & Heierhorst, J. (2008). Location-specific functions of the two forkhead-associated domains in Rad53 checkpoint kinase signaling. *Biochemistry*, 47, 3912-6.

[41] Yaffe, M. B., & Smerdon, S. J. (2004). The use of in vitro peptide-library screens in the analysis of phosphoserine/threonine-binding domain structure and function. *Annu Rev Biophys Biomol Struct.*, 33, 225-44.

[42] Hammet, A., Pike, B. L., Mc Nees, C. J., Conlan, L. A., Tenis, N., & Heierhorst, J. (2003). FHA domains as phospho-threonine binding modules in cell signaling. *IUBMB Life.*, 55, 23-7.

[43] Durocher, D., Henckel, J., Fersht, A. R., & Jackson, S. P. (1999). The FHA domain is a modular phosphopeptide recognition motif. *Mol Cell.*, 4, 387-94.

[44] Durocher, D., & Jackson, S. P. (2002). The FHA domain. *FEBS Lett.*, 513, 58-66.

[45] Mahajan, A., Yuan, C., Pike, B. L., Heierhorst, J., Chang, C. F., & Tsai, M. D. (2005). FHA domain-ligand interactions: importance of integrating chemical and biological approaches. *J Am Chem Soc.*, 127, 14572-3.

[46] Liang, X., & Van Doren, S. R. (2008). Mechanistic insights into phosphoprotein-binding FHA domains. *Acc Chem Res.*, 41, 991-9.

[47] Wang, P., Byeon, I. J., Liao, H., Beebe, K. D., Yongkiettrakul, S., Pei, D., & Tsai, M. D. (2000). II. Structure and specificity of the interaction between the FHA2 domain of Rad53 and phosphotyrosyl peptides. *J Mol Biol.*, 302, 927-40.

[48] Pennell, S., Westcott, S., Ortiz-Lombardia, M., Patel, D., Li, J., Nott, T. J., Mohammed, D., Buxton, R. S., Yaffe, M. B., Verma, C., & Smerdon, S. J. (2010). Structural and functional analysis of phosphothreonine-dependent FHA domain interactions. *Structure.*, 18, 1587-95.

[49] Lee, H., Yuan, C., Hammet, A., Mahajan, A., Chen, E. S. W., Wu, M. R., Su, M. I., Heierhorst, J., & Tsai, M. D. (2008). Diphosphothreonine-specific interaction between an SQ/TQ cluster and an FHA domain in the Rad53-Dun1 kinase cascade. *Mol Cell.*, 30, 767-78.

[50] Byeon, I. J., Li, H., Song, H., Gronenborn, A. M., & Tsai, M. D. (2005). Sequential phosphorylation and multisite interactions characterize specific target recognition by the FHA domain of Ki67. *Nat Struct Mol Biol.*, 12, 987-93.

[51] Durocher, D., Taylor, I. A., Sarbassova, D., Haire, L. F., Westcott, S. L., Jackson, S. P., Smerdon, S. J., & Yaffe, M. B. (2000). The molecular basis of FHA domain:phospho-peptide binding specificity and implications for phospho-dependent signaling mechanisms. *Mol Cell.*, 6, 1169-82.

[52] Li, J., Williams, B. L., Haire, L. F., Goldberg, M., Wilker, E., Durocher, D., Yaffe, M. B., Jackson, S. P., & Smerdon, S. J. (2002). Structural and functional versatility of the FHA domain in DNA-damage signaling by the tumor suppressor kinase Chk2. *Mol Cell.*, 9, 1045-54.

[53] Hirokawa, N., & Noda, Y. (2008). Intracellular transport and kinesin superfamily proteins, KIFs: structure, function, and dynamics. *Physiol Rev.*, 88, 1089-118.

[54] Tong, Y., Tempel, W., Wang, H., Yamada, K., Shen, L., Senisterra, G. A., Mac Kenzie, F., Chishti, A. H., & Park, H. W. (2010). Phosphorylation-independent dual-site binding of the FHA domain of KIF13 mediates phosphoinositide transport via centaurin alpha1. *Proc Natl Acad Sci, U S A*, 107, 20346-51.

[55] Lee, J. R., Shin, H., Choi, J., Ko, J., Kim, S., Lee, H. W., Kim, K., Rho, S. H., Lee, J. H., Song, H. E., Eom, S. H., & Kim, E. (2004). An intramolecular interaction between the FHA domain and a coiled coil negatively regulates the kinesin motor KIF1A. *EMBO J.*, 23, 1506-15.

[56] Danilenko, V. N., Osolodkin, D. I., Lakatosh, S. A., Preobrazhenskaya, M. N., & Shtil, A. A. (2011). Bacterial eukaryotic type serine-threonine protein kinases: from structural biology to targeted anti-infective drug design. *Curr Top Med Chem.*, 11, 1352-69.

[57] Av-Gay, Y., & Everett, M. (2000). The eukaryotic-like Ser/Thr kinases of Mycobacterium tuberculosis. *Trends Microbiol.*, 8, 238-44.

[58] Pallen, M., Chaudhuri, R., & Khan, A. (2002). Bacterial FHA domains: neglected players in the phospho-threonine signalling game? *Trends Microbiol.*, 10, 556-63.

[59] Nott, T. J., Kelly, G., Stach, L., Li, J., Westcott, S., Patel, D., Hunt, D. M., Howell, S., Buxton, R. S., O'Hare, H. M., & Smerdon, S. J. (2009). An intramolecular switch regulates phosphoindependent FHA domain interactions in Mycobacterium tuberculosis. *Sci Signal.*, 2, ra12.

[60] Sheng, Z. Z., Zhao, Y. Q., & Huang, J. F. (2011). Functional Evolution of BRCT Domains from Binding DNA to Protein. *Evol Bioinform Online.*, 7, 87-97.

[61] Glover, J. N., Williams, R. S., & Lee, M. S. (2004). Interactions between BRCT repeats and phosphoproteins: tangled up in two. *Trends Biochem Sci.*, 29, 579-85.

[62] Zhang, W., Zhang, J., Zhang, X., Xu, C., & Tu, X. (2011). Solution structure of Rap1 BRCT domain from Saccharomyces cerevisiae reveals a novel fold. *Biochem Biophys Res Commun.*, 404, 1055-9.

[63] Gong, Z., Kim, J. E., Leung, C. C., Glover, J. N., & Chen, J. (2010). BACH1/FANCJ acts with TopBP1 and participates early in DNA replication checkpoint control. *Mol Cell.*, 37, 438-46.

[64] Leung, C. C., Gong, Z., Chen, J., & Glover, J. N. (2011). Molecular basis of BACH1/ FANCJ recognition by TopBP1 in DNA replication checkpoint control. *J Biol Chem.*, 286, 4292-301.

[65] Leung, C. C., Kellogg, E., Kuhnert, A., Hanel, F., Baker, D., & Glover, J. N. (2010). Insights from the crystal structure of the sixth BRCT domain of topoisomerase IIbeta binding protein 1. *Protein Sci.*, 19, 162-7.

[66] Koonin, E. V., Altschul, S. F., & Bork, P. (1996). BRCA1 protein products... Functional motifs. *Nat Genet.*, 13, 266-8.

[67] Glover, J. N. (2006). Insights into the molecular basis of human hereditary breast cancer from studies of the BRCA1 BRCT domain. *Fam Cancer.*, 5, 89-93.

[68] Williams, R. S., Lee, M. S., Hau, D. D., & Glover, J. N. (2004). Structural basis of phosphopeptide recognition by the BRCT domain of BRCA1. *Nat Struct Mol Biol.*, 11, 519-25.

[69] Williams, R. S., Green, R., & Glover, J. N. (2001). Crystal structure of the BRCT repeat region from the breast cancer-associated protein BRCA1. *Nature Structural Biology*, 8, 838-42.

[70] Lofblom, J., Feldwisch, J., Tolmachev, V., Carlsson, J., Stahl, S., & Frejd, F. Y. (2010). Affibody molecules: engineered proteins for therapeutic, diagnostic and biotechnological applications. *FEBS Lett.*, 584, 2670-80.

[71] Leung, C. C., & Glover, J. N. (2011). BRCT domains: easy as one, two, three. *Cell Cycle.*, 10, 2461-70.

[72] Cescutti, R., Negrini, S., Kohzaki, M., & Halazonetis, T. D. (2010). TopBP1 functions with 53BP1 in the G1 DNA damage checkpoint. *EMBO J.*, 29, 3723-32.

[73] Derbyshire, D. J., Basu, B. P., Serpell, L. C., Joo, W. S., Date, T., Iwabuchi, K., & Doherty, A. J. (2002). Crystal structure of human 53BP1 BRCT domains bound to p53 tumour suppressor. *EMBO J.*, 21, 3863-72.

[74] Joo, W. S., Jeffrey, P. D., Cantor, S. B., Finnin, M. S., Livingston, D. M., & Pavletich, N. P. (2002). Structure of the 53BP1 BRCT region bound to p53 and its comparison to the Brca1 BRCT structure. *Genes Dev.*, 16, 583-93.

[75] Du, L. L., Moser, B. A., & Russell, P. (2004). Homo-oligomerization is the essential function of the tandem BRCT domains in the checkpoint protein Crb2. *J Biol Chem.*, 279, 38409-14.

[76] Rappas, M., Oliver, A. W., & Pearl, L. H. (2011). Structure and function of the Rad9-binding region of the DNA-damage checkpoint adaptor TopBP1. *Nucleic Acids Res.*, 39, 313-24.

[77] Martin-Garcia, R., de Leon, N., Sharifmoghadam, M. R., Curto, M. A., Hoya, M., Bustos-Sanmamed, P., & Valdivieso, M. H. (2011). The FN3 and BRCT motifs in the exomer component Chs5p define a conserved module that is necessary and sufficient for its function. *Cell Mol Life Sci.*, 68, 2907-17.

[78] Lloyd, J., Chapman, R. J., Clapperton, J. A., Haire, L. F., Hartsuiker, E., Li, J. J., Carr, A. M., Jackson, S. P., & Smerdon, S. J. (2009). A supramodular FHA/BRCT-repeat architecture mediates Nbs1 adaptor function in response to DNA damage. *Cell.*, 139, 100-11.

[79] Williams, R. S., Dodson, G. E., Limbo, O., Yamada, Y., Williams, J. S., Guenther, G., Classen, S., Glover, J. N., Iwasaki, H., Russell, P., & Tainer, J. A. (2009). Nbs1 flexibly tethers Ctp1 and Mre11-Rad50 to coordinate DNA double-strand break processing and repair. *Cell.*, 139, 87-99.

[80] Wu, P. Y., Frit, P., Meesala, S., Dauvillier, S., Modesti, M., Andres, S. N., Huang, Y., Sekiguchi, J., Calsou, P., Salles, B., & Junop, MS. (2009). Structural and functional interaction between the human DNA repair proteins DNA ligase IV and XRCC4. *Mol Cell Biol.*, 29, 3163-72.

[81] Loeffler, P. A., Cuneo, M. J., Mueller, G. A., DeRose, E. F., Gabel, S. A., & London, R. E. (2011). Structural studies of the PARP-1 BRCT domain. *BMC Struct Biol.*, 11, 37.

[82] de Groote, F. H., Jansen, J. G., Masuda, Y., Shah, D. M., Kamiya, K., de Wind, N., & Siegal, G. (2011). The Rev1 translesion synthesis polymerase has multiple distinct DNA binding modes. *DNA Repair (Amst).*, 10, 915-25.

[83] Matthews, L. A., Duong, A., Prasad, A. A., Duncker, B. P., & Guarne, A. (2009). Crystallization and preliminary X-ray diffraction analysis of motif N from Saccharomyces cerevisiae Dbf4. *Acta Crystallogr Sect F Struct Biol Cryst Commun.*, 65, 890-4.

[84] Johnston, L. H., Masai, H., & Sugino, A. (1999). First the CDKs, now the DDKs. *Trends Cell Biol.*, 9, 249-52.

[85] Ogi, H., Wang, C. Z., Nakai, W., Kawasaki, Y., & Masumoto, H. (2008). The role of the Saccharomyces cerevisiae Cdc7-Dbf4 complex in the replication checkpoint. *Gene*, 414, 32-40.

[86] Stead, B. E., Brandl, C. J., Sandre, M. K., & Davey, M. J. (2012). Mcm2 phosphorylation and the response to replicative stress. *BMC Genet.*, 13, 36.

[87] Masai, H., & Arai, K. (2000). Dbf4 motifs: conserved motifs in activation subunits for Cdc7 kinases essential for S-phase. *Biochem Biophys Res Commun.*, 275, 228-32.

[88] Varrin, A. E., Prasad, A. A., Scholz, R. P., Ramer, M. D., & Duncker, B. P. (2005). A mutation in Dbf4 motif M impairs interactions with DNA replication factors and confers increased resistance to genotoxic agents. *Mol Cell Biol.*, 25, 7494-504.

[89] Pike, B. L., Yongkiettrakul, S., Tsai, M. D., & Heierhorst, J. (2004). Mdt1, a novel Rad53 FHA1 domain-interacting protein, modulates DNA damage tolerance and G(2)/M cell cycle progression in Saccharomyces cerevisiae. *Mol Cell Biol.*, 24, 2779-88.

Mechanisms that Protect Chromosome Integrity During DNA Replication

Preserving the Replication Fork in Response to Nucleotide Starvation: Evading the Replication Fork Collapse Point

Sarah A. Sabatinos and Susan L. Forsburg

Additional information is available at the end of the chapter

1. Introduction

Replication fork progression is blocked by a variety impediments including DNA damage, aberrant DNA structures, or nucleotide depletion [1-3]. The response to replication fork stalling varies according the type of replication inhibition, the number of stalled forks and the duration of the treatment [3-7]. Stalled replication forks are at increased risk for DNA damage, which can lead to mutation or cell death [7-13]. The cell relies on the Intra-S phase checkpoint and DNA damage response proteins to preserve fork structure to allow recovery and resumption of the cell cycle [5, 10, 14-19]. Thus, the mechanisms that maintain replication fork structure are crucial for genome maintenance, and form a primary barrier to malignant transformation [20, 21].

The drug hydroxyurea (HU) induces a reversible early S-phase arrest by causing deoxynucleotide triphosphate (dNTP) depletion [22-24]. HU is a venerable chemotherapeutic, used for its ability to inhibit cell proliferation, but also because it predisposes proliferating cells to genome instability. The loss of replication fork stability and its associated DNA damage following HU treatment is loosely termed "replication fork collapse". Changes in dNTP pool levels through other mechanisms (*e.g.* exogenous thymidine or 5-bromo-2'deoxyuridine treatment) are known to cause point mutations [25-27], plasmid instability [28] and polyploidization [29]. Further, dNTP pool changes in human cells may cause hypersensitivity to secondary treatment with alkylating agents [30, 31].

Wild type cells recover from HU arrest and complete S-phase once drug is removed from the culture medium. Alternatively, some cultures may recover from HU arrest prior to its removal by up-regulating nucleotide synthesis and overcoming HU replication inhibition to slow-ly complete S-phase [17, 32, 33]. The ability to recover stalled replication out of an HU arrest requires restoration of replication forks, restart of DNA synthesis and completion of S-phase.

Whether a replication fork successfully recovers, or collapses with DNA damage, de-pends in part on the Intra-S phase checkpoint pathway. Cells lacking the checkpoint suf-fer fork collapse and death. Notably, cells that do not trigger the Intra-S phase checkpoint continue to synthesize DNA despite the presence of HU. Continued synthesis in the pres-ence of low dNTP pools leads to reduced replication rates and increased single stranded DNA (ssDNA) [33-37]. This is a fragile state of "open" DNA that is prone to double strand breaks (DSBs) [38-40]. Further, altered dNTP levels during DNA replication enhance point mutations, in which the base inserted shifts towards that of the dominant pool or away from the lowest pool [25, 41-44]. This explains why the replication checkpoint is a crucial barrier to genome instability.

Thus, replication fork collapse in checkpoint mutants does not occur immediately after HU treatment, detection of decreased dNTP levels, or failure to mount a checkpoint response. Instead, replication fork collapse across a population of forks, within a culture of cells, is a consequence of continued fork activity. The signs and symptoms of replication fork collapse represent a new execution point, the Replication Fork Collapse Point. This metric describes the time at which the majority of replication forks in a cell population become non-function-al. In this review, we describe the causes and symptoms of the Replication Fork Collapse Point, with particular regard to the Intra S-phase checkpoint.

2. Replication Fork Structure is Maintained During Stalling

The replication fork describes a region of denatured DNA where DNA synthesis is actively occurring, resembling a two-tined fork. The replisome encompasses the forked DNA, and the entire complex is large and dynamic, coupling DNA unwinding and polymerization [45-47]. Unwinding is performed by a conserved hexameric helicase (MCM) and its associat-ed proteins Cdc45 and GINS. The processive helicase produces single strand DNA (ssDNA) which becomes transiently coated with replication protein A (RPA), a ssDNA binding pro-tein homologue. ssDNA is the substrate for leading- (polε) and lagging-strand (polδ and polα-primase) polymerases.

These functions must be linked to facilitate DNA synthesis. Coupling generation of ssDNA with its use in replication is particularly important, because ssDNA is vulnerable to form-ing secondary structures, which leads to DNA damage [40, 48, 49], and recombination [50-52]. Thus, fork proteins limit the amount of DNA unwinding and ssDNA [39, 53]. In normal conditions, synthesis may occur rapidly and the goal of minimizing ssDNA production (<200bp) is easily accomplished [54]. However, if either the leading or lagging strand poly-merases become stalled or arrested in a slow zone, the helicase must also be slowed down to prevent it from generating excessive ssDNA and potentially dissociating entirely from the replisome.

Helicase and polymerases are linked by the replication Fork Protection Complex (FPC), which contributes to replication fidelity and later chromosome segregation. Tim1 (*S. pombe* Swi1) and Tipin (*Sp*Swi3) are evolutionarily conserved core components of the FPC that are essential for fork stability [11, 55-58](Table 1). This core is joined by AND1 (*Sp*Mcl1) and CLASPIN (*Sp*Mrc1), two proteins that bridge the helicase and polymerases. AND1 links the lagging strand primase (polα) [59-62], while CLASPIN connects the leading strand polymerase (polε) [58, 63].

Because of its role maintaining replisome structure, the FPC promotes replication fork efficiency and speed, particularly during fork stalling or pausing. While not essential for DNA replication [58, 64-66], the FPC contributes to processivity [67-70], and has additional roles in response to replication stalling [55, 71, 72], and facilitating sister chromatid cohesion, which is essential for faithful chromosome segregation [73-75].

Human FPC component	Homologues	References
TIMELESS (TIM)	Tim1 (*M. musculus, X. laevis*) Tof1 (*S. cerevisiae*) Swi1 (*S. pombe*)	[56, 57, 63, 65, 72]
TIPIN (TIP)	Tipin (*M. musculus, X. laevis*) Csm3 (*S. cerevisiae*) Swi3 (*S. pombe*)	[11, 56-58, 63, 75, 75-77]
CLASPIN	Claspin (*M. musculus, X. laevis*) Mrc1 (*S. cerevisiae, S. pombe*)	[14, 63, 78-82]
AND1	And1 (*M. musculus, X. laevis*) Ctf4 (*S. cerevisiae*) Mcl1 (*S. pombe*)	[59, 61, 73, 83]

Table 1. Fork Protection Complex Proteins in Various Species.

3. Causes of Replication Fork Stalling

DNA replication occurs in a short period during the cell cycle. In yeasts, replication of the ~12 Mb genome occurs within 20 to 30 minutes out of a 2.5 to 3h cell cycle. Human cells require several hours, a fraction of a full cell cycle, to replicate a substantially larger genome. The rate-limiting factor is replication fork velocity at 1–2 kb/min. This is an astonishing rate, considering secondary and tertiary structure of the genome packaged into higher order chromatin domains. The tight links between helicase, polymerase and FPC promote highly processive replication. Importantly, they also contribute to replication fidelity. Disruption of any one component (if not already lethal) leads to significant disruptions in processivity and/or fidelity. This is particularly true when impediments to replication are encountered.

Replication pausing and stalling is caused by both natural barriers and external factors [3, 84, 85]. Some regions of DNA cause replication fork stalling through sequence elements (*e.g.* DNA secondary structure in repetitive elements), or protein interference (*i.e.* transcription). Replication "slow zones" have been described in many model organisms, and these may contribute to genome instability and chromosome fragility. One characteristic shared by many "difficult templates" is the presence of repetitive sequence elements that cause fork stalling [86-89].

A replication termination sequence (RTS1) at the mating locus of fission yeast also promotes unidirectional fork progression by binding the replication termination factor 1 (Rtf1) [88, 90-95]. Unidirectional DNA replication is required to establish an imprint that directs mating type switching. RTS1 replication fork pausing is polar, meaning that forks approaching the barrier from one direction will be affected; forks from the opposite direction continue replication [93, 96].

Similarly, ribosomal DNA (rDNA) arrays are an example of a natural, repetitive element that is at risk for fork pausing. Each of the rDNA repeats contains a polar replication terminator, which ensures that forks proceed unidirectionally through each element [86, 97-100]. This occurs as a response to the binding of a fork arrest protein. For example, in fission yeast the Reb1 protein binds the replication termination element Ter3, which promotes long-range DNA interactions with other chromosomal Ter sequences [101, 102]. Localized to the nucleolus, this may nucleate a zone for replication termination [103]. Based on similarity to prokaryotic replication terminators, Reb1-Ter binding may stop the MCM helicase from creating more ssDNA leading to fork pausing and stalling. Pausing of the fork at this site also depends on FPC proteins Swi1 and Swi3.

Replication termination at rDNA is also seen in budding yeast and mammals. In *S.cerevisiae*, the FOB1 protein binds to ribosomal fork barrier elements and arrests progression of the replication fork so that replication is in concert with rDNA transcription [104, 105]. In mice, transcription termination factor 1 (TTF1) binds to termination sites in the rDNA and causes fork arrest [106-108]. It is suggested that Reb1/FOB1/TTF1 binding to their specific rDNA elements blocks the replicative MCM helicase and arrests forks.

The rDNA elements define one type of genomic sequence that causes replication slowing or pausing sites. Other regions of the genome may also cause fragile sites, which are broadly characterized as replication slow zones that are prone to forming DNA breaks [38, 40, 109, 110]. These may be dependent upon the chromatin context, transcriptional activity, or impairment of the fork by external agents, such as HU [111].

HU inhibits the activity of ribonucleotide reductase, which causes a reduction of dNTP pools [112]. HU is frequently used as a reversible early-S phase block reagent in cultured cells. In this sense, HU response is similar to excess thymidine treatment, which changes dNTP pools and induces an early S-phase arrest in metazoan cells [35]. The size of dNTP pools is intimately linked to cell cycle and checkpoint responses [24, 32, 113-115]. Critically, checkpoint proficiency allows cells to survive HU arrest, hold forks stable, and efficiently restart during release.

4. Intra-S Checkpoint: keeping things connected

The Intra-S phase checkpoint is a kinase cascade that responds to HU treatment. It serves to stabilize replication forks and arrest replication until dNTP pools recover. The checkpoint also prevents DNA damage from forming, particularly DNA double strand breaks, by restricting endonucleases such as Mus81 that can act on stalled fork structures [9, 10]. In addition, the Intra-S checkpoint regulates recombination enzymes (*e.g.* *Sp*Rqh1/*Sc*Sgs1/*Hs*BLM, Rad60), to preserve stalled forks in a state competent for restart without loss of genetic information [18, 116].

The remainder of this review will focus on the effects of the checkpoint on the replisome itself. During checkpoint activation, the helicase is restrained and stabilized, to prevent excessive unwinding and allow the fork to restart when HU is removed or bypassed. DNA synthesis is also restrained, preventing mutations that may occur during replication in the presence of altered dNTP pools. Late replication origins are prohibited from firing, conserving these "second-chance" origins for later replication restart. These activities help to stabilize established forks after HU treatment, later allowing them to restart. Alternatively, new forks may be established from the late origins in restart to rescue collapsed forks and complete DNA synthesis.

Wild type cells are actively inhibited from DNA synthesis during HU block [10, 17, 36, 58, 65, 117, 118]. That is, the forks do not cease synthesis because they run out of nucleotides. Rather, the checkpoint ensures that the forks are slowed or stopped before such starvation occurs, saving them from the mutagenic effects of dNTP imbalance [34, 42, 119]. These observations are consistent with depletion, rather than exhaustion of specific dNTP pools [22], and extremely slow residual synthesis [33]. This fork arrest is accompanied by inhibition of the helicase [15, 53, 54, 65], which reduces ssDNA accumulation and concomitant RPA binding until HU is removed.

The Intra-S phase checkpoint is a key component of the response to HU and actively restrains forks during replication stress. The initial signal to activate the checkpoint is provided by increased ssDNA created as replication forks stall [39, 53, 54]. ssDNA is bound by RPA and recruits the Rad9-Rad1-Hus1 (9-1-1) complex and ATR family kinases to stalled forks [120, 121]. Thus, the symptom of slow or stalled forks (generation of ssDNA) initiates the checkpoint [4, 120-125]. However, the FPC and checkpoint together ensure that the helicase cannot generate too much ssDNA, which provides one defense against replication fork collapse during HU stalling.

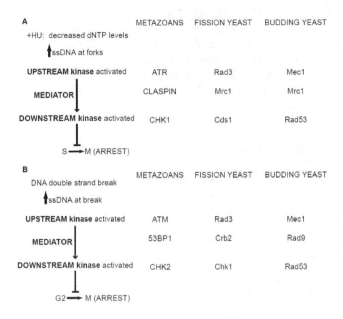

Figure 1. The Intra-S phase checkpoint across species. Key components of the checkpoint and their names are described for metazoan (Human, mouse, *Xenopus*), budding yeast (*S. cerevisiae*) and fission yeast (*S. pombe*). A) The replication checkpoint signals the presence of replication fork stalling and ssDNA accumulation through the upstream kinase
(ATR/Rad3/Mec1), which activates the downstream effector kinase (CHK1/Cds1/Rad53). CLASPIN/Mrc1 is the mediator of the replication checkpoint and is responsible for efficient dimerization and activation of the downstream kinase.
B) Similarities to the DNA damage response checkpoint. When DSBs are generated in G2, ssDNA is created around the
break, which recruits the upstream kinase (ATM/Rad3/Mec1). Through the 53BP1/Crb2/Rad9 mediator, the upstream
kinase phosphorylates and activates the downstream kinase (CHK2/Chk1/Rad53), which arrests the cell cycle and allows time for repair.

Checkpoint activation is also coupled to the FPC proteins, particularly CLASPIN and its
yeast equivalent, Mrc1 [118, 126, 127]. In fission yeast, Mrc1 is phosphorylated by the upstream Rad3/ATR kinase to a checkpoint-active form [128]. This activation recruits the
downstream Cds1 kinase to the stalled replication fork and is essential to signal amplification and transmission by activated Cds1. This pathway is conserved: in humans and budding yeast, respectively, Chk1/Rad53 is recruited to stalled forks by CLASPIN/Mrc1 and
ATR/Mec1 kinase [6, 14, 16, 124, 129-131].

This S phase checkpoint has a parallel structure to the DNA double strand break (DSB) response: Mrc1 is a replication-specific version of the *Hs*53BP1/*Sc*RAD9/*Sp*Crb2 mediator,
which brings together master kinases (ATM/ATR, Mec1 and Rad3) with an effector kinase
(CHK2, Rad53 or Chk1) for DSB response [14, 129] (Figure 1).

Fission yeast *cds1Δ* and *mrc1Δ* cells rapidly die in HU [117, 118, 127]. These cells lack the
Intra-S phase checkpoint and cannot restrain late origin firing or nuclease activity at stalled
forks [34, 36, 51, 58, 132]. In contrast to wild type cells, however, these mutants continue

DNA synthesis during HU block [53, 133]. The replication forks develop extensive ssDNA which can be observed by RPA binding. However, the fork proteins do not remain together, suggesting that the link between helicase and polymerase components is lost. Upon release from HU, *cds1Δ* or *mrc1Δ* cells manage a limited amount of further DNA synthesis. Their DNA synthesis proceeds slowly, but cells never achieve a fully replicated amount of DNA. Thus, the forks are failing as they reach the Replication Fork Collapse point, which results in S-phase failure and ultimately, cell death.

5. The Rules of Replisome Restraint and Restart,1: Fork Movement

Considering the phenotype of checkpoint mutants, we infer that an active mechanism restrains the helicase during HU treatment. Genome-wide studies in budding yeast show accumulation of single stranded DNA occurs in checkpoint mutants, adjacent to replication forks upon treatment with HU [39, 58, 133]. Similarly, in fission yeast checkpoint mutants, large masses of RPA can be visualized in whole cells treated with HU, which depend upon the MCM helicase [53].

A simple interpretation is that the helicase becomes uncoupled from the stalling polymerase and unwinds DNA ahead of it. This excessive unwinding generates ssDNA that is prone to breakage, which generates a characteristic DSB marker, phosphorylated histone H2A(X) [15, 39]. In many cases, the RPA signal is associated with markers of DNA synthesis, such as incorporation of the nucleotide analogue BrdU [53], or proximity to replication fork proteins [133]. Importantly, this uncoupling and unwinding occurs at the same time as DNA synthesis during both HU block and release. This suggests a more subtle effect in which leading and lagging strand synthesis is uncoupled, which leads to simultaneous accumulation of ssDNA and markers of synthesis, either because they are in the same region or because the ssDNA is a functional template.

6. Rules of Restraint and Restart, 2: Synthesis

The second key to restraint and successful restart is modulating the DNA polymerases. Wild type cells incorporate minimal amounts of nucleoside analogue in the presence of HU. Forks slow but remain stable [7, 17, 34, 54]. The rate of nucleotide analogue incorporation decreases, and DNA content does not increase significantly [36, 117, 134, 135]. In the yeast system, studies suggest that early replication forks extend about 5kb from the origin in the presence of HU before stopping [134, 136]. Decreased dNTP pools slow replication elongation during HU arrest. However, ectopic expansion of dNTPs by expressing ribonucleotide reductase from a plasmid can increase fork velocity even in HU [34]. Upon release from HU, replication rapidly restarts, whether from new origins or reactivation of existing forks, which results in rapid completion of DNA synthesis before cell division.

Budding yeast dNTP metabolism is quite robust and resistant to challenge, sensitive only to high levels of HU or significant NTP imbalance. In contrast, fission yeast [137] and metazo-

an cells are sensitive to low levels of HU, or modest dNTP imbalance, both which are sufficient to provoke replication arrest [7]. In all systems, there is an intimate connection to the
Intra-S phase checkpoint.

Surprisingly, checkpoint mutants do not block DNA synthesis in HU, indicating that they
are not actually starved for nucleotides, but rather lacking the ability to monitor pool levels [53]. Fission yeast cds1Δ mutants continue to synthesize DNA and incorporate analogue. In analysis of chromatin fibers, these can be visualized as extended tracts of newly
synthesized DNA despite the presence of HU. Upon release from the drug, cds1Δ mutants
continue to incorporate some analogue before reaching a plateau, by which time they have
accumulated approximately 66% the total amount of DNA incorporated in wild type (continuous labeling, block and release) [53]. These differences can be measured by detection of
analogue incorporation, but are obscured by total DNA content analysis, which is prone to
artifact [22, 36, 134].

The difference between the two situations is that much of this synthesis occurs *during* HU
treatment in the mutant, and only *after release* in wild type [53]. Thus, it is not until the recovery period that the majority of cds1Δ forks break down and can no longer synthesize
DNA. We define the point at which synthesis ceases as the Replication Fork Collapse Point
(see discussion of the Collapse Point in section "The Collapse Point: A Metric for Fork Stability"). Importantly, this is an extended window of time where there is a stochastic probability
of forks arresting and suffering collapse. Our data show that regions of DNA synthesis upon
HU release have high levels of RPA [53], which indicates that fork collapse is accompanied
by accumulation of ssDNA. This may reflect a burden of damage, incurred in HU, which is
remembered during release at the Collapse Point.

Polymerase ε is coupled to the helicase by Mrc1 and the FPC proteins Tof1 (Swi1) and Csm1
(Swi3) [63, 65, 132, 138]. This is thought to stabilize leading strand components at stalled
forks in HU. Asynchronously growing mrc1Δ cells lack this connection, which leads to intrinsic damage and a higher level of basal Cds1 phosphorylation even without added replication stress [128]. This essentially uncouples leading and lagging strands. mrc1Δ cells
treated with HU incorporate more nucleotide analogue but in shorter DNA fiber tracts. This
is consistent with a role for Mrc1 in modulating origin firing, as well as rate [139-141]. The
increase in DNA synthesis is only slightly higher after release, which could be attributed to
slower forks or polε uncoupling as in *S. cerevisiae* [81, 141]. These data suggest that in the
absence of Mrc1, forks continue to synthesize a low, steady level of DNA and this is independent of Cds1.

Mrc1 brings Cds1 and Rad3 together to phosphorylate Cds1 on threonine 11 [128, 142]. Subsequently, Cds1 activation is amplified by dimerization and autophosphorylation, setting in
motion the full Intra-S phase checkpoint [128]. HU treatment induces little Cds1-T11 phosphorylation in mrc1Δ; instead, the damage response kinase Chk1 is activated. This suggests
that there is conversion of stalled synthesis into DNA damage. Consistent with this, phos-
pho-H2A accumulates in mrc1Δ nuclei and replicated fiber tracts after HU release [53]. The
damage signal is frequently coincident with areas of synthesis, but often distinct from RPA-

heavy areas. Thus, unwinding and synthesis are likely uncoupled and distinct in *mrc1Δ* with HU treatment.

7. Rules of Restraint and Restart, 3: the late origins

An additional function of the Intra-S phase checkpoint is to restrain late origins from firing. Upon release from HU, these origins become competent for replication, and establish "rescue forks" that ensure completion of DNA replication [33, 36, 143-145]. Could these origins explain the post-release DNA synthesis observed in the checkpoint mutants? While late origin firing must contribute to some of the synthesis after release, we suggest that much of the post-release DNA synthesis does not occur from late origin firing, for the following reasons.

First, origin firing is de-regulated in HU blocked checkpoint mutants, which suggests that many late origins have already fired at the time of release, and are not available for this further synthesis. Recent work on dNTP pools in budding yeast suggests that >200 additional origins are fired in a *rad53-11* mutant compared to wild type in HU [34]. Deleting the ribonucleotide reductase inhibitor Sml1 results in activation of late origins in HU, while a *rad53-11sml1Δ* double mutant shows increased replication tracts in HU [34]. Sml1 is regulated by the replication checkpoint [114], and *sml1Δ* cells have increased dNTP pools [113, 146]. *sml1Δ* mutant backgrounds are frequently used in replication checkpoint studies because they overcome the lethal effects of *rad53Δ* or *mec1Δ*, but this makes direct comparison of these double mutants with other organisms, which retain controls of NTP pools, difficult. However, HU has also been shown to arrest replication without completely exhausting dNTP pools [22, 34], which suggests that cells sense small dNTP changes. Perhaps the Intra-S phase checkpoint also contributes to fork slowing and stalling, and is not limited to signal transduction at stalled forks.

Second, it is likely that late origins that fire in checkpoint mutants after HU release are incapable of synthesizing more than a short tract length, due to lack of nucleotides [137]. More analogue is incorporated in *cds1Δ* compared to wild type for the first 30 minutes after HU release, suggesting that start-up replication is both different and faster in the mutant cells. Additionally, more origins fire in mutants during HU block [140, 143, 147], suggesting that fewer origins remain to be activated after release. These observations imply that for forks established during HU arrest in *cds1Δ* and *mrc1Δ*, synthesis cannot proceed past a point of increased fork collapse and template damage [9, 15, 116, 148-150]. The role of origin repression, using mutant cells that impede origin firing, is required to confirm the degree to which DNA synthesis occurring in checkpoint mutants after HU release is dependent on late origins.

Together, these observations from multiple systems suggest that wild type cells survive HU block and release through coordination of several mechanisms: control of late origin firing, maintenance of existing replication forks, and later restart of the stabilized forks. Wild type cells do not encounter the Replication Fork Collapse Point because forks are maintained, replication is successfully restarted, and DNA synthesis completed.

8. Converting Stalled Forks to Restart

After HU is removed from culture medium, stabilized replication forks are returned to competence for DNA synthesis. In theory, immediate restart from a stabilized fork may be possible if all components are in place, having been protected from disassembly during HU arrest. In many cases this is likely to involve recombination pathways and the Rad51 recombinase. Rad51 binds to replisome components in HU, and around damaged replication forks [7, 15, 151]. Rad51 binds to ssDNA and promotes homologous recombination by allowing broken DNA to invade a homologous region for repair [52, 152, 153]. Checkpoint mutants have additional ssDNA, and experience "branch migration" of the fork structure [7, 52, 94, 154]. The resulting "chicken foot" structure is at risk for becoming a break or collapsed fork. Alternatively, the cruciform structure can be resolved by exonuclease Exo1, but leads to a partially replicated structure that cannot be replicated without *de novo* polymerase recruitment or break-induced replication [154-158].

The amount of time in HU until release has different effects in yeast and metazoan cultures. Both budding and fission yeast begin to arrest in HU within the first hour of HU exposure (e.g. [53, 134, 144, 159]). After a few hours at normal growth temperature, adaptation occurs, probably through changes in ribonucleotide reductase activity. In budding yeast, long-term HU exposure causes normal replication profiles to proceed at a glacial pace [33]. Similarly, human cells show increased sensitivity to HU over time, where fewer forks are observed with extended HU dose [7]. Peterman *et.al.* (2010) demonstrated that restarting long-HU treatment forks depends on both Rad51 and the repair protein XRCC3. Thus, protracted HU exposure may change that repair pathway used in fork repair, slowing the entire replication program [7, 153]. Slightly later than ssDNA accumulation, we detect Rad52 foci in live cells, increasing during HU block and release but lagging behind RPA accumulation [53]. Intriguingly, MRN/MRX- components co-localize at the replication fork and are important for fork stability (e.g. [35, 149, 160-162]), pointing at the essential role of recombination repair in restarting stalled replication forks.

9. The Collapse Point: A Metric for Fork Stability

The concept of replication fork "collapse" encompasses the observations that DNA damage and broken forks lead to loss of replication. ssDNA accumulates at susceptible forks and is a marker of increased risk of collapse [39, 54, 133, 137]. The DNA damage created at a stalled fork at or before collapse may not simply be DSBs. In fact, single strand breaks may form an important part in the damage process, converted to DSBs either during fork regression or in a second S-phase [163-165].

We propose the Collapse Point as the time when the balance between replication fork processivity and instability tips toward disaster. The time when the majority of forks in a cell have irreversibly, irrecoverably failed and replication will not be completed. Ongoing synthesis in checkpoint mutants during HU treatment sets the forks on a course to destruction, but actual collapse does not occur until the attempt to recover. We suggest some replication forks retain activity and undergo a shortened replication restart after HU release. This is consistent with data in fission and budding yeasts that fork components are retained and move during HU arrest in checkpoint deficient cells [53, 133].

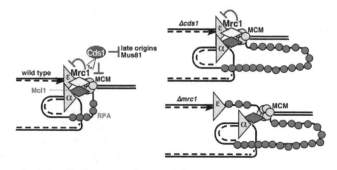

Figure 2. Model of the Replication Fork in HU before the Collapse Point.

Model of how replication fork architecture is affected by HU treatment in wild type, cds1Δ or mrc1Δ fission yeast cells. Left, wild type forks are stabilized through the Fork Protection Complex of Mrc1 (yellow diamond) and Mcl1 (green diamond), and the Intra-S phase checkpoint. As part of the FPC, Mrc1 forms a link between polymerase epsilon (ε), while Mcl1 links the helicase to polymerase alpha (α). A small loop of ssDNA forms at the fork in response to HU. This signals through the mediator function of Mrc1, activating Cds1 kinase. Cds1 kinase stabilizes and stalls the helicase activity of the replisome. Cds1 also inhibits Mus81 endonucleolytic activity and late origin firing, maintaining wild type cells in a state where replication can be restarted and/or finished from late origins after HU is removed. Right, top; when Cds1 is absent, a larger ssDNA loop forms from failing to slow the helicase. ssDNA is coated with replication protein A (RPA, blue circles) which form large foci in nuclei during HU block and release. Mrc1 is present, likely stabilizing polε, but the ssDNA may serve as template for lagging strand synthesis by pol α. Note that a ssDNA loop is presumed on the cds1Δ leading strand, but is omitted for clarity. The Intra-S phase checkpoint is not activated, late origins fire, and nucleases are not regulated.

Right, bottom; in the absence of Mrc1 the helicase is deregulated and potentially detached from polymerases. Large amounts of ssDNA form at individual forks and are coated by RPA. Although there is a great deal of ssDNA template, the length of mrc1Δ tracts in HU suggests that replication is slow, and that ssDNA areas are used as template during block and release up to a point where damage is encountered and forks collapse. In both cds1Δ and mrc1Δ cases, loss of replicative activity correlates with increased ssDNA foci, which build during HU block and release, suggesting that both mutants reach the Replication Fork Collapse Point during release.

The Replication Fork Collapse Point has no meaning for an individual fork; instead, it is the emergent property of the sum behavior of forks in a cell. The Collapse Point will generally be *estimated* by ensemble averaging across all cells in a culture.

While forks in *cds1Δ* and *mrc1Δ* retain synthesis activity, they are not necessarily the same as wild type (Figure 2). The amount of ssDNA and DNA damage signal (phospho-H2A(X)) is increased proximal to *cds1Δ* and *mrc1Δ* replication forks [39, 53], which could represent uncoupling of leading and lagging strand synthesis in advance of replication fork collapse. Together, these observations suggest that replication fork activity in checkpoint mutants shapes their stability. The inability of the mutant cells to restrain replication during HU and throughout release contributes to extensive ssDNA, DNA damage, and eventual collapse.

These results point to the *cds1Δ* and *mrc1Δ* Replication Fork Collapse Points occurring later than previously expected, and largely during HU release as cells attempt to resume the cell cycle. Thus, forks do not immediately collapse, but instead retain synthesis activity. The ends of new synthesis in release bear the marks of ssDNA and DNA damage. We conclude that fork collapse for these mutants is delayed, but its seeds are sown during HU block, only coming into full effect during release. The Replication Fork Collapse Point may be used as a descriptor for other genotypes to describe both how and when the majority of replication forks are destroyed in a population of cells. This is an execution point: while the cells are not viable by this time, the Collapse Point signals the time at which decay leading to death is fully established. We anticipate that the Collapse Point will be much later for *rad51Δ* or other repair-deficient mutants, which collapse by failing to properly restart. In contrast, Fork Protection Complex mutants may show an intermediate timing, or an incomplete Fork Collapse Point.

In turn, these studies prompt further questions. Do dNTP pools recover after release in *cds1Δ* and *mrc1Δ* cells? If this is the case, as suggested by the increase in replication after release in *cds1Δ* and *mrc1Δ* cells, what are the additional defining features of replication fork collapse?

10. Conclusions

Monitoring replication competency, accumulation of ssDNA and DNA damage signals around replication forks permits modeling to determine how replication forks respond to HU arrest and recovery. This, in turn, indicates what role checkpoint proteins Cds1 and Mrc1 play in fork stability and effective restart. The Replication Fork Collapse Point incorporates the signs and symptoms of fork collapse and attempts to put a time to when the majority of replication forks undergo collapse. This is likely different for different genetic backgrounds missing key components of checkpoint signal, fork stabilization and replication restart. Future work will dissect replication fork proteins in HU and release, and take the genome-wide data from microarray and sequencing, moving into monitoring patterns at individual replication forks. Since replication stability and fidelity is a key barrier to malignancy, defining when and how replication forks collapse in the absence of checkpoint will allow insights into the development and prevention of cancer.

Acknowledgements

We thank Forsburg and Aparicio lab members (USC) for helpful discussions, and Marc Green and Ruben Petreaca for manuscript comments. This work funded by a National Institutes of Health grant R01 GM059321 to SLF.

Author details

Sarah A. Sabatinos[1] and Susan L. Forsburg[1*]

*Address all correspondence to: forsburg@usc.edu

1 Department of Molecular and Computational Biology, University of Southern California, USA

References

[1] Torres-Rosell, J., De Piccoli, G., Cordon-Preciado, V., Farmer, S., Jarmuz, A., Machin, F., et al. (2007). Anaphase onset before complete DNA replication with intact checkpoint responses. *Science,* 315(5817), 1411-1415.

[2] Rothstein, R., Michel, B., & Gangloff, S. (2000). Replication fork pausing and recombination or "gimme a break". *Genes Dev,* 14(1), 1-10.

[3] Labib, K., & Hodgson, B. (2007). Replication fork barriers: pausing for a break or stalling for time? *EMBO Rep,* 8(4), 346-353.

[4] Lucca, C., Vanoli, F., Cotta-Ramusino, C., Pellicioli, A., Liberi, G., Haber, J., et al. (2004). Checkpoint-mediated control of replisome-fork association and signalling in response to replication pausing. *Oncogene,* 23(6), 1206-1213.

[5] Meister, P., Taddei, A., Vernis, L., Poidevin, M., Gasser, S. M., & Baldacci, G. (2005). Temporal separation of replication and recombination requires the intra-S checkpoint. *J Cell Biol,* 168(4), 537-544.

[6] Tourriere, H., & Pasero, P. (2007). Maintenance of fork integrity at damaged DNA and natural pause sites. *DNA Repair (Amst),* 6(7), 900-913.

[7] Petermann, E., Orta, M. L., Issaeva, N., Schultz, N., & Helleday, T. (2010). Hydroxyurea-stalled replication forks become progressively inactivated and require two different RAD51-mediated pathways for restart and repair. *Mol Cell,* 37(4), 492-502.

[8] Mao, N., Kojic, M., & Holloman, W. K. (2009). Role of Blm and collaborating factors in recombination and survival following replication stress in Ustilago maydis. *DNA Repair (Amst),* 8(6), 752-9.

[9]   Froget, B., Blaisonneau, J., Lambert, S., & Baldacci, G. (2008). Cleavage of stalled
      forks by fission yeast Mus81/Eme1 in absence of DNA replication checkpoint. *Mol Bi-
      ol Cell*, 19(2), 445-456.

[10] Kai, M., Boddy, M. N., Russell, P., & Wang, T. S. (2005). Replication checkpoint kin-
      ase Cds1 regulates Mus81 to preserve genome integrity during replication stress.
      Genes Dev, 19(8), 919-932.

[11] Noguchi, E., Noguchi, C., Du, L. L., & Russell, P. (2003). Swi1 prevents replication
      fork collapse and controls checkpoint kinase Cds1. *Mol Cell Biol*, 23(21), 7861-7874.

[12] Bryant, H. E., Petermann, E., Schultz, N., Jemth, A. S., Loseva, O., & Issaeva, N.
      (2009). PARP is activated at stalled forks to mediate Mre11-dependent replication re-
      start and recombination. *Embo J*, 28(17), 2601-2615.

[13] Bernstein, K. A., Shor, E., Sunjevaric, I., Fumasoni, M., Burgess, R. C., Foiani, M., et
      al. (2009). Sgs1 function in the repair of DNA replication intermediates is separable
      from its role in homologous recombinational repair. *Embo J*, 28(7), 915-925.

[14] Alcasabas, A. A., Osborn, A. J., Bachant, J., Hu, F., Werler, P. J., Bousset, K., et al.
      (2001). Mrc1 transduces signals of DNA replication stress to activate Rad53. *Nat Cell
      Biol*, 3(11), 958-965.

[15] Bailis, J. M., Luche, D. D., Hunter, T., & Forsburg, S. L. (2008). Minichromosome
      maintenance proteins interact with checkpoint and recombination proteins to pro-
      mote s-phase genome stability. *Molecular and cellular biology*, 28(5), 1724-1738.

[16] Branzei, D., & Foiani, M. (2007). Interplay of replication checkpoints and repair pro-
      teins at stalled replication forks. *DNA Repair (Amst)*, 6(7), 994-1003.

[17] Lopes, M., Cotta-Ramusino, C., Pellicioli, A., Liberi, G., Plevani, P., Muzi-Falconi, M.,
      et al. (2001). The DNA replication checkpoint response stabilizes stalled replication
      forks. *Nature*, 412(6846), 557-561.

[18] Marchetti, M. A., Kumar, S., Hartsuiker, E., Maftahi, M., Carr, A. M., & Freyer, G. A.
      (2002). A single unbranched S-phase DNA damage and replication fork blockage
      checkpoint pathway. *Proc Natl Acad Sci U S A*, 99(11), 7472-7477.

[19] Tsang, E., & Carr, A. M. (2008). Replication fork arrest, recombination and the main-
      tenance of ribosomal DNA stability. *DNA Repair (Amst)*, 7(10), 1613-1623.

[20] Bartkova, J., Horejsi, Z., Koed, K., Kramer, A., Tort, F., & Zieger, K. (2005). DNA
      damage response as a candidate anti-cancer barrier in early human tumorigenesis.
      Nature, 434(7035), 864-870.

[21] Bartkova, J., Rezaei, N., Liontos, M., Karakaidos, P., Kletsas, D., Issaeva, N., et al.
      (2006). Oncogene-induced senescence is part of the tumorigenesis barrier imposed by
      DNA damage checkpoints. *Nature*, 444(7119), 633-637.

[22] Koc, A., Wheeler, L. J., Mathews, C. K., & Merrill, G. F. (2004). Hydroxyurea arrests DNA replication by a mechanism that preserves basal dNTP pools. *J Biol Chem*, 279(1), 223-30.

[23] Matsumoto, M., Rey, D. A., & Cory, J. G. (1990). Effects of cytosine arabinoside and hydroxyurea on the synthesis of deoxyribonucleotides and DNA replication in L1210 cells. *Adv Enzyme Regul*, 30-47.

[24] Bianchi, V., Pontis, E., & Reichard, P. (1986). Changes of deoxyribonucleoside triphosphate pools induced by hydroxyurea and their relation to DNA synthesis. *J Biol Chem*, 261(34), 16037-16042.

[25] Goodman, M. F., Hopkins, R. L., Lasken, R., & Mhaskar, D. N. (1985). The biochemical basis of 5-bromouracil- and 2-aminopurine-induced mutagenesis. *Basic Life Sci*, 31, 409-423.

[26] Hakansson, P., Dahl, L., Chilkova, O., Domkin, V., & Thelander, L. (2006). Thelander L. The Schizosaccharomyces pombe replication inhibitor Spd1 regulates ribonucleotide reductase activity and dNTPs by binding to the large Cdc22 subunit. *The Journal of biological chemistry*, 281(3), 1778-1783.

[27] Kunz, BA, Kang, X. L., & Kohalmi, L. (1991). The yeast rad18 mutator specifically increases G.C----T.A transversions without reducing correction of G-A or C-T mismatches to G.C pairs. *Molecular and cellular biology*, 11(1), 218-225.

[28] Kohalmi, S. E., Haynes, R. H., & Kunz, B. A. (1988). Instability of a yeast centromere plasmid under conditions of thymine nucleotide stress. *Mutat Res*, 207(1), 13-16.

[29] Potter, C. G. (1971). Induction of polyploidy by concentrated thymidine. *Exp Cell Res*, 68(2), 442-448.

[30] Meuth, M. (1981). Role of deoxynucleoside triphosphate pools in the cytotoxic and mutagenic effects of DNA alkylating agents. *Somatic Cell Genet*, 7(1), 89-102.

[31] Meuth, M. (1983). Deoxycytidine kinase-deficient mutants of Chinese hamster ovary cells are hypersensitive to DNA alkylating agents. *Mutat Res*, 110(2), 383-391.

[32] Mulder, K. W., Winkler, G. S., & Timmers, H. T. (2005). DNA damage and replication stress induced transcription of RNR genes is dependent on the Ccr4-Not complex. *Nucleic Acids Res*, 33(19), 6384-6392.

[33] Alvino, G. M., Collingwood, D., Murphy, J. M., Delrow, J., Brewer, B. J., & Raghuraman, M. K. (2007). Replication in hydroxyurea: it's a matter of time. *Molecular and cellular biology*, 27(18), 6396-6406.

[34] Poli, J., Tsaponina, O., Crabbe, L., Keszthelyi, A., Pantesco, V., & Chabes, A. (2012). dNTP pools determine fork progression and origin usage under replication stress. *The EMBO journal*, 31(4), 883-894.

[35] Bolderson, E., Scorah, J., Helleday, T., Smythe, C., & Meuth, M. (2004). ATM is required for the cellular response to thymidine induced replication fork stress. *Hum Mol Genet*, 13(23), 2937-2945.

[36] Feng, W., Collingwood, D., Boeck, M. E., Fox, L. A., Alvino, G. M., Fangman, W. L., et al. (2006). Genomic mapping of single-stranded DNA in hydroxyurea-challenged yeasts identifies origins of replication. *Nature cell biology*, 8(2), 148-55.

[37] Vassin, V. M., Anantha, R. W., Sokolova, E., Kanner, S., & Borowiec, J. A. (2009). Human RPA phosphorylation by ATR stimulates DNA synthesis and prevents ssDNA accumulation during DNA-replication stress. *J Cell Sci*, Pt 22, 4070-4080.

[38] Letessier, A., Millot, G. A., Koundrioukoff, S., Lachages, A. M., Vogt, N., Hansen, R. S., et al. (2011). Cell-type-specific replication initiation programs set fragility of the FRA3B fragile site. *Nature*, 470(7332), 120-123.

[39] Feng, W., Di Rienzi, S. C., Raghuraman, M. K., & Brewer, B. J. (2011). Replication stress-induced chromosome breakage is correlated with replication fork progression and is preceded by single-stranded DNA formation. *G3 (Bethesda)*, 1(5), 327-35.

[40] Durkin, S. G., & Glover, T. W. (2007). Chromosome fragile sites. *Annu Rev Genet*, 41-169.

[41] Chabes, A., Georgieva, B., Domkin, V., Zhao, X., Rothstein, R., & Thelander, L. (2003). Survival of DNA damage in yeast directly depends on increased dNTP levels allowed by relaxed feedback inhibition of ribonucleotide reductase. *Cell*, 112(3), 391-401.

[42] Davidson, M. B., Katou, Y., Keszthelyi, A., Sing, T. L., Xia, T., Ou, J., et al. (2012). Endogenous DNA replication stress results in expansion of dNTP pools and a mutator phenotype. *The EMBO journal*, 31(4), 895-907.

[43] Fasullo, M., Tsaponina, O., Sun, M., & Chabes, A. (2010). Elevated dNTP levels suppress hyper-recombination in Saccharomyces cerevisiae S-phase checkpoint mutants. *Nucleic acids research*, 38(4), 1195-1203.

[44] Kumar, D., Abdulovic, A. L., Viberg, J., Nilsson, A. K., Kunkel, T. A., & Chabes, A. (2011). Mechanisms of mutagenesis in vivo due to imbalanced dNTP pools. *Nucleic acids research*, 39(4), 1360-1371.

[45] Muzi-Falconi, M., Giannattasio, M., Foiani, M., & Plevani, P. (2003). The DNA polymerase alpha-primase complex: multiple functions and interactions. *ScientificWorldJournal*, 3-21.

[46] Langston, L. D., Indiani, C., & O'Donnell, M. (2009). Whither the replisome: emerging perspectives on the dynamic nature of the DNA replication machinery. *Cell Cycle*, 8(17), 2686-2691.

[47] Hubscher, U. (2009). DNA replication fork proteins. *Methods Mol Biol*, 521-19.

[48] Lopez-Contreras, A. J., & Fernandez-Capetillo, O. (2010). The ATR barrier to replication-born DNA damage. *DNA repair*, 9(12), 1249-1255.

[49] Glover, T. W., Arlt, M. F., Casper, A. M., & Durkin, S. G. (2005). Mechanisms of common fragile site instability. *Hum Mol Genet* [2], R197-R205.

[50] Wang, X., & Haber, J. E. (2004). Role of Saccharomyces single-stranded DNA-binding protein RPA in the strand invasion step of double-strand break repair. *PLoS Biol*, E21.

[51] Alabert, C., Bianco, J. N., & Pasero, P. (2009). Differential regulation of homologous recombination at DNA breaks and replication forks by the Mrc1 branch of the S-phase checkpoint. *Embo J*, 28(8), 1131-1141.

[52] Sugiyama, T., & Kantake, N. (2009). Dynamic regulatory interactions of rad51, rad52, and replication protein-a in recombination intermediates. *J Mol Biol*, 390(1), 45-55.

[53] Sabatinos, S. A., Green, M. D., & Forsburg, S. L. (2012). Continued DNA synthesis in replication checkpoint mutants leads to fork collapse. *submitted*.

[54] Sogo, J. M., Lopes, M., & Foiani, M. (2002). Fork reversal and ssDNA accumulation at stalled replication forks owing to checkpoint defects. *Science*, 297(5581), 599-602.

[55] Mc Farlane, R. J., Mian, S., & Dalgaard, J. Z. (2010). The many facets of the Tim-Tipin protein families' roles in chromosome biology. *Cell Cycle*, 9(4), 700-705.

[56] Gotter, A. L., Suppa, C., & Emanuel, BS. (2007). Mammalian TIMELESS and Tipin are evolutionarily conserved replication fork-associated factors. *J Mol Biol*, 366(1), 36-52.

[57] Noguchi, E., Noguchi, C., Mc Donald, W. H., Yates, J. R. 3rd, & Russell, P. (2004). Swi1 and Swi3 are components of a replication fork protection complex in fission yeast. *Mol Cell Biol*, 24(19), 8342-8355.

[58] Katou, Y., Kanoh, Y., Bando, M., Noguchi, H., Tanaka, H., Ashikari, T., et al. (2003). S-phase checkpoint proteins Tof1 and Mrc1 form a stable replication-pausing complex. *Nature*, 424(6952), 1078-1083.

[59] Williams, D. R., & McIntosh, J. R. (2002). mcl1+, the Schizosaccharomyces pombe homologue of CTF4, is important for chromosome replication, cohesion, and segregation. *Eukaryot Cell*, 1(5), 758-773.

[60] Miles, J., & Formosa, T. (1992). Evidence that POB1, a Saccharomyces cerevisiae protein that binds to DNA polymerase alpha, acts in DNA metabolism in vivo. *Mol Cell Biol*, 12(12), 5724-6735.

[61] Gambus, A., van Deursen, F., Polychronopoulos, D., Foltman, M., Jones, R. C., & Edmondson, R. D. (2009). A key role for Ctf4 in coupling the MCM2-7 helicase to DNA polymerase alpha within the eukaryotic replisome. *Embo J*, 28(19), 2992-3004.

[62] Im, J. S., Ki, S. H., Farina, A., Jung, Hurwitz. J., & Lee, J. K. (2009). Assembly of the Cdc45-Mcm2-7-GINS complex in human cells requires the Ctf4/And-1, RecQL4, and Mcm10 proteins. *Proc Natl Acad Sci U S A*, 106(37), 15628-15632.

[63] Bando, M., Katou, Y., Komata, M., Tanaka, H., Itoh, T., Sutani, T., et al. (2009). Csm3, Tof1, and Mrc1 form a heterotrimeric mediator complex that associates with DNA replication forks. *J Biol Chem*, 284(49), 34355-34365.

[64] Gambus, A., Jones, R. C., Sanchez-Diaz, A., Kanemaki, M., van Deursen, F., & Edmondson, R. D. (2006). GINS maintains association of Cdc45 with MCM in replisome progression complexes at eukaryotic DNA replication forks. *Nat Cell Biol*, 8(4), 358-366.

[65] Nedelcheva, M. N., Roguev, A., Dolapchiev, L. B., Shevchenko, A., Taskov, H. B., Shevchenko, A., et al. (2005). Uncoupling of unwinding from DNA synthesis implies regulation of MCM helicase by Tof1/Mrc1/Csm3 checkpoint complex. *J Mol Biol*, 347(3), 509-21.

[66] Calzada, A, Hodgson, B, Kanemaki, M, Bueno, A, & Labib, K. (2005). Molecular anatomy and regulation of a stable replisome at a paused eukaryotic DNA replication fork. *Genes Dev*, 19(16), 1905-1919.

[67] Hamdan, S. M., Johnson, D. E., Tanner, N. A., Lee, J. B., Qimron, U., Tabor, S., et al. (2007). Dynamic DNA helicase-DNA polymerase interactions assure processive replication fork movement. *Molecular cell*, 27(4), 539-549.

[68] Kim, S., Dallmann, H. G., Mc Henry, C. S., & Marians, K. J. (1996). Coupling of a replicative polymerase and helicase: a tau-DnaB interaction mediates rapid replication fork movement. *Cell*, 84(4), 643-650.

[69] Stano, N. M., Jeong, Y. J., Donmez, I., Tummalapalli, P., Levin, M. K., & Patel, S. S. (2005). DNA synthesis provides the driving force to accelerate DNA unwinding by a helicase. *Nature*, 435(7040), 370-373.

[70] Tougu, K., & Marians, K. J. (1996). The interaction between helicase and primase sets the replication fork clock. *The Journal of biological chemistry*, 271(35), 21398-21405.

[71] Unsal-Kacmaz, K., Chastain, P. D., Qu, P. P., Minoo, P., Cordeiro-Stone, M., Sancar, A., et al. (2007). The human Tim/Tipin complex coordinates an Intra-S checkpoint response to UV that slows replication fork displacement. *Mol Cell Biol*, 27(8), 3131-3142.

[72] Yoshizawa-Sugata, N., & Masai, H. (2007). Human Tim/Timeless-interacting protein, Tipin, is required for efficient progression of S phase and DNA replication checkpoint. *J Biol Chem*, 282(4), 2729-2740.

[73] Errico, A., Cosentino, C., Rivera, T., Losada, A., Schwob, E., Hunt, T., et al. (2009). Tipin/Tim1/And1 protein complex promotes Pol alpha chromatin binding and sister chromatid cohesion. *Embo J*, 28(23), 3681-3692.

[74] Tanaka, H., Kubota, Y., Tsujimura, T., Kumano, M., Masai, H., & Takisawa, H. (2009). Replisome progression complex links DNA replication to sister chromatid cohesion in Xenopus egg extracts. *Genes Cells*, 14(8), 949-963.

[75] Leman, A. R., Noguchi, C., Lee, C. Y., & Noguchi, E. (2010). Human Timeless and Ti-
      pin stabilize replication forks and facilitate sister-chromatid cohesion. *J Cell Sci*, Pt 5,
      660-670.

[76] Errico, A., Costanzo, V., & Hunt, T. (2007). Tipin is required for stalled replication
      forks to resume DNA replication after removal of aphidicolin in Xenopus egg ex-
      tracts. *Proc Natl Acad Sci U S A*, 104(38), 14929-34.

[77] Tanaka, T., Yokoyama, M., Matsumoto, S., Fukatsu, R., You, Z., & Masai, H. (2010).
      Fission yeast Swi1-Swi3 complex facilitates DNA binding of Mrc1. *The Journal of bio-
      logical chemistry*, 285(51), 39609-22.

[78] Kumagai, A., & Dunphy, W. G. (2000). Claspin, a novel protein required for the acti-
      vation of Chk1 during a DNA replication checkpoint response in Xenopus egg ex-
      tracts. *Mol Cell*, 6(4), 839-849.

[79] Kumagai, A., Kim, S. M., & Dunphy, W. G. (2004). Claspin and the activated form of
      ATR-ATRIP collaborate in the activation of Chk1. *J Biol Chem*, 279(48), 49599-45608.

[80] Lee, J., Gold, D. A., Shevchenko, A., Shevchenko, A., & Dunphy, W. G. (2005). Roles
      of replication fork-interacting and Chk1-activating domains from Claspin in a DNA
      replication checkpoint response. *Mol Biol Cell*, 16(11), 5269-5282.

[81] Lou, H., Komata, M., Katou, Y., Guan, Z., Reis, C. C., Budd, M., et al. (2008). Mrc1
      and DNA polymerase epsilon function together in linking DNA replication and the S
      phase checkpoint. *Molecular cell*, 32(1), 106-117.

[82] Osborn, A. J., & Elledge, S. J. (2003). Mrc1 is a replication fork component whose
      phosphorylation in response to DNA replication stress activates Rad53. *Genes & de-
      velopment*, 17(14), 1755-1767.

[83] Williams, D. R., & McIntosh, J. R. (2005). Mcl1p is a polymerase alpha replication ac-
      cessory factor important for S-phase DNA damage survival. *Eukaryotic cell*, 4(1),
      166-177.

[84] Arlt, M. F., Mulle, J. G., Schaibley, V. M., Ragland, R. L., Durkin, S. G., & Warren, S.
      T. (2009). Replication stress induces genome-wide copy number changes in human
      cells that resemble polymorphic and pathogenic variants. *Am J Hum Genet*, 84(3),
      339-350.

[85] Mirkin, E. V., & Mirkin, S. M. (2007). Replication fork stalling at natural impedi-
      ments. *Microbiol Mol Biol Rev*, 71(1), 13-35.

[86] Krings, G., & Bastia, D. (2004). swi1- and swi3-dependent and independent replica-
      tion fork arrest at the ribosomal DNA of Schizosaccharomyces pombe. *Proc Natl Acad
      Sci U S A*, 101(39), 14085-90.

[87] Noguchi, C., & Noguchi, E. (2007). Sap1 promotes the association of the replication
      fork protection complex with chromatin and is involved in the replication checkpoint
      in Schizosaccharomyces pombe. *Genetics*, 175(2), 553-566.

[88] Dalgaard, J. Z., & Klar, A. J. (2000). swi1 and swi3 perform imprinting, pausing, and termination of DNA replication in S. pombe. *Cell*, 102(6), 745-751.

[89] Razidlo, D. F., & Lahue, R. S. (2008). Mrc1, Tof1 and Csm3 inhibit CAG.CTG repeat instability by at least two mechanisms. *DNA Repair (Amst)*, 7(4), 633-640.

[90] Ahn, J. S., Osman, F., & Whitby, M. C. (2005). Replication fork blockage by RTS1 at an ectopic site promotes recombination in fission yeast. *The EMBO journal*, 24(11), 2011-2023.

[91] Eydmann, T., Sommariva, E., Inagawa, T., Mian, S., Klar, A. J., & Dalgaard, J. Z. (2008). Rtf1-mediated eukaryotic site-specific replication termination. *Genetics*, 180(1), 27-39.

[92] Codlin, S., & Dalgaard, J. Z. (2003). Complex mechanism of site-specific DNA replication termination in fission yeast. *The EMBO journal*, 22(13), 3431-3440.

[93] Dalgaard, J. Z., & Klar, A. J. (2001). A DNA replication-arrest site RTS1 regulates imprinting by determining the direction of replication at mat1 in S. pombe. *Genes & development*, 15(16), 2060-2068.

[94] Lambert, S., Mizuno, K., Blaisonneau, J., Martineau, S., Chanet, R., Freon, K., et al. (2010). Homologous recombination restarts blocked replication forks at the expense of genome rearrangements by template exchange. *Molecular cell*, 39(3), 346-359.

[95] Vengrova, S., Codlin, S., & Dalgaard, J. Z. (2002). RTS1-an eukaryotic terminator of replication. *Int J Biochem Cell Biol*, 34(9), 1031-1034.

[96] Lee, B. S., Grewal, S. I., & Klar, A. J. (2004). Biochemical interactions between proteins and mat1 cis-acting sequences required for imprinting in fission yeast. *Molecular and cellular biology*, 24(22), 9813-9822.

[97] Coulon, S, Noguchi, E, Noguchi, C, Du, LL, Nakamura, TM, & Russell, P. (2006). Rad22Rad52-dependent repair of ribosomal DNA repeats cleaved by Slx1-Slx4 endonuclease. *Mol Biol Cell*, 17(4), 2081-2090.

[98] Kaplan, D. L., & Bastia, D. (2009). Mechanisms of polar arrest of a replication fork. *Mol Microbiol*, 72(2), 279-285.

[99] Krings, G., & Bastia, D. (2005). Sap1p binds to Ter1 at the ribosomal DNA of Schizosaccharomyces pombe and causes polar replication fork arrest. *J Biol Chem*, 280(47), 39135-39142.

[100] Maric, C., Levacher, B., & Hyrien, O. (1999). Developmental regulation of replication fork pausing in Xenopus laevis ribosomal RNA genes. *J Mol Biol*, 291(4), 775-788.

[101] Biswas, S., & Bastia, D. (2008). Mechanistic insights into replication termination as revealed by investigations of the Reb1-Ter3 complex of Schizosaccharomyces pombe. *Mol Cell Biol*, 28(22), 6844-6857.

[102] Zhao, A., Guo, A., Liu, Z., & Pape, L. (1997). Molecular cloning and analysis of Schizosaccharomyces pombe Reb1p: sequence-specific recognition of two sites in the far upstream rDNA intergenic spacer. *Nucleic acids research*, 25(4), 904-910.

[103] Singh, S. K., Sabatinos, S., Forsburg, S., & Bastia, D. (2010). Regulation of replication termination by Reb1 protein-mediated action at a distance. *Cell*, 142(6), 868-78.

[104] Bochman, M. L., Sabouri, N., & Zakian, V. A. (2010). Unwinding the functions of the Pif1 family helicases. *DNA Repair (Amst)*, 9(3), 237-49.

[105] Bairwa, N. K., Zzaman, S., Mohanty, B. K., & Bastia, D. (2010). Replication fork arrest and rDNA silencing are two independent and separable functions of the replication terminator protein Fob1 of Saccharomyces cerevisiae. *The Journal of biological chemistry*, 285(17), 12612-9.

[106] Evers, R., & Grummt, I. (1995). Molecular coevolution of mammalian ribosomal gene terminator sequences and the transcription termination factor TTF-I. *Proceedings of the National Academy of Sciences of the United States of America*, 92(13), 5827-31.

[107] Gerber, J. K., Gogel, E., Berger, C., Wallisch, M., Muller, F., Grummt, I., et al. (1997). Termination of mammalian rDNA replication: polar arrest of replication fork movement by transcription termination factor TTF-I. *Cell*, 90(3), 559-567.

[108] Langst, G., Becker, P. B., & Grummt-I, I. (1998). TTF-I determines the chromatin architecture of the active rDNA promoter. *The EMBO journal*, 17(11), 3135-45.

[109] Arlt, M. F., Durkin, S. G., Ragland, R. L., & Glover, T. W. (2006). Common fragile sites as targets for chromosome rearrangements. *DNA repair*, 5(9-10), 1126-1135.

[110] Howlett, N. G., Taniguchi, T., Durkin, S. G., D'Andrea, A. D., & Glover, T. W. (2005). The Fanconi anemia pathway is required for the DNA replication stress response and for the regulation of common fragile site stability. *Hum Mol Genet*, 14(5), 693-701.

[111] Bermejo, R., Capra, T., Gonzalez-Huici, V., Fachinetti, D., Cocito, A., Natoli, G., et al. (2009). Genome-organizing factors Top2 and Hmo1 prevent chromosome fragility at sites of S phase transcription. *Cell*, 138(5), 870-884.

[112] Yarbro, J. W. (1992). Mechanism of action of hydroxyurea. *Semin Oncol*, (9), 1-10.

[113] Tsaponina, O., Barsoum, E., Astrom, S. U., & Chabes, A. (2011). Ixr1 is required for the expression of the ribonucleotide reductase Rnr1 and maintenance of dNTP pools. *PLoS Genet*, 7(5), e1002061.

[114] Zhao, X., Chabes, A., Domkin, V., Thelander, L., & Rothstein, R. (2001). Thelander L, and Rothstein R. The ribonucleotide reductase inhibitor Sml1 is a new target of the Mec1/Rad53 kinase cascade during growth and in response to DNA damage. *The EMBO journal*, 20(13), 3544-3553.

[115] Huang, A., Fan, H., Taylor, W. R., & Wright, J. A. (1997). Ribonucleotide reductase R2 gene expression and changes in drug sensitivity and genome stability. *Cancer Res*, 57(21), 4876-4881.

[116] Miyabe, I., Morishita, T., Shinagawa, H., & Carr, A. M. (2009). Schizosaccharomyces pombe Cds1Chk2 regulates homologous recombination at stalled replication forks through the phosphorylation of recombination protein Rad60. *J Cell Sci*, Pt 20, 3638-3643.

[117] Lindsay, H. D., Griffiths, D. J., Edwards, R. J., Christensen, P. U., Murray, J. M., Osman, F., et al. (1998). S-phase-specific activation of Cds1 kinase defines a subpathway of the checkpoint response in Schizosaccharomyces pombe. *Genes & development*, 12(3), 382-395.

[118] Zhao, H., Tanaka, K., Nogochi, E., Nogochi, C., & Russell, P. (2003). Replication checkpoint protein Mrc1 is regulated by Rad3 and Tel1 in fission yeast. *Mol Cell Biol*, 23(22), 8395-8403.

[119] Kumar, D., Viberg, J., Nilsson, A. K., & Chabes, A. (2010). Highly mutagenic and severely imbalanced dNTP pools can escape detection by the S-phase checkpoint. *Nucleic acids research*, 38(12), 3975-3983.

[120] Zou, L., Liu, D., & Elledge, S. J. (2003). Replication protein A-mediated recruitment and activation of Rad17 complexes. *Proc Natl Acad Sci U S A*, 100(24), 13827-13832.

[121] Kanoh, Y., Tamai, K., & Shirahige, K. (2006). Different requirements for the association of ATR-ATRIP and 9-1-1 to the stalled replication forks. *Gene*, 377-388.

[122] Kemp, M., & Sancar, A. (2009). DNA distress: just ring 9-1-1. *Curr Biol*, 19(17), R733-R734.

[123] Yan, S., & Michael, W. M. (2009). TopBP1 and DNA polymerase-alpha directly recruit the 9-1-1 complex to stalled DNA replication forks. *J Cell Biol*, 184(6), 793-804.

[124] Branzei, D., & Foiani, M. (2006). The Rad53 signal transduction pathway: Replication fork stabilization, DNA repair, and adaptation. *Exp Cell Res*, 312(14), 2654-2659.

[125] Yan, S., & Michael, W. M. (2009). TopBP1 and DNA polymerase alpha-mediated recruitment of the 9-1-1 complex to stalled replication forks: implications for a replication restart-based mechanism for ATR checkpoint activation. *Cell Cycle*, 8(18), 2877-2884.

[126] Tanaka, K., & Russell, P. (2001). Mrc1 channels the DNA replication arrest signal to checkpoint kinase Cds1. *Nat Cell Biol*, 3(11), 966-972.

[127] Tanaka, K., & Russell, P. (2004). Cds1 phosphorylation by Rad3-Rad26 kinase is mediated by forkhead-associated domain interaction with Mrc1. *J Biol Chem*, 279(31), 32079-32086.

[128] Xu, Y. J., Davenport, M., & Kelly, T. J. (2006). Two-stage mechanism for activation of the DNA replication checkpoint kinase Cds1 in fission yeast. *Genes Dev*, 20(8), 990-1003.

[129] Osborn, A. J., Elledge, S. J., & Zou, L. (2002). Checking on the fork: the DNA-replication stress-response pathway. *Trends Cell Biol*, 12(11), 509-516.

[130] Naylor, M. L., Li, J. M., Osborn, A. J., & Elledge, S. J. (2009). Mrc1 phosphorylation in response to DNA replication stress is required for Mec1 accumulation at the stalled fork. *Proc Natl Acad Sci U S A*, 106(31), 12765-12770.

[131] Schleker, T., Nagai, S., & Gasser, S. M. (2009). Posttranslational modifications of repair factors and histones in the cellular response to stalled replication forks. *DNA Repair (Amst)*, 8(9), 1089-100.

[132] Tourriere, H., Versini, G., Cordon-Preciado, V., Alabert, C., & Pasero, P. (2005). Mrc1 and Tof1 promote replication fork progression and recovery independently of Rad53. *Mol Cell*, 19(5), 699-706.

[133] De Piccoli, G., Katou, Y., Itoh, T., Nakato, R., Shirahige, K., & Labib, K. (2012). Replisome stability at defective DNA replication forks is independent of S phase checkpoint kinases. *Molecular cell*, 45(5), 696-704.

[134] Kim, S. M., & Huberman, J. A. (2001). Regulation of replication timing in fission yeast. *The EMBO journal*, 20(21), 6115-6126.

[135] Feng, W., Bachant, J., Collingwood, D., Raghuraman, M. K., & Brewer, B. J. (2009). Centromere replication timing determines different forms of genomic instability in Saccharomyces cerevisiae checkpoint mutants during replication stress. *Genetics*, 183(4), 1249-1260.

[136] Lengronne, A., Pasero, P., Bensimon, A., & Schwob, E. (2001). Monitoring S phase progression globally and locally using BrdU incorporation in TK(+) yeast strains. *Nucleic acids research*, 29(7), 1433-1442.

[137] Sabatinos, S. A., Mastro, T. L., & Forsburg, S. L. (2012). Nucleoside analogues create DNA damage and sensitivity in fission yeast. *Eukaryotic Cell*, submitted.

[138] Chin, J. K., Bashkirov, V. I., Heyer, WD, & Romesberg, FE. (2006). Esc4/Rtt107 and the control of recombination during replication. *DNA repair*, 5(5), 618-628.

[139] Koren, A., Soifer, I., & Barkai, N. (2010). MRC1-dependent scaling of the budding yeast DNA replication timing program. *Genome Res*, 20(6), 781-790.

[140] Hayano, M, Kanoh, Y, Matsumoto, S, & Masai, H. (2011). Mrc1 marks early-firing origins and coordinates timing and efficiency of initiation in fission yeast. *Molecular and cellular biology*, 31(12), 2380-2391.

[141] Szyjka, S. J., Viggiani, C. J., & Aparicio, O. M. (2005). Mrc1 is required for normal progression of replication forks throughout chromatin in S. cerevisiae. *Molecular cell*, 19(5), 691-697.

[142] Xu, Y. J., & Kelly, T. J. (2009). Autoinhibition and autoactivation of the DNA replication checkpoint kinase Cds1. *The Journal of biological chemistry*, 284(23), 16016-16027.

[143] Hayashi, M., Katou, Y., Itoh, T., Tazumi, A., Yamada, Y., Takahashi, T., et al. (2007). Genome-wide localization of pre-RC sites and identification of replication origins in fission yeast. *The EMBO journal*, 26(5), 1327-1339.

[144] Santocanale, C., & Diffley, J. F. (1998). A Mec1- and Rad53-dependent checkpoint controls late-firing origins of DNA replication. *Nature*, 395(6702), 615-618.

[145] Heichinger, C., Penkett, C. J., Bahler, J., & Nurse, P. (2006). Genome-wide characterization of fission yeast DNA replication origins. *The EMBO journal*, 25(21), 5171-5179.

[146] Chabes, A., & Stillman, B. (2007). Constitutively high dNTP concentration inhibits cell cycle progression and the DNA damage checkpoint in yeast Saccharomyces cerevisiae. *Proceedings of the National Academy of Sciences of the United States of America*, 104(4), 1183-8.

[147] Mickle, K. L., Ramanathan, S., Rosebrock, A., Oliva, A., Chaudari, A., Yompakdee, C., et al. (2007). Checkpoint independence of most DNA replication origins in fission yeast. *BMC Mol Biol*, 8, 112.

[148] Hanada, K., Budzowska, M., Davies, S. L., van Drunen, E., Onizawa, H., Beverloo, H. B., et al. (2007). The structure-specific endonuclease Mus81 contributes to replication restart by generating double-strand DNA breaks. *Nat Struct Mol Biol*, 14(11), 1096-1104.

[149] Robison, J. G., Elliott, J., Dixon, K., & Oakley, G. G. (2004). Replication protein A and the Mre11.Rad50.Nbs1 complex co-localize and interact at sites of stalled replication forks. *J Biol Chem*, 279(33), 34802-34810.

[150] Schlacher, K., Christ, N., Siaud, N., Egashira, A., Wu, H., & Jasin, M. (2011). Double-strand break repair-independent role for BRCA2 in blocking stalled replication fork degradation by MRE11. *Cell*, 145(4), 529-42.

[151] Ouyang, K. J., Woo, L. L., Zhu, J., Huo, D., Matunis, M. J., & Ellis, N. A. (2009). SUMO modification regulates BLM and RAD51 interaction at damaged replication forks. *PLoS Biol*, 7(12), e1000252.

[152] Kurokawa, Y., Murayama, Y., Haruta-Takahashi, N., Urabe, I., & Iwasaki, H. (2008). Reconstitution of DNA strand exchange mediated by Rhp51 recombinase and two mediators. *PLoS Biol*, 6(4), e88.

[153] Wray, J., Liu, J., Nickoloff, J. A., & Shen, Z. (2008). Distinct RAD51 associations with RAD52 and BCCIP in response to DNA damage and replication stress. *Cancer Res*, 68(8), 2699-2707.

[154] Lambert, S., Froget, B., & Carr, A. M. (2007). Arrested replication fork processing: interplay between checkpoints and recombination. *DNA Repair (Amst)*, 6(7), 1042-1061.

[155] Aggarwal, M., Sommers, J. A., Morris, C., & Brosh, R. M. , Jr. (2010). Delineation of WRN helicase function with EXO1 in the replicational stress response. *DNA Repair (Amst)*, 9(7), 765-776.

[156] Tinline-Purvis, H., Savory, A. P., Cullen, J. K., Dave, A., Moss, J., Bridge, W. L., et al. (2009). Failed gene conversion leads to extensive end processing and chromosomal rearrangements in fission yeast. *The EMBO journal*, 28(21), 3400-3412.

[157] Tran, P. T., Fey, J. P., Erdeniz, N., Gellon, L., Boiteux, S., & Liskay, R. M. (2007). A mutation in EXO1 defines separable roles in DNA mismatch repair and post-replication repair. *DNA Repair (Amst)*, 6(11), 1572-1583.

[158] Lambert, S., Watson, A., Sheedy, D. M., Martin, B., & Carr, A. M. (2005). Gross chromosomal rearrangements and elevated recombination at an inducible site-specific replication fork barrier. *Cell*, 121(5), 689-702.

[159] Meister, P., Taddei, A., Ponti, A., Baldacci, G., & Gasser, S. M. (2007). Replication foci dynamics: replication patterns are modulated by S-phase checkpoint kinases in fission yeast. *The EMBO journal*, 26(5), 1315-1326.

[160] Willis, N., & Rhind, N. (2010). The fission yeast Rad32(Mre11)-Rad50-Nbs1 complex acts both upstream and downstream of checkpoint signaling in the S-phase DNA damage checkpoint. *Genetics*, 184(4), 887-897.

[161] Hashimoto, Y., Puddu, F., & Costanzo, V. (2012). RAD51- and MRE11-dependent reassembly of uncoupled CMG helicase complex at collapsed replication forks. *Nature structural & molecular biology*, 19(1), 17-24.

[162] Brugmans, L., Verkaik, N. S., Kunen, M., van Drunen, E., Williams, B. R., & Petrini, J. H. (2009). NBS1 cooperates with homologous recombination to counteract chromosome breakage during replication. *DNA repair*, 8(12), 1363-1370.

[163] Kuzminov, A. (2001). Single-strand interruptions in replicating chromosomes cause double-strand breaks. *Proceedings of the National Academy of Sciences of the United States of America*, 98(15), 8241-8246.

[164] Caldecott, K. W. (2007). Mammalian single-strand break repair: mechanisms and links with chromatin. *DNA repair*, 6(4), 443-453.

[165] Hutchinson, F. (1993). Induction of large DNA deletions by persistent nicks: a new hypothesis. *Mutat Res*, 299(3-4), 211-218.

Chapter 9

The Role of WRN Helicase/Exonuclease in DNA Replication

Lynne S. Cox and Penelope A. Mason

Additional information is available at the end of the chapter

1. Introduction

1.1. WRN is a RecQ helicase/exonuclease required for genome stability and to prevent premature ageing

1.1.1. Clinical phenotype of Werner's syndrome

Humans possess five distinct RecQ helicases (see Figure 1), all of which possess a hallmark RecQ helicase domain. Mutation or loss in any one of three human RecQ helicases give rise to genetic instability syndromes: WRN mutation gives Werner's syndrome (WS), BLM loss results in Bloom syndrome (BS), and Rothmund-Thomson syndrome (RTS) is caused by mutation of RECQL4[1]. WRN has come to prominence because its loss of function results in human Werner's syndrome, a segmental progeria (premature ageing) characterised by many signs and symptoms of normal ageing at both the organismal and cellular levels, with shortened lifespan (median age of death 47 years [2]). In particular WS patients suffer from osteoporosis, athero-and arterio-sclerosis and a high cancer incidence (particularly sarcoma) together with metabolic disorders normally associated with increased age, especially type II diabetes and lipodystrophy. Furthermore, patients show outwardly recognisable signs of ageing such as cataracts, greying hair and skin wrinkling, while female WS patients suffer premature menopause and both sexes show hypogonadism, with decreased fertility (reviewed in ref. [2]).

---

1 RTS is found in a subset of patients with RECQL4 mutation; different mutations in the same gene give rise to RAPA-DILLINO syndrome [1]

1.2. Cellular phenotype on WRN loss

This premature ageing phenotype is also observed at the cellular level: fibroblasts from WS patients undergo highly premature replicative senescence in culture, failing to proliferate after only 9-11 population doublings, compared with the 50-60 doublings characteristic of wild type fibroblasts [3]. Transcriptomic studies have demonstrated that >90% gene expression changes associated with normal ageing are seen in young WS cells [4], while glycosylation of blood albumin (a biomarker of ageing) in young WS patients is equivalent to levels detected in normal centenarians [5]. Importantly, loss of function of WRN is associated with significant genome instability with a high frequency of chromosomal translocations and deletions [6, 7], which is thought to contribute to the increased cancer risk. Genome instability is a hallmark of defective S phase checkpoint proteins (reviewed in ref. [8]), suggesting either than WRN is directly involved in the checkpoint, or that it normally serves downstream of the checkpoint such that its loss prevents correct execution of the arrest and recovery pathways. Notably, it is not only WS patients who are more susceptible to cancer on WRN loss: epigenetic inactivation by methylation of CpG islands in the WRN gene promoter has been reported in epithelial and mesenchymal cancers with value in prognosis in colorectal cancer [9], while specific WRN SNPs have been correlated with breast cancer incidence [10], even though such genetic changes do not alter the helicase or exonuclease activities of the protein or modulate the levels expressed. WRN is therefore of interest not only to those attempting to understand the molecular basis of human ageing, but also to cancer biologists – indeed WRN knockdown is likely to promote cancer cell death and hypersensitise cells to current chemotherapeutic agents such as camptothecin that impact on DNA replication [9, 11, 12]. Small molecules that specifically inhibit WRN but not other RecQ helicases are therefore likely to have therapeutic potential [13].

1.3. WRN protein

The wide range of ageing-associated phenotypes in WS patients and their cells indicates a fundamental role for WRN in preventing premature ageing, but how can loss of one protein lead to the pleiotropic outcomes of human ageing? The most important clue came from cloning the WRN gene [14], which showed for the first time that the human WRN gene encodes a large protein of 1432 amino acid (~162kDa) with an amino terminal exonuclease domain conserved with proteins of the DnaQ family, and a central helicase domain of the RecQ family. In addition, DNA binding (RQC) and protein interaction (HRDC) domains exist distal to the helicase domain (Figure 1A). Immunofluorescence and mutational studies have demonstrated that WRN is a nuclear protein with both NLS and NoLS sequences situated at the C terminus [15]), that appears to be sequestered in the nucleolus [16] except during S phase or upon DNA damage, when it is redistributed to sites of DNA replication or repair ([17-19].

Of the five human RecQ proteins, WRN is the only one to possess exonuclease activity [20]. Acting in a 3'-5' direction (as shown using 3'- or 5'-end labelled substrates), WRN exonuclease has been demonstrated to bind onto overhanging 5' ends of the guide strand of duplex DNA and cleave the target strand sequentially, though with relatively low processivity [21]. While it cannot cleave blunt ended substrates, nor those where ends are blocked by bulky lesions [22], WRN exonuclease degrades substrates that are likely to be found both during DNA repair and as intermediates in DNA replication, including forks and bubble substrates [23] (see Table 1). Despite early reports of lack of activity on short single-stranded DNA (e.g. [21]), WRN exonuclease can digest single stranded oligonucleotides over 50 bases in length [24, 25].

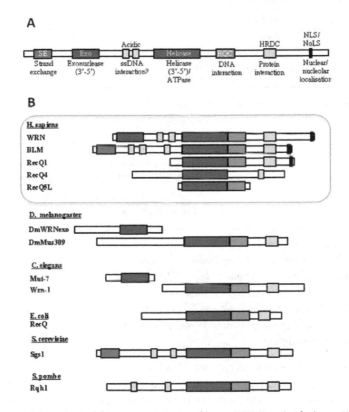

Figure 1. The RecQ helicase family. (A) Domain organization of human WRN. Note that for human WRN, the RQC serves in DNA binding and the HRDC is probably involved in protein-protein interaction, though these roles may be reversed in other RecQs. (B) Humans have 5 RecQ helicases (boxed), named after the archetypal RecQ of *E. coli*. Human WRN is unique in the family in possessing an exonuclease domain. In invertebrates such as *Drosophila* and *C. elegans*, the exonuclease (red) and helicase (blue) activities are encoded by separate genes.

Helicase substrates	Exonuclease substrates
Holliday junction	Holliday junction
Bubble duplex	Bubble duplex
3'-recessed duplex	3'-recessed duplex

Helicase substrates	Exonuclease substrates
D-loop duplex	D-loop duplex
Duplex with 5'-flaps	Looped duplex
G-quadruplex	Nicked looped duplex
Partial duplex	
Forked duplex	
Triplex	
NOT blunt duplex	
NOT 5'-recessed duplex	

Table 1. Substrates unwound by WRN helicase or degraded by WRN exonuclease. (Note that BLM also unwinds the same substrates as WRN helicase)

The helicase activity of WRN is highly conserved with other RecQ helicase family members, acting 3'-5'to unwind duplex DNA in an ATP-dependent manner [26]. Within the helicase domain are seven conserved motifs characteristic of the RecQ family. In general, RecQ helicases are adept at unwinding unusual DNA structures that can inhibit the course of normal DNA replication. Examples are tailed and forked duplexes, small gaps and flaps (commonly found as DNA repair and recombination intermediates), bubble substrates and displacement-loop triplex and Holliday junctions (common at telomeres and during recombinational repair and sister chromatid exchange), and G-quadruplexes which are often found at tracts rich in guanine such as at the telomere (e.g. [27], reviewed in ref. [28], see Table 1). It is important to note that the helicase and exonuclease activities do not simply act as independent entities in cells, but that their actions are almost certainly co-ordinated and interlinked. For example, co-operation between them is required during telomere maintenance ([29]; see section 4 below for more detail).

WRN helicase template specificity requires DNA binding that is probably mediated through the conserved RQC domain. X-ray crystallographic analysis has shown some unusual features, in that binding of WRN to DNA does not occur through a standard 'recognition helix', but instead through a beta wing of the RQC domain that inserts like a wedge between the terminal bases of blunt duplex DNA to unwind one base even in the absence of ATP [30]. How this binding correlates with WRN's lack of unwinding of blunt ended substrates remains to be determined. In addition to binding to DNA, WRN binds to many different proteins at the replication fork, the telomere and during fork recovery after stalling. Protein interaction with WRN may occur through the helicase-and-ribonuclease D/C-terminal (HRDC) domain; while this region is through to be important for DNA binding in *E. coli* RecQ and yeast Sgs1, the conserved for DNA interaction surface is lacking in human WRN, and the domain is unable to bind DNA *in vitro*, but that reveals many exposed alpha helicases that are likely to bind to protein partners [31].

1.4. WRN orthologues

While the exonuclease and helicase activities are both encoded by the same gene in verte-
brates, giving rise to one multifunctional protein, the enzymes are encoded by separate genetic loci in plants, invertebrates and prokaryotes (Figure 1B, reviewed in ref. [32]), with
physical and/or functional interaction between the helicase and nuclease proposed *in vivo*.
(Figure 1, reviewed in ref. [32, 33] For example, in the fruit fly *Drosophila melanogaster*, we
have cloned and characterised the orthologue of human WRN exonuclease encoded by the
fly locus *CG7670* [34]. *Drosophila* WRN exonuclease (DmWRNexo) is a 3'-5' exonuclease [35]
that shows remarkable substrate conservation with human WRN exonuclease and utilises
conserved residues at the active site for nucleic acid cleavage [36]. Flies homozygous for a
strong hypomorphic mutation in *CG7670* have greatly elevated levels of recombination that
appears to occur through reciprocal exchange, and are hypersensitive to the topoisomerase
poison camptothecin, that leads to replication fork collapse [37]. Hence loss of only the
WRN exonuclease activity in flies results in many features characteristic of human WS, suggesting a key role for the exonuclease in preventing premature ageing. We consider the possible role(s) of WRN exonuclease in replication fidelity, restart of stalled forks and telomere
maintenance in more detail below (see sections 2.2, 3.4 and 4 below).

A limitation to studying WRN in flies is the lack of a fully characterised WRN helicase orthologue. However, the nematode worm *C. elegans* has a highly conserved WRN-like helicase, encoded by the *wrn-1* gene, and two candidate exonucleases, at loci *ZK1098.8* (*mut-7*
[38]) and adjacent *ZK1098.3*. RNAi knockdown of *wrn-1* results in shortened lifespan [39]
and perturbation of the S phase checkpoint via ATM/R kinases [40], suggesting both that
WRN is important during DNA replication, and that its role is critical in maintaining normal
longevity of the organism. These outcomes are of particular interest since they so closely
echo the findings in humans, but in a genetically tractable and short-lived lower eukaryotic
model organism. In plants, WRN has been most studied in *Arabidopsis*, where physical and
function interaction has been described between the exonuclease (AtWEX) and helicase
(AtWRN) orthologues [41]. In budding yeast and fission yeast, there is only one RecQ helicase (Sgs1 and Rqh1, respectively); whether these proteins interacts directly with an exonuclease to reconstitute human WRN-like activity is yet to be determined, though genetic
interaction between Rqh1 and Mus81/Eme1 has been reported [42].

Because of the phenotypes resulting from WRN loss or mutation, it has been implicated in
many aspects of DNA metabolism, including transcription, DNA repair, recombination and
telomere maintenance. Its role in DNA replication will be discussed in this chapter, including not only a direct role in normal processive DNA replication, and replication of the telomeres, but also in preventing replication fork stalling or assisting fork recovery after arrest.

1.5. S phase defects in WS cells

Fibroblasts and lymphoblastoid cells from Werner's syndrome patients show a defect in progression through S phase [17, 43]. FACS analysis demonstrates both a longer duration of S
phase and an overall significant increase in cell cycle time in primary fibroblasts from WS patients ([17] and in normal primary fibroblasts in which WRN was depleted by shRNAi by

80-90% [44]. Early studies on replication rates in WS fibroblasts used alkaline sucrose gradients to detect the size of nascent DNA, demonstrating slower replication in WS cells compared with normal controls [45]. The ability of WS cells to incorporate Texas-red-dUTP into nascent DNA is also significantly impaired [46]. Interestingly, while acute shRNAi-mediated WRN depletion in SV40 T antigen-transformed cells had no impact on cell cycle progression in the absence of imposed replication stress, primary fibroblasts depleted of WRN did show an S phase delay [44]. Hence it appears that loss of WRN protein results in an S phase phenotype.

WRN has been isolated within a large multi-protein replication complex [47]and found to interact *in vitro* with purified PCNA. The binding region has been localised to a PIP-like motif on WRN towards the amino terminus [18], which is likely to bind within the hydrophobic pocket of PCNA, as described for other PIP-containing proteins (see section 2.1). Studies on *Xenopus* egg cell-free extracts depleted of the frog orthologue of WRN, called FFA-1 (focus-forming activity-1) initially suggested that the protein was required for establishment of replication foci and thus served a central role in DNA synthesis [48]. (Note however that immunoprecipitation from *Xenopus* egg extracts is fraught with difficulties and accidental removal of other components such as membranes may inadvertently lead to loss of replication capacity). Subsequently, FFA-1 was shown to localise to sites of DNA synthesis coincident with RPA, and expression of a dominant negative GST-FFA-1 fusion protein blocked replication activity [49]. Similar immunofluorescence studies in both HeLa cells and primary human fibroblasts, supported by high-resolution immuno-electron microscopy, also showed WRN present at a subset (~60%) of replication foci, colocalising with PCNA [18]. This localisation is in the absence of replication stress, while on HU arrest, the majority of WRN relocates from the nucleolus to RPA-containing foci that are suggested to represent stalled forks [19]. Hence WRN is present at replication sites, and in its absence, cell cycle and DNA synthesis phenotypes are consistent with a replication defect.

2. WRN at the replication fork

In order to appreciate where WRN acts during DNA replication, it is necessary to understand the core structure of the DNA replication fork during the elongation stage of DNA replication. During elongation, processive polymerisation of the leading strand is carried out by DNA polymerase epsilon (pol ε) and the leading strand by DNA polymerase delta (pol δ) (based on mutational studies of the proof-reading domains of each in yeast) [50-52]. The replicative polymerases are tethered to the template by association with the homotrimeric sliding clamp protein PCNA (proliferating cell nuclear antigen) [53]. Co-ordination between leading and lagging strands may be achieved through the action of the GINS/ Cdc45 complex that has been proposed to act as a replisome progression complex (RPC) [54]. On the lagging strand, repeated cycles of priming by DNA pol α-primase results in synthesis of 7-10 nucleotide of RNA primer followed by ~20 nucleotides of initiator DNA (with error rates of 10^{-2} and 10^{-4} respectively), followed by switching to the higher fidelity and more processive DNA pol δ on the lagging strand and pol ε on the leading strand. This switch occurs through a multistep loading process essentially requiring recognition of the primer-template junction (where RPA is bound to the unwound single-stranded parental DNA) by RFC, an AAA+ ATPase that serves to load the sliding clamp PCNA. Pol δ is then recruited to PCNA through its p66 subunit to synthesise approximately 200 nucleotides of the Okazaki fragment. (For a more detailed discussion of fork establishment, see ref. [55]).

2.1. Okazaki fragment processing

Because of the low fidelity of pol α-primase, it is essential to remove both the RNA primer and iDNA during Okazaki fragment processing (OFP) This is coincident with continued synthesis of nascent DNA on the lagging strand; processive replication by pol δ results in displacement of the RNA-iDNA primer as a 5' flap and its removal by one of a range of postulated pathways involving RNase H1, FEN1, Dna2 (on long RPA-coated flaps) and other helicases/nucleases including Pif1 and possibly a RecQ helicase (Sgs1 in yeast, WRN in humans) (reviewed in ref. [56]). Pol δ synthesises DNA to fill the gap and DNA ligase seals the nick in the phosphodiester backbone. These steps in Okazaki fragment processing (OFP) may be co-ordinated through differential binding of the separate enzymes to PCNA, which has been suggested to act as a molecular 'toolbelt' in OFP [57]. Association of the OFP proteins[2] with PCNA occurs through a conserved PCNA-interacting peptide (PIP) of the general motif QxxL/M/IxxFF to the hydrophobic pocket of PCNA formed at the interdomain connector loop (e.g. [58, 60], reviewed in ref. [61]). Each PIP is likely to bind by an induced fit mechanism, since the crystal structures of PCNA bound by its various partners shows variation in this loop region [62]. Notably, WRN has a conserved PIP, and peptide ELISA studies showed that this region is sufficient for PCNA binding *in vitro* [18]. Additionally, WRN binds to and stimulates the nuclease activity of Fen1, which may contribute to efficiency of Okazaki fragment processing [63]; as WRN binds to Fen1 immediately adjacent to its PCNA binding site, it is likely that there is some interplay between the three proteins [64] that may be important in Okazaki fragment processing, though this has not been fully explored.

2.2. Proof-reading during processive DNA synthesis

DNA replication overall has an extremely low error rate of 10^{-9}, achieved in part by the very high fidelity of the processive replicative polymerase ε and δ, and also by additional 'extrinsic' proofreading activities together with mismatch repair (MMR) to remove incorrectly incorporated bases. The high fidelity DNA polymerases ε and δ achieve an error rate of ~2 $x10^{-5}$ (reviewed in ref. [65]) through two key structural features. Firstly, the active site is only fully formed upon acceptance of the correct incoming dNTP to create a solvent–inaccessible site that is partially specified by correct helical geometry of duplex DNA, thus increasing enthalpy and decreasing entropy for correct nucleotides and allowing high discrimination over incorrect nucleotides. Secondly, these polymerases each possesses a 3'-5' exonuclease active site whereby the nascent DNA swings through ~40° to present to this site [66], and where incorrect nucleotides are removed by hydrolysis of the phosphodiester backbone just created. X-ray crystal structures of the isolated WRN exonuclease domain have shown that

---

2 Many other proteins also bind to PCNA in this manner – some regulate PCNA's activity (e.g. p21) [58] while others are regulated by such binding (e.g. Cdt1 degradation is PIP-dependent) [59].

WRN shares structural homology with exonuclease domains of the high fidelity DnaQ family of replication polymerases, suggesting a possible role for WRN in editing DNA, either
during DNA synthesis or in processing free ends, in collaboration with and stimulated by
the end-binding protein Ku [67]. Very recently, it has been shown that WRN assists pol δ
(possibly on the lagging strand during Okazaki fragment synthesis) by removing 3' mismatches, thus allowing the polymerase to extend primers [68]. This supports a direct role for
WRN in Okazaki fragment synthesis.

3. Replication fork stalling – the role of WRN

3.1. High rates of replication fork stalling in WS

Early electron microscopy studies of ^{3}H-T labelled DNA in fibre autoradiographs suggested
a problem with replication origin spacing in WS [69, 70], though subsequent higher resolution studies using fluorescent antibodies to halogenated nucleotides suggest rather that it is
replication fork rate, not inter-origin distance, which is abnormal in WS cells [17, 44]. Indeed, these DNA combing studies, that analyse individual DNA molecules labelled during
replication, have demonstrated a problem with replication fork progression in WS cells, resulting in a high degree of replication fork asymmetry from what should be bidirectional
origins [17]. Such studies led to the proposal that replication forks stall at high frequency in
cells lacking WRN protein. Why should WS cells be particularly prone to fork stalling?

3.2. Causes of fork stalling

The replication fork encounters barriers during normal replication, such as unusual DNA
structures arising at G-rich regions (G4-quadruplex) or fragile sites. These structures must
be unwound to present a single stranded template suitable for copying; a high incidence of
replication fork stalling is likely if the normal mechanisms for tackling the unusual structures is lacking. Alteration in nucleotide pools through treatment with hydroxyurea (HU),
or polymerase inhibition with the dCTP mimic aphidicolin results in replication fork arrest
in the absence of template abnormalities or lesions. In addition, exogenous agents can cause
formation of lesions in the DNA that the replication fork cannot easily pass over – for example, methylated or oxidized bases.

Replication fork pausing or stalling is therefore likely to be a common occurrence, and the cell
has mechanisms to stabilise the fork, deal with the unusual structure or repair the damaged region, and allow fork restart. Where DNA synthesis pauses but the MCM replicative helicases
proceed to unwind the duplex template, regions of single stranded parental DNA arise, that
are rapidly coated with RPA. This forms a signal to the S phase checkpoint machinery, particularly the kinase ATR, that, together with other checkpoint kinases such as Mec1, Chk1 and
Chk2 (Rad53) and mediator Mrc1, leads both to recruitment of proteins to deal with the particular fork progression barrier, and to stabilisation of the replisome at the stalled fork, reviewed
in ref. [8]. Indeed, DNA pol ε has been shown to stay associated with stalled forks in yeast [71]

under the influence of Rad53 signalling. Replication fork restart then occurs once the damage has been resolved and the checkpoint lifted. More serious to the cell is the collapse of replication forks as they traverse regions of the template containing single strand breaks – single-stranded breaks are converted to double-strand breaks (DSBs) by the passage of the replication fork, forming highly cytotoxic and potentially recombinogenic lesions. Hence surveillance and rescue mechanisms must exist in the cell to deal both with stalled and collapsed forks. The RecQ helicase family has been implicated as key in this mechanism.

3.3. Dealing with unusual structures before they arrest the fork

The most efficient mode of replication involves the removal of barriers to fork progression before they lead to fork stalling. Importantly, WRN has been shown to be required by DNA pol δ (but not α or ε) to unwind G4 DNA [72], bubbles and D loops [68] to allow pol δ-mediated synthesis over such template sequences without leading to fork stalling. In addition, the helicase activity of WRN is also required to limit the formation of single stranded DNA regions and gaps during replication of common fragile sites (CFS) [73, 74] and enhances processivity of DNA pol δ on fragile site FRA16D over hairpins and microsatellite regions, requiring either the helicase or DNA binding activities of WRN [75]. Hence one important role of WRN in DNA replication is to present the replisome with a template that is easy to replicate, but does it act at any other point to ensure efficient replication?

3.4. Is WRN involved in fork restart or progression following restart?

Where replication forks have stalled, replication restart can occur in one of a number of ways: (i) the block may be repaired (or removed); (ii) it may be bypassed using error-prone translesional synthesis (TLS), or (iii) it may be avoided by using an alternative template (e.g. the newly synthesised region on the opposite strand, resulting from fork regression or generated by recombination). The first option is usually the easiest and the least likely to have mutational consequences; translesional synthesis is inherently more likely to cause mutation (pol iota (ι), for example, has an error rate of 0.72 i.e. it incorporates nucleotides almost at random, irrespective of the template sequence [76, 77]), whilst recombination requires a suitable donor template that is not always available. The type of lesion, whether it is on the leading or lagging strand, and the surrounding environment all contribute to how the replication block is dealt with. For example, nucleotide depletion following HU treatment imposes replication stress and can lead to fork stalling, but such stalling may be 'seen' differently by the checkpoint and restart machinery to forks that stall at physical barriers caused by damaging agents such as MMS.

It appears that RecQ helicases may aid in pathway 'choice', although the mechanisms that dictate which pathway is utilised are not fully understood. For instance, yeast complementation studies in *rad50* mutants have demonstrated that BLM is important in resistance to ionising radiation that causes double-strand breaks [78], while WRN confers resistance to drugs such as MMS that lead to replication fork stalling [79, 80]. In human cells, dual labelling of DNA before and after either HU or MMS treatment and analysis by fibre spreading (DNA combing) has shown that cells acutely depleted of WRN using shRNAi were still able to pre-serve replisome integrity upon HU- or MMS-induced fork stalling, though following recov-ery, replication fork rates were slower in WRN-depleted cells than controls, as evidenced by much shorter tracts of labelled DNA post-treatment compared with those synthesised before treatment [44]. It has been proposed [44] that WRN leads to rapid elimination of single-stranded DNA tracts by promoting recombination (using the sister chromatid as template), by enhancing translesion polymerase-mediated gap filling, or by removing DNA immedi-ately after fork passage. It has therefore been suggested that the genome instability in WS results from a defective response to stalled replication forks.

3.5. Error-prone translesional synthesis to relieve the replication block

Some lesions such as those caused by MMS or 4NQO present an insurmountable barrier to templating for the high fidelity B family DNA polymerases, but error-prone replication through these small lesions is often less costly for the cell than replication pausing and re-cruitment of repair complexes. Such error-prone synthesis is conducted by the Y family translesion DNA polymerases (TLS pols). These can pair nucleotides opposite modified and unusual bases, but at the cost of fidelity (ranging from error rates of ~6 x 10^{-3} for pol kappa (κ), through 3.5 x 10^{-2} for pol eta (η) to the essentially random 0.72 error rate for pol ι [76, 77, 81, 82]). The active site of such polymerasis is much larger than that of the proofreading pol-ymerasis, allowing for unusual base pairing geometry, helical distortion of the template DNA, and solvent access [83]. Consistent with an important role for WRN in replication fork progression after pausing, WRN has been found to promote the processivity of Y-family TLS pols on a wide range of substrates including oxidized bases, abasic sites, and thymine dimers [84]. This activity is specific to WRN, and appears to increase the apparent V_{max} of polymerisation. This does not require either catalytic activity of WRN, as proteins with point mutations that ablate both helicase and exonuclease activities can still promote pol η poly-merisation, although neither catalytically-active BLM nor RecQ5 can substitute [84].

3.6. WRN suppresses illegitimate recombination at stalled forks

Whilst the experiments described above strongly support the assertion that WRN is re-quired for fork progression after restart, others have suggested that WRN is itself required to promote restart, possibly through preventing either the accumulation of recombinogenic substrates or in suppressing recombination itself. High levels of spontaneous Rad51 foci in WS cells indicate the presence of an increased number of DNA double-strand breaks (DSBs) and elevated recombination when WRN is absent, supporting the assertion that WRN blocks excessive and illegitimate recombination. Indeed, stalled forks are thought to regress to 'chicken foot; structures with 4-way Holiday junctions that can either be removed by exo-nuclease degradation of the free ends, by branch migration to a point at which replication can simply restart, or by recombination at the junction (see Figure 2). WRN is likely to sup-press the recombinational route, as shown by partial complementation of yeast cells defec-tive in Sgs1 by expression of human WRN. Accumulation and persistence of Holliday junctions is likely, since ectopic expression of the bacterial RusA resolvase allows WS cells to proliferate as rapidly as control cells, and to resist treatment with CPT or 4NQO (fork collapse and fork stalling agents) to which WS are normally hypersensitive [46].

WRN helicase may branch migrate the chicken foot to 'fold back' the regressed form and thus re-establish a normal fork structure (Figure 2). Indeed, fork regression by WRN on RPA-coated DNA has recently been reported [85]. Alternatively, WRN exonuclease may degrade regions of the chicken foot and allow reformation of a normal replication fork. In addition to its own exonuclease activity, WRN associates with human Exonuclease 1 (Exo1), stimulating its activity [86]. It may therefore be the case that the two nuclease activities combine to remove regressed forks. It has been suggested that in the absence of WRN, the recombinational route is used to process the accumulated HJs, and that this requires the action of the nuclease Mus81; fission yeast Rqh1 suppresses Mus81 mutation [42] and human WRN suppresses Mus81-mediated recombination [87].

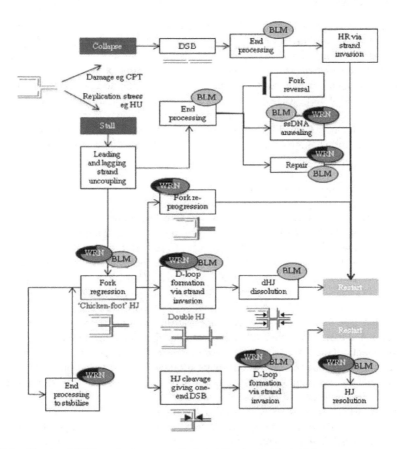

Figure 2. Possible roles of WRN in replication restart after fork stalling (see text for details)

3.7. Template switching at stalled forks

Leading strand blockage often uncouples the replicative helicases from the rest of the repli-some, allowing significant unwinding to form long tracts of single-stranded DNA, with lag-ging strand synthesis continuing for a distance [88]. The accumulated long single stranded loop of leading strand DNA is highly susceptible to damage. Replication fork restart on the leading strand might simply utilise new priming by RPA-mediated recruitment of pol α-pri-mase to the region of transition between singe stranded and duplex DNA (i.e. where the previous polymerase ceased synthesis), in much the same way that it normally reassociates with the primer-template junction in Okazaki fragment synthesis. Alternatively, regression of the replication fork may permit annealing to the new lagging strand using 'template switching' to give a Holliday junction that can then be reversed past the lesion [89, 90]. In bacteria, this can be done by RecQ helicase, with RecJ exonuclease to remove the protruding lagging strand flap [91].

In mammals this is likely to require WRN and the flap endonuclease activity of FEN-1 [92]. WRN (and BLM) can induce fork regression over the lesion by local unwinding, and can lead to the formation of the chicken-foot. WRN can also reverse a regressed fork. Both BLM and WRN helicase activities can also catalyse branch migration of the DNA leading to recov-ery of the template daughter strand annealing via Rad51 [93], formation of a double Holli-day junction and strand exchange. If the product here is a hemicatenenes, it can be resolved into either a chicken foot or a HJ and processed the same way. Ultimately, functional repli-cation forks may be reformed [94]. Alternatively, the Holliday junction can then be cleaved by a resolvase and DSB repair as before. See Figure 2 (above) for a schematic of replication fork restart.

3.8. How is WRN recruited to stalled forks?

Stalling of replication forks initiates the caffeine-sensitive S phase checkpoint, mediated by RPA, ATR and Rad53. WRN recruitment to, or retention at, stalled forks may be direct through binding to RPA [85], but it also appears to require phosphorylation by the check-point kinase ATR [95]. When such phosphorylation is prevented, WRN cannot accumulate at repair sites and DNA strand breaks are detected [73]. That WRN is an *in vivo* as well as *in vitro* target of ATR has been confirmed by phosphoproteomic studies [96]. However, it is still the subject of research and debate as to whether WRN is an upstream sensor or down-stream effector in the S phase checkpoint that responds to replication stress or stalled forks. For example, shRNAi-mediated WRN knockdown abrogated the S phase checkpoint on CPT treatment but did not affect checkpoint induction on HU exposure [97], suggesting that WRN may be an important 'sensor' of collapsed but not stalled forks, although the mecha-nism has yet to been defined. Perhaps fork collapse (e.g. upon CPT treatment) requires ATM, with its double-strand break sensing activity through recruitment by Ku and activa-tion by DNA-PKcs (DNA-dependent protein kinase catalytic subunit), while fork stalling (e.g. on HU) uses the ATR pathway. This is consistent with differential regulation of WRN by the two kinases [73], and with a requirement for WRN not only in replication fork pro-gression after stalling (see above) but also in directing recombination in concert with RAD51 and RAD54 [93]. Recently, it has been shown that WRN also interacts with the repair sliding clamp 9-1-1 (homologous structurally and functionally to PCNA, though acting in repair rather than replication), and that upon fork arrest, the 9-1-1 complex recruits TopBP1 that in turn recruits ATR which phosphorylates WRN [98]. Perhaps the initial type of damage that leads to fork arrest is therefore a deciding factor in the pathways of WRN recruitment and post-translational modification.

3.9. Role of WRN at stalled forks on the lagging strand

Lagging strand blocks do not uncouple the replication fork; rather, lagging strand polymerase merely stutters to the next primer to restart synthesis of the next Okazaki fragment [99, 100]. The resulting single-stranded gap is repaired by translesional synthesis as above (which may be error-prone) or by homologous recombination with the sister chromatid (which is more likely to retain fidelity). In *E. coli*, this requires formation of a double Holliday junction and resolution via non-crossover [101]. In mammals, BLM has the ability to mobilise double Holliday junctions and the resulting catenated DNA is resolved by topoisomerase III without crossover [102]. WRN does not interact with TopoIII and cannot migrate a double Holliday junction [103], although structures involved in intermediate formation (D-loops, G-quadruplex) might require either WRN or BLM. WRN can process a mobile D-loop [104] using the co-ordinate action of both helicase and exonuclease.

However, WRN is also linked to the functionality of the lagging strand polymerase, pol δ. WRN stimulates the base incorporation of pol δ (but not α or ε) even in the absence of PCNA [105]. Pol δ is slowed at fragile sites and repetitive runs likely to cause hairpin or bubble structures, but this can be alleviated by the helicase functionality of WRN [75]. Like WRN, pol δ has a 3′-5′ exonuclease capability which it can use to proofread bases after insertion [106]. WRN can substitute at this proofreader, and cells with low levels of WRN show increased mutation of the lagging strand [107]. Interestingly, the exonuclease activity of pol δ is active on WRN-preferred DNA substrates such as Holliday junctions, D-loops and bubble duplex, and can form a complex with WRN [107] that increases the degradation of these substrates. WRN exonuclease is blocked by many common lesions [22, 108]; it will be interesting to find out whether the nuclease activity of pol δ is complementary to this, and might suggest why the two would functionally substitute within the lagging strand complex.

Ultimately, fork restart requires proximal repositioning of the replication complex; this remodelling may make use of WRN nuclease activities to further process DNA ends and allow removal of damage. Interactions with PCNA and either strict (pol δ, pol α) or promiscuous (TLS pathway) repair polymerases and FEN1 flap removal activity can allow bypass of nicks and modified DNA bases at the same time as restart positioning, allowing many lesions to be handled.

4. Involvement of WRN in telomere maintenance

4.1. Telomere structure and replication

Mammalian telomeres consist of a few kilobases of repetitive non-coding G-rich sequence (the human sequence is (TTAGGG)n) which must be 'capped' rather like a bootlace in order to stop the DNA end being recognised as a DSB via p53/p21 signalling [109] and instigating profligate double-strand break (DSB) repair [110]. Functional capping forms a lasso-like structure [111] called the telomere-loop (T-loop) where the repetitive telomere sequence folds back upon itself to displace a short segment of proximal sequence with a 3' single stranded end to give a displacement-loop (the D-loop)[112]. The proteins that make up the telosome (or core shelterin complex [112]) include TRF1 and TRF2 [113], which bind and sta-bilise telomeric duplex DNA at the T-loop [114], and POT1 [115], a DNA-binding protein which coats and protects the tracts of single stranded telomeric sequence that occur at the telomeric D-loop and during telomeric replication and processing. Figure 3 shows the T and D loop structure with associated proteins.

Telomeres are replicated by passage of a replication fork that initiated upstream of the chro-mosome end: obviously it is not possible to load the replisome or prime DNA synthesis be-yond the end of the chromosome. At each round of replication, the telomeric sequence is unwound from the D (and possibly also the T) loops, and passively replicated by an incom-ing fork. While early reports suggested that priming on the lagging strand was defective at the very end of the chromosome, it has become apparent that both leading and lagging strands are normally replicated but that regeneration of a 3' overhang for strand invasion to form the D loop involves end resection of the leading strand, thus removing sequence infor-mation and shortening the telomere at each round of replication.

4.2. Telomere shortening leads to replicative senescence and genome instability

Telomere shortening acts as a counting mechanism to indicate the number of cell divisions a somatic cell has passed through, and normal fibroblasts generally arrest at the Hayflick limit of 55-60 population doublings [3] under the influence of this telomere attrition. Hence cellu-lar ageing is in a large part caused by progressive telomere loss – cells that lose telomeres more rapidly senesce more quickly that those with long telomeres, and people with prema-turely short telomeres (e.g. mothers of chronically sick children[116], carers of partners with dementia and low paid workers experiencing
work-related stress) age prematurely [117, 118]. (Note that this is not the case in mice, where lab strains have extremely long telomeres and cells senesce prior to telomeres reaching a critical length).

To overcome this cellular ageing, it is vital that immortal cells such as those of the germline have a mechanism to restore telomeric DNA at every round of replication. Such cells express active telomerase, a reverse transcriptase which utilises its endogenous RNA template to re-generate telomeric sequence [119], but telomerase levels are extremely low or absent in most somatic cells [120]. Notably, immortalisation of cancer cells is accompanied by re-expression of telomerase [121] in about 85% of all human cancers, while the remaining 15% are able to maintain their telomere lengths in the absence of telomerase, by alternatives mechanisms, reviewed in ref. [122] (see section 4.6).

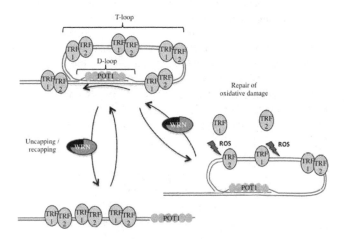

Figure 3. The structure of the telomere, showing the large telomere (T) loop and the smaller displacement (D) loop. Proteins TRF1, TRF2 and POT1 are critically important in stabilising the telomeric structure. WRN binds to all of these proteins

Dysfunctional telomeres that become uncapped are liable to degradation or immediate repair by homologous recombination (HR) or non-homologous end-joining (NHEJ), the latter causing chromosome fusions that are usually catastrophic for the cell. However, the tightly capped telomere cannot serve as a template during replication, so regulated disassembly of the shelterin complex and unwinding of the D (and possibly T) loop is necessary for efficient copying of telomeric regions. The transient uncapping that occurs during replication is recognised by repair proteins as DNA damage [123], and the correct reformation of the T-loop requires correct handling and processing by repair enzymes. Uncontrolled uncapping is therefore a powerful cause of genomic instability, and loss of telomeres shortens replicative lifespan; both are hallmarks of WS.

4.3. Are telomeres defective in WS?

The major clinical characteristics of WS are premature ageing, presumably resulting from the highly premature replicative senescence, and elevated cancer risk, which is caused by excess genome instability. Since replicative senescence is caused, at least in part, by telomere shortening, and chromosome fusions result from telomere loss, it has been of major importance to determine whether telomeres are indeed defective in WS cells, and whether WRN plays any role in telomere maintenance. Human WS cells in culture show elevated rates of telomere loss [124]. Contradictory to this, however, are data from single telomere length analysis (STELA) that suggest WS cells do not experience exceptional rates of telomere shortening, at least in clonal populations, though in bulk cultures of WS fibroblast, telomere loss ranges from a normal 99bp/PD to a four fold increase at 355 bp/PD [125].

Support for the importance of telomeric dysfunction in WS replicative senescence comes from studies of mouse models that are null for WRN. However, mice lacking WRN do not exhibit the premature ageing symptoms seen in humans [126] because laboratory mouse strains possess much longer telomeres than humans (40-80kb compared to 2-10kb) and detectable levels of telomerase even in somatic cells [127]. When mice deficient in telomerase are bred for several generations to reduce their telomere lengths to that approaching the normal human mean, removal of WRN gives similar premature ageing characteristics to those seen in human WS [128, 129]. Crucially, later-generation telomerase-null mice that still retain longer telomeres do not show this phenotype even though premature senescence is seen in their littermates that have short telomeres. Hence short telomeres combined with lack of WRN results in premature ageing.

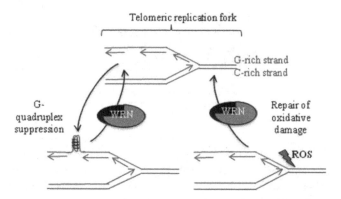

Figure 4. Roles of WRN at the telomere include unwinding of G4 DNA, that would otherwise lead to replication fork stalling, and repair of oxidative damage to which the telomeric DNA is exquisitely sensitive.

4.4. WRN helicase and exonuclease co-operate at the telomere

The repetitive nature of telomeric DNA arises as a consequence of the short RNA template within telomerase; this, combined with the G-rich nature leads to these sequences forming secondary structures called G-quadruplexes, which stall replication machinery much as any bulky lesion or DNA gap or break will. It is therefore essential for cells to unwind telomeric DNA ahead of the replication fork to prevent stalling, or worse, collapse. As discussed above, D-loops, recombination intermediates and G-quadruplexes may all require WRN and other RecQ helicases to remove these blockages (Figure 4). Under experimental conditions *in vitro*, WRN localises to a sub-set of telomeres during S-phase without the induction of stress [29], and is enriched when cells are subjected to damaging agents that cause replication stress such as CPT. Thus WRN catalysis is needed to police both endogenous replication fork blocks and induced damage.

WRN interacts with many of the proteins making up the shelterin complex or telosome [130-133] (see also Figure 3, above). Such interactions are likely to have functional consequences: for example, POT1 stimulates WRN helicase activity on linear and D-loop structures *in vitro* [134], whilst the presence of TRF1 and TRF2 can modulate their activity. TRF2 recruits WRN to D-loops and therefore stimulates unwinding [134], but it inhibits the helicase activity of WRN if binding to telomeric HJ substrates [135].

There are fewer pathways for replication fork recovery at telomeric ends because of the lack of downstream origins [136]. This obviously increases the need for proteins such as WRN that can dissolve or resolve replication blocks and promote fork progression before irreversible fork collapse occurs. One such block is G-quadruplex DNA: it has been shown to stall the major replicative polymerase δ [72]. G-quadruplex structures can arise spontaneously in single-stranded telomeric sequence [137] and can be suppressed by the binding of POT1 to release single stranded telomeric sequence during uncapping [138]. WRN preferentially unwinds G-quadruplex DNA [139] and its presence will suppress polymerase δ stalling [72], suggesting it is a good candidate for this role.

Interestingly, WRN – the only human RecQ helicase to also have exonuclease activity – unwinds D-loops *in vitro* in the absence of other proteins, using co-ordinate activity of both its helicase and exonuclease functions (RecQ helicase activity on these substrates is not particularly processive without stimulation for example by RPA [140]). The catalytic subunit of DNA-PK has also been shown to interact with WRN at telomeres [141], acting to suppress its exonuclease function and allow longer tracts to be unwound by the helicase activity. Therefore in the presence of DNA-PKcs, WRN processing of telomeric DNA does not shorten telomeric ends.

4.5. WRN acts on the lagging strand during telomere replication

Despite these detailed studies, the exact catalytic role(s) of WRN in telomere maintenance are still not fully defined. There is good evidence that cells lacking WRN have defective lagging strand synthesis at the telomere [142], as metaphase chromosomes in WRN helicase-deficient cells show a characteristic (if low-level) loss of telomeric sequence on one but not both sister chromatids. This is called sister telomere loss (STL), and suggests dysfunctional processing of one strand of the telomere during replication. The sister telomere lost is always the one resulting from lagging strand synthesis [142] This phenotype is thought to arise because in the absence of WRN activity, G-quadruplexes accumulate in the G-rich template strand and cause failure of lagging-telomere replication. Expression of active (but not inactive) telomerase suppresses STLs in cells lacking WRN [142], suggesting that sister telomere loss occurs during WRN-dependent processing of telomeres at times other than normal S phase when telomeres are uncapped for replication elongation.

The low levels of STL that occur (if the experimental data from chromosomal FISH reflect the underlying levels) suggest that the events that cause the telomere loss might be difficult to process or close to irreparable, or are merely rare; they might be alternatively-processed in a pathway that does not induce loss. Conversely, the catalytic activity supplied by WRN might be substituted by other enzymes – its helicase role by one of the other RecQs, or its exonuclease function by another appropriate 3'-5' exonuclease such as ExoI [143]. Whilst this has not yet been determined, however it is notable that cells deficient in BLM also show telomeric defects, although these are not the end-fusions arising from DSB repair as seen with WRN, but seem to be catenated associations possibly from aberrant HR [131]. BLM may thus have a role in resolving late-replicating DNA intermediates at telomeres distinct from WRN, as the rate of telomere dysfunction seen in cells with either single-null genotype is exacerbated in a double null [144].

Since the processing of Okazaki fragments during lagging strand synthesis gives rise to regions of ssDNA, and G-rich sequences have a tendency to form G-quadruplex structures spontaneously, at the telomere there is increased likelihood of G-quadruplex formation in the single-stranded tracts. POT1 cannot actively dissociate the structure by binding, strongly suggesting that the G-quadruplex must first be dissociated before POT1 can bind and protect the telomeric sequence, and implicating a role for WRN in removing the replication block before problems arise (see review [145]). Interestingly, the available levels of POT1 may modulate the coupling of the leading and lagging strands at telomeres in the absence of WRN, allowing uncoupled synthesis of leading strand without processing of the lagging strand block [146].

Supporting this hypothesis, recent research in yeast suggests an alternative protein that may function to suppress G-quadruplex formation at telomeres, but this time on the leading strand. Pif1 is a 5'-3' helicase that negatively regulates telomere length [147]. Loss of Pif1 leads to slow replication fork progression, and in vitro Pif1 can unwind replication substrates [148]. Recently it was shown that cells without Pif1 have chromosome breakage at sites of G-tracts, and Pif1 can unwind G-quadruplex DNA that forms in the leading strand [149]. The higher eukaryote C. elegans also possesses a helicase (DOG-1) that is able to inhibit loss of guanine tracts, presumably by suppression of G-quadruplex structures [150]. It is tempting to speculate that genome surveillance utilises Pif1 on the leading strand and WRN on the lagging strand to suppress G-quadruplex formation and subsequent replication fork blockage at sites of high guanine content such as fragile sites and telomeric sequence.

The loss of WRN in this putative mechanism inherently implies loss specifically of lagging-strand DNA at the telomeres. In this model, the rarity of STL may be explained by a low rate of G-quadruplex formation at single-stranded telomeric tracts during Okazaki fragment replication, the ability of BLM (or another RecQ) to substitute for WRN, or the specific need for WRN in a small subset of these events – perhaps because exonuclease processing is also required. The WRN exonuclease activity is itself specifically implicated in processing of the 3'-end of the telomere, although other nucleases such as ExoI or perhaps FEN-1 [151] might possess the capability to substitute for WRN. Addition of exogenous DNA oligonucleotides homologous to the 3'-overhang structure of an uncapped telomeric end to cells lacking WRN results in an increase in DNA damage responses and ultimately cell senescence [152].

The loss of WRN in telomerase-positive cells in vivo causes the generation of extrachromosomal telomeric structures [153], [154] and this requires both helicase and exonuclease activities. WRN has exonuclease activity here that requires telomeric sequence in both double-stranded and single-stranded portions, and shows a characteristic limited degradation pattern [155]. TRF2 recruits WRN to telomeric sequence and *in vitro* it synergistically enhances the ability of WRN to degrade the G-rich 3'-overhangs of telomeric D-loops substrates [132, 156]. POT1 inhibits WRN exonuclease activity here [155]. TRF2 or WRN alone exhibit little or no stimulation on these substrates. Non-telomeric substrates show similarly little WRN-dependent degradation, presumably because TRF2 does not bind/recruit and stimulate WRN exonuclease, whilst TRF2 bound to telomeric sequence completely inhibits the activity of other nucleases such as ExoIII [132]. WRN helicase and exonuclease, together with TRF2, POT1, and Ku therefore probably act together to prevent telomeric free ends from becoming substrates for HR or other aberrant pathways. Taken together, these results support the specificity of WRN exonuclease in reducing the length of the telomeric 3'-end to the optimal length for regeneration of the T-loop after replication, and suppression of extrachromosomal telomeric circles.

4.6. WRN may be important in ALT

Telomeres can be lengthened without the use of telomerase using recombination to generate the template DNA needed. In yeast, this ALT[3] pathway requires Sgs1 (the RecQ homologue in *S. cerevisiae*), for which WRN and BLM may both partially substitute [131, 157, 158]. Both WRN and BLM have also been seen to interact with telomeric DNA in human cells that utilize the ALT pathway [29, 130, 131], albeit only a small proportion. Although the ALT pathways are not yet elucidated, most models suggest recombinational mechanisms where strand invasion into telomeric DNA of the same (or different) chromosome or chromatid is utilized as template for resynthesis (e.g. see [159, 160]). BLM-deficient cells show elevated rates of sister chromatid exchange [161] that were not detected in cells lacking WRN, however finer resolution experiments suggest that WS cells do show elevated SCE, but only at telomeres [162, 163]. The WS mouse models with shortened telomeres (described in section 4.3 above) show elevated levels of this telomere-specific SCE [164], as do cells deficient in POT1, or Ku and TRF2 together [165]. Ku stimulates both helicase and exonuclease activities of WRN [166, 167], and suppresses telomeric recombination brought on by the absence of TRF2 and consequent telomeric uncapping [110]. Taken together, these data suggest that WRN is prominent in a pathway that specifically suppresses telomeric recombination or dissolves junctions, and it is at least partially distinct from the role of BLM.

4.7. Telomeric DNA is hypersensitive to oxidative lesions – a further role for WRN

The G-rich nature of telomeric sequence means it is a rich target for oxidative damage[4] [168], and oxidative stress and mitochondrial dysfunction often give rise to concomitant telomeric dysfunction [169], which can be reduced using antioxidants. Notably, artificial replicative senescence can be induced with a burst of oxidative damage [170, 171]. Oxidation of telomeric bases can disrupt DNA binding of TRF1 and TRF2, and presumably therefore telo-

---

3 ALT = alternative lengthening of telomeres

4 8-oxoG is a common product of oxidative attack of DNA

some and T-loop assembly [166], whilst over-expression of TRF2 protect cells with shortened telomeres from early senescence [172].

WRN is a central component of base excision repair (BER) of oxidative lesions, interacting with most of the key proteins in the pathway such as pol beta (β) and FEN1 [173, 174]. Consistent with an important role for WRN in removing oxidative lesions, WS cells show increased oxidative damage [175, 176].

It has been shown that D-loops containing oxidised bases can be bound by POT1 and are a preferred substrate for WRN [177]. The strand-displacement activity of pol β, the repair polymerase in BER, is also stimulated by TRF2 [178], and TRF1, TRF2 and POT1 can enhance all the constituent steps of long patch BER [179]. As previously mentioned, WRN itself can also stimulate TLS pols to replicate past an oxidative block [84]. This suggests active recruitment and stimulation of anti-oxidative damage processes at telomeres involving RecQ helicases. These findings partly illustrate how the activities of RecQ helicases are tightly controlled by the surrounding milieu in order to differentiate their roles in replication and repair.

Although the wider significance of all these data is yet to be determined, it is obvious that WRN is active at multiple points in telomere replication and repair.

5. Conclusions

The helicase/exonuclease WRN has been shown to be critically important in DNA replication, acting to enhance fidelity, regulate template unwinding to prevent fork stalling at unusual structures, assist with replication fork restart and/or enhance processivity post-restart, aid translesion synthesis over otherwise unreplicatable lesions, promote regression of stalled replication forks to allow error-free restart, modulate recombination at collapsed replication forks, and aid telomere replication. It is recruited to sites of DNA synthesis, possibly through association with the sliding clamp PCNA, and to sites of stalled/collapsed forks probably by RPA in concert with the S phase checkpoint kinase ATR and its downstream effectors and mediators Chk1, Rad53, Mec1 and Mrc1. Loss of WRN results in high levels of chromosomal instability and elevated cancer risk, and the defects in DNA replication on WRN loss also results in premature onset of replicative senescence with concomitant organismal ageing, manifest as progeroid Werner's syndrome. While much has been discovered as to WRN's mode of action, there is still an enormous amount to learn as to how its activities are co-ordinated with the cell during DNA replication.

Acknowledgements

We thank Hayley Lees (Department of Biochemistry, University of Oxford) for critical reading of the manuscript. We gratefully acknowledge support from the Economic and Social Sciences Research Council of Great Britain (ESRC) grant [ES/G037086/1] under the cross-council New Dynamics of Ageing initiative.

Author details

Lynne S. Cox* and Penelope A. Mason

*Address all correspondence to: Lynne.cox@bioch.ox.ac.uk

Department of Biochemistry, University of Oxford, South Parks Road, Oxford

References

[1] Larizza, L., Roversi, G., & Volpi, L. Rothmund-Thomson syndrome. *Orphanet J Rare Dis*, vol. 5, 2.

[2] Goto, M. (2001). Clinical characteristics of Werner syndrome and other premature aging syndromes: pattern of aging in progeroid syndromes *From premature gray hair to helicase- Werner syndrome: implications for aging and cancer, M. Goto and R. W. Miller eds., Tokyo: Japan Scientific Societies Press*, 27-39.

[3] Hayflick, L. (1979). The cell biology of aging. *J Invest Dermatol*, 73(1), 8-14.

[4] Kyng, K. J., May, A., Kolvraa, S., & Bohr, V. A. (2003). Gene expression profiling in Werner syndrome closely resembles that of normal aging. *Proc Natl Acad Sci U S A*, 100(21), 12259-12264.

[5] Vanhooren, V., Desmyter, L., Liu, X. E., Cardelli, M., Franceschi, C., Federico, A., Libert, C., Laroy, W., Dewaele, S., Contreras, R., & Chen, C. (2007). N-glycomic changes in serum proteins during human aging. *Rejuvenation Res*, 10(4), 521-531a.

[6] Fukuchi, K., Martin, G. M., & Monnat, R. J. (1989). Mutator phenotype of Werner syndrome is characterized by extensive deletions. *Proc Natl Acad Sci U S A*, 86(15), 5893-5897.

[7] Scappaticci, S., Cerimele, D., & Fraccaro, M. (1982). Clonal structural chromosomal rearrangements in primary fibroblast cultures and in lymphocytes of patients with Werner's Syndrome,. *Hum Genet*, 62(1), 16-24.

[8] Segurado, M., & Tercero, J. A. (2009). The S-phase checkpoint: targeting the replication fork. *Biol Cell*, 101(11), 617-627.

[9] Agrelo, R., Cheng, W. H., Setien, F., Ropero, S., Espada, J., Fraga, M. F., Herranz, M., Paz, M. F., Sanchez-Cespedes, M., Artiga, M. J., Guerrero, D., Castells, A., von, C., Kobbe, V. A., & Esteller, M. (2006). Epigenetic inactivation of the premature aging Werner syndrome gene in human cancer. *Proc Natl Acad Sci U S A*, 103(23), 8822-8827.

[10] Ding, S. L., Yu, J. C., Chen, S. T., Hsu, G. C., & Shen, C. Y. (2007). Genetic variation in the premature aging gene WRN: a case-control study on breast cancer susceptibility. *Cancer Epidemiol Biomarkers Prev*, 16(2), 263-269.

[11] Bird, J. L., Jennert-Burston, K. C., Bachler, M. A., Mason, P. A., Lowe, J. E., Heo, S. J., Campisi, J., Faragher, R. G., & Cox, L. S. (2012). Recapitulation of Werner syndrome sensitivity to camptothecin by limited knockdown of the WRN helicase/exonuclease. *Biogerontology*, 13(1), 49-62.

[12] Futami, K., Takagi, M., Shimamoto, A., Sugimoto, M., & Furuichi, Y. (2007). Increased chemotherapeutic activity of camptothecin in cancer cells by siRNA-induced silencing of WRN helicase. *Biol Pharm Bull*, 30(10), 1958-1961.

[13] Aggarwal, M., Sommers, J. A., Shoemaker, R. H., & Brosh, R. M. (2011). Inhibition of helicase activity by a small molecule impairs Werner syndrome helicase (WRN) function in the cellular response to DNA damage or replication stress. *Proc Natl Acad Sci U S A*, 108(4), 1525-1530.

[14] Yu, C. E., Oshima, J., Fu, Y. H., Wijsman, E. M., Hisama, F., Alisch, R., Matthews, S., Nakura, J., Miki, T., Ouais, S., Martin, G. M., Mulligan, J., & Schellenberg, G. D. (1996). Positional cloning of the Werner's syndrome gene,. *Science*, 272(5259), 258-262.

[15] von Kobbe, C., & Bohr, V.A. (2002). A nucleolar targeting sequence in the Werner syndrome protein resides within residues 949-1092. *J Cell Sci*, 115, 3901-3907.

[16] Marciniak, R. A., Lombard, D. B., Johnson, F. B., & Guarente, L. (1998). Nucleolar localization of the Werner syndrome protein in human cells. *Proc Natl Acad Sci U S A*, 95(12), 6887-6892.

[17] Rodriguez-Lopez, A. M., Jackson, D. A., Iborra, F., & Cox, L. S. (2002). Asymmetry of DNA replication fork progression in Werner's syndrome. *Aging Cell*, 1(1), 30-39.

[18] Rodriguez-Lopez, A. M., Jackson, D. A., Nehlin, J. O., Iborra, F., Warren, A. V., & Cox, L. S. (2003). Characterisation of the interaction between WRN, the helicase/exonuclease defective in progeroid Werner's syndrome, and an essential replication factor, PCNA. *Mech Ageing Dev*, 124(2), 167-174.

[19] Constantinou, A., Tarsounas, M., Karow, J. K., Brosh, R. M., Bohr, V. A., Hickson, I. D., & West, S. C. (2000). Werner's syndrome protein (WRN) migrates Holliday junctions and co-localizes with RPA upon replication arrest. *EMBO Rep*, 1(1), 80-84.

[20] Huang, S., Li, B., Gray, M. D., Oshima, J., Mian, I. S., & Campisi, J. (1998). The premature ageing syndrome protein, WRN, is a 3'-->5' exonuclease. *Nat Genet*, 20(2), 114-116.

[21] Huang, S., Beresten, S., Li, B., Oshima, J., Ellis, N. A., & Campisi, J. (2000). Characterization of the human and mouse WRN 3'-->5' exonuclease. *Nucleic Acids Res*, 28(12), 2396-2405.

[22] Harrigan, J. A., Fan, J., Momand, J., Perrino, F. W., Bohr, V. A., & Wilson, D. M. (2007). WRN exonuclease activity is blocked by DNA termini harboring 3′ obstructive groups. *Mech Ageing Dev*, 128(3), 259-266.

[23] Ozgenc, A., & Loeb, L. A. (2005). Current advances in unraveling the function of the Werner syndrome protein. *Mutat Res*, 577(1-2), 237-251.

[24] Xue, Y., Ratcliff, G. C., Wang, H., Davis-Searles, P. R., Gray, M. D., Erie, D. A., & Redinbo, M. R. (2002). A minimal exonuclease domain of WRN forms a hexamer on DNA and possesses both 3′- 5′ exonuclease and 5′-protruding strand endonuclease activities. *Biochemistry*, 41(9), 2901-2912.

[25] Machwe, A., Xiao, L., & Orren, D. K. (2006). Length-dependent degradation of single-stranded 3′ ends by the Werner syndrome protein (WRN): implications for spatial orientation and coordinated 3′ to 5′ movement of its ATPase/helicase and exonuclease domains. *BMC Mol Biol*, 7, 6.

[26] Gray, M. D., Shen, J. C., Kamath-Loeb, A. S., Blank, A., Sopher, B. L., Martin, G. M., Oshima, J., & Loeb, L. A. (1997). The Werner syndrome protein is a DNA helicase. *Nat Genet*, 17(1), 100-103.

[27] Brosh, R. M., Waheed, J., & Sommers, J. A. (2002). Biochemical characterization of the DNA substrate specificity of Werner syndrome helicase. *J Biol Chem*, 277(26), 23236-23245.

[28] Shen, J., & Loeb, L. A. (2001). Unwinding the molecular basis of the Werner syndrome. *Mech Ageing Dev*, 122(9), 921-944.

[29] Opresko, P. L., Otterlei, M., Graakjaer, J., Bruheim, P., Dawut, L., Kolvraa, S., May, A., Seidman, M. M., & Bohr, V. A. (2004). The Werner syndrome helicase and exonuclease cooperate to resolve telomeric D loops in a manner regulated by TRF1 and TRF2. *Mol Cell*, 14(6), 763-774.

[30] Kitano, K., Kim, S. Y., & Hakoshima, T. (2010). Structural basis for DNA strand separation by the unconventional winged helix domain of RecQ helicase WRN. *Structure*, 18(2), 177-187.

[31] Kitano, K., Yoshihara, N., & Hakoshima, T. (2007). Crystal structure of the HRDC domain of human Werner syndrome protein, WRN. *J Biol Chem*, 282(4), 2717-2728.

[32] Hartung, F., & Puchta, H. (2006). The RecQ gene family in plants. *J Plant Physiol*, 163(3), 287-296.

[33] Opresko, P. L., Cheng, W. H., & Bohr, V. A. (2004). Junction of RecQ helicase biochemistry and human disease. *J Biol Chem*, 279(18), 18099-18102.

[34] Cox, L. S., Clancy, D. J., Boubriak, I., & Saunders, R. D. (2007). Modeling Werner Syndrome in Drosophila melanogaster: hyper-recombination in flies lacking WRN-like exonuclease. *Ann N Y Acad Sci*, 1119, 274-288.

[35] Boubriak, I., Mason, P. A., Clancy, D. J., Dockray, J., Saunders, R. D., & Cox, L. S. (2009). DmWRNexo is a 3'-5' exonuclease: phenotypic and biochemical characterization of mutants of the Drosophila orthologue of human WRN exonuclease. *Biogerontology*, 10(3), 267-277.

[36] Mason, P. A., Boubriak, I., Robbins, T., Lasala, R., Saunders, R., & Cox, L. S. (2012). The Drosophila orthologue of progeroid human WRN exonuclease, DmWRNexo, cleaves replication substrates but is inhibited by uracil or abasic sites : Analysis of DmWRNexo activity in vitro. *Age (Dordr)*.

[37] Saunders, R. D., Boubriak, I., Clancy, D. J., & Cox, L. S. (2008). Identification and characterization of a Drosophila ortholog of WRN exonuclease that is required to maintain genome integrity. *Aging Cell*, 7(3), 418-425.

[38] Ketting, R. F., Haverkamp, T. H., van Luenen, H. G., & Plasterk, R. H. (1999). Mut-7 of C. elegans, required for transposon silencing and RNA interference, is a homolog of Werner syndrome helicase and RNaseD. *Cell*, 99(2), 133-141.

[39] Lee, S. J., Yook, J. S., Han, S. M., & Koo, H. S. (2004). A Werner syndrome protein homolog affects C. elegans development, growth rate, life span and sensitivity to DNA damage by acting at a DNA damage checkpoint. *Development*, 131(11), 2565-2575.

[40] Lee, S. J., Gartner, A., Hyun, M., Ahn, B., & Koo, H. S. (2010). The Caenorhabditis elegans Werner syndrome protein functions upstream of ATR and ATM in response to DNA replication inhibition and double-strand DNA breaks. *PLoS Genet*, 6(1), e1000801.

[41] Hartung, F., Plchova, H., & Puchta, H. (2000). Molecular characterisation of RecQ homologues in Arabidopsis thaliana. *Nucleic Acids Res*, 28(21), 4275-4282.

[42] Doe, C. L., Ahn, J. S., Dixon, J., & Whitby, M. C. (2002). Mus81-Eme1 and Rqh1 involvement in processing stalled and collapsed replication forks. *J Biol Chem*, 277(36), 32753-32759.

[43] Poot, M., Hoehn, H., Runger, T. M., & Martin, G. M. (1992). Impaired S-phase transit of Werner syndrome cells expressed in lymphoblastoid cell lines. *Exp Cell Res*, 202(2), 267-273.

[44] Sidorova, J. M., Li, N., Folch, A., & Monnat, R. J. (2008). The RecQ helicase WRN is required for normal replication fork progression after DNA damage or replication fork arrest. *Cell Cycle*, 7(6), 796-807.

[45] Fujiwara, Y., Higashikawa, T., & Tatsumi, M. (1977). A retarded rate of DNA replication and normal level of DNA repair in Werner's syndrome fibroblasts in culture. *J Cell Physiol*, 92(3), 365-374.

[46] Rodriguez-Lopez, A. M., Whitby, M. C., Borer, C. M., Bachler, M. A., & Cox, L. S. (2007). Correction of proliferation and drug sensitivity defects in the progeroid Werner's Syndrome by Holliday junction resolution. *Rejuvenation Res*, 10(1), 27-40.

[47] Lebel, M., Spillare, E. A., Harris, C. C., & Leder, P. (1999). The Werner syndrome gene product co-purifies with the DNA replication complex and interacts with PCNA and topoisomerase I. *J Biol Chem*, 274(53), 37795-37799.

[48] Yan, H., & Newport, J. (1995). FFA-1, a protein that promotes the formation of replication centers within nuclei. *Science*, 269(5232), 1883-1885.

[49] Chen, C. Y., Graham, J., & Yan, H. (2001). Evidence for a replication function of FFA-1, the Xenopus orthologue of Werner syndrome protein. *J Cell Biol*, 152(5), 985-996.

[50] Garg, P., & Burgers, P. M. (2005). DNA polymerases that propagate the eukaryotic DNA replication fork. *Crit Rev Biochem Mol Biol*, 40(2), 115-128.

[51] Pursell, Z. F., Isoz, I., Lundstrom, E. B., Johansson, E., & Kunkel, T. A. (2007). Yeast DNA polymerase epsilon participates in leading-strand DNA replication. *Science*, 317(5834), 127-130.

[52] McCulloch, S.D., & Kunkel, T.A. (2008). The fidelity of DNA synthesis by eukaryotic replicative and translesion synthesis polymerases. *Cell Res*, 18(1), 148-161.

[53] Podust, V. N., Tiwari, N., Stephan, S., & Fanning, E. (1998). Replication factor C disengages from proliferating cell nuclear antigen (PCNA) upon sliding clamp formation, and PCNA itself tethers DNA polymerase delta to DNA. *J Biol Chem*, 273(48), 31992-31999.

[54] Gambus, A., Jones, R. C., Sanchez-Diaz, A., Kanemaki, M., van Deursen, F., Edmondson, R. D., & Labib, K. (2006). GINS maintains association of Cdc45 with MCM in replisome progression complexes at eukaryotic DNA replication forks. *Nat Cell Biol*, 8(4), 358-366.

[55] Cox, L.S. (2009). Molecular Themes in DNA Replication in Book Molecular Themes in DNA Replication. *Series Molecular Themes in DNA Replication*, Editor ed.^eds., City: Royal Society of Chemistry, 2009, 443.

[56] Budd, M. E., Cox, L. S., & Campbell, J. L. (2009). Coordination of nucleases and helicases during DNA replication and double-strand break repair. *Molecular Themes in DNA Replication, L. S. Cox ed., Cambridge: Royal Society of Chemistry*, 112-155.

[57] Beattie, T.R., & Bell, S.D. (2012). Coordination of multiple enzyme activities by a single PCNA in archaeal Okazaki fragment maturation. *EMBO J*, 31(6), 1556-1567.

[58] Warbrick, E., Lane, D. P., Glover, D. M., & Cox, L. S. (1995). A small peptide inhibitor of DNA replication defines the site of interaction between the cyclin-dependent kinase inhibitor p21WAF1 and proliferating cell nuclear antigen. *Curr Biol*, 5(3), 275-282.

[59] Guarino, E., Shepherd, M. E., Salguero, I., Hua, H., Deegan, R. S., & Kearsey, S. E. (2011). Cdt1 proteolysis is promoted by dual PIP degrons and is modulated by PCNA ubiquitylation. *Nucleic Acids Res*, 39(14), 5978-5990.

[60] Warbrick, E., Lane, D. P., Glover, D. M., & Cox, L. S. (1997). Homologous regions of Fen1 and p21Cip1 compete for binding to the same site on PCNA: a potential mechanism to co-ordinate DNA replication and repair. *Oncogene*, 14(19), 2313-2321.

[61] Cox, L. S. (1997). Who binds wins: Competition for PCNA rings out cell-cycle changes. *Trends Cell Biol*, 7(12), 493-498.

[62] Cox, L. S., & Kearsey, S. (2009). Ring structures and six-fold symmetry in DNA replication. *Molecular Themes in DNA Replication, L. S. Cox ed., Cambridge: Royal Society of Chemistry*, 47-85.

[63] Brosh, R. M., von Kobbe, C., Sommers, J. A., Karmakar, P., Opresko, P. L., Piotrowski, J., Dianova, I., Dianov, G. L., & Bohr, V. A. (2001). Werner syndrome protein interacts with human flap endonuclease 1 and stimulates its cleavage activity. *EMBO J*, 20(20), 5791-5801.

[64] Sharma, S., Sommers, J. A., Gary, R. K., Friedrich-Heineken, E., Hubscher, U., & Brosh, R. M. (2005). The interaction site of Flap Endonuclease-1 with WRN helicase suggests a coordination of WRN and PCNA. *Nucleic Acids Res*, 33(21), 6769-6781.

[65] Nick, S. A., Mc Elhinny, Z. F., Pursell, S., & Kunkel, T. A. (2009). Mechanisms for high fidelity DNA replication. *Molecular Themes in DNA Replication, L. S. Cox ed., Cambridge, UK: Royal Society of Chemistry*, 86-111.

[66] Swan, M. K., Johnson, R. E., Prakash, L., Prakash, S., & Aggarwal, A. K. (2009). Structural basis of high-fidelity DNA synthesis by yeast DNA polymerase delta. *Nat Struct Mol Biol*, 16(9), 979-986.

[67] Perry, J. J., Yannone, S. M., Holden, L. G., Hitomi, C., Asaithamby, A., Han, S., Cooper, P. K., Chen, D. J., & Tainer, J. A. (2006). WRN exonuclease structure and molecular mechanism imply an editing role in DNA end processing. *Nat Struct Mol Biol*, 13(5), 414-422.

[68] Kamath-Loeb, A. S., Shen, J. C., Schmitt, M. W., & Loeb, L. A. (2012). The Werner syndrome exonuclease facilitates DNA degradation and high fidelity DNA polymerization by human DNA polymerase delta. *J Biol Chem*, 287(15), 12480-12490.

[69] Takeuchi, F., Hanaoka, F., Goto, M., Akaoka, I., Hori, T., Yamada, M., & Miyamoto, T. (1982). Altered frequency of initiation sites of DNA replication in Werner's syndrome cells. *Hum Genet*, 60(4), 365-368.

[70] Hanaoka, F., Yamada, M., Takeuchi, F., Goto, M., Miyamoto, T., & Hori, T. (1985). Autoradiographic studies of DNA replication in Werner's syndrome cells. *Adv Exp Med Biol*, 190, 439-457.

[71] Foiani, M., Ferrari, M., Liberi, G., Lopes, M., Lucca, C., Marini, F., Pellicioli, A., Muzi, M., Falconi, , & Plevani, P. (1998). S-phase DNA damage checkpoint in budding yeast. *Biol Chem*, 379(8-9), 1019-1023.

[72] Kamath-Loeb, A. S., Loeb, L. A., Johansson, E., Burgers, P. M., & Fry, M. (2001). Interactions between the Werner syndrome helicase and DNA polymerase delta specifi-

cally facilitate copying of tetraplex and hairpin structures of the d(CGG)n trinucleotide repeat sequence. *J Biol Chem*, 276(19), 16439-16446.

[73] Ammazzalorso, F., Pirzio, L. M., Bignami, M., Franchitto, A., & Pichierri, P. (2010). ATR and ATM differently regulate WRN to prevent DSBs at stalled replication forks and promote replication fork recovery. *EMBO J*, 29(18), 3156-3169.

[74] Murfuni, I., De Santis, A., Federico, M., Bignami, M., Pichierri, P., & Franchitto, A. (2012). Perturbed replication induced genome-wide or at common fragile sites is differently managed in the absence of WRN. *Carcinogenesis*, 33(9), 1655-63.

[75] Shah, S. N., Opresko, P. L., Meng, X., Lee, M. Y., & Eckert, K. A. (2010). DNA structure and the Werner protein modulate human DNA polymerase delta-dependent replication dynamics within the common fragile site FRA16D,. *Nucleic Acids Res*, 38(4), 1149-1162.

[76] Johnson, R. E., Washington, M. T., Haracska, L., Prakash, S., & Prakash, L. (2000). Eukaryotic polymerases iota and zeta act sequentially to bypass DNA lesions. *Nature*, 406(6799), 1015-1019.

[77] Choi, J.Y., & Guengerich, F.P. (2006). Kinetic evidence for inefficient and error-prone bypass across bulky N2-guanine DNA adducts by human DNA polymerase iota. *J Biol Chem*, 281(18), 12315-12324.

[78] Nimonkar, A. V., Ozsoy, A. Z., Genschel, J., Modrich, P., & Kowalczykowski, S. C. (2008). Human exonuclease 1 and BLM helicase interact to resect DNA and initiate DNA repair. *Proc Natl Acad Sci U S A*, 105(44), 16906-16911.

[79] Aggarwal, M., Sommers, J. A., Morris, C., & Brosh, R. M. (2010). Delineation of WRN helicase function with EXO1 in the replicational stress response. *DNA Repair (Amst)*, 9(7), 765-776.

[80] Aggarwal, M., & Brosh, R. M. (2010). Genetic mutants illuminate the roles of RecQ helicases in recombinational repair or response to replicational stress. *Cell Cycle*, 9(16), 3139-3141.

[81] Matsuda, T., Bebenek, K., Masutani, C., Rogozin, I. B., Hanaoka, F., & Kunkel, T. A. (2001). Error rate and specificity of human and murine DNA polymerase eta. *J Mol Biol*, 312(2), 335-346.

[82] Zhang, Y., Yuan, F., Xin, H., Wu, X., Rajpal, D. K., Yang, D., & Wang, Z. (2000). Human DNA polymerase kappa synthesizes DNA with extraordinarily low fidelity. *Nucleic Acids Res*, 28(21), 4147-4156.

[83] Vasquez Del, R., Carpio, T. D., Silverstein, S., Lone, M. K., Swan, J. R., Choudhury, R. E., Johnson, S., Prakash, L., & Aggarwal, A. K. (2009). Structure of human DNA polymerase kappa inserting dATP opposite an 8-OxoG DNA lesion. *PLoS One*, 4(6), e5766.

[84] Kamath-Loeb, A. S., Lan, L., Nakajima, S., Yasui, A., & Loeb, L. A. (2007). Werner syndrome protein interacts functionally with translesion DNA polymerases. *Proc Natl Acad Sci U S A*, 104(25), 10394-10399.

[85] Machwe, A., Lozada, E., Wold, M. S., Li, G. M., & Orren, D. K. (2011). Molecular cooperation between the Werner syndrome protein and replication protein A in relation to replication fork blockage. *J Biol Chem*, 286(5), 3497-3508.

[86] Sharma, S., Sommers, J. A., Driscoll, H. C., Uzdilla, L., Wilson, T. M., & Brosh, R. M. (2003). The exonucleolytic and endonucleolytic cleavage activities of human exonuclease 1 are stimulated by an interaction with the carboxyl-terminal region of the Werner syndrome protein. *J Biol Chem*, 278(26), 23487-23496.

[87] Franchitto, A., Pirzio, L. M., Prosperi, E., Sapora, O., Bignami, M., & Pichierri, P. (2008). Replication fork stalling in WRN-deficient cells is overcome by prompt activation of a MUS81-dependent pathway. *J Cell Biol*, 183(2), 241-252.

[88] Byun, T. S., Pacek, M., Yee, M. C., Walter, J. C., & Cimprich, K. A. (2005). Functional uncoupling of MCM helicase and DNA polymerase activities activates the ATR-dependent checkpoint. *Genes Dev*, 19(9), 1040-1052.

[89] Singleton, M. R., Scaife, S., Raven, N. D., & Wigley, D. B. (2001). Crystallization and preliminary X-ray analysis of RecG, a replication-fork reversal helicase from Thermotoga maritima complexed with a three-way DNA junction. *Acta Crystallogr D Biol Crystallogr*, 57(11), 1695-1696.

[90] Singleton, M. R., Scaife, S., & Wigley, D. B. (2001). Structural analysis of DNA replication fork reversal by RecG. *Cell*, 107(1), 79-89.

[91] Courcelle, J., Donaldson, J. R., Chow, K. H., & Courcelle, C. T. (2003). DNA damage-induced replication fork regression and processing in Escherichia coli. *Science*, 299(5609), 1064-1067.

[92] Sharma, S., Otterlei, M., Sommers, J. A., Driscoll, H. C., Dianov, G. L., Kao, H. I., Bambara, R. A., & Brosh, R. M. (2004). WRN helicase and FEN-1 form a complex upon replication arrest and together process branchmigrating DNA structures associated with the replication fork. *Mol Biol Cell*, 15(2), 734-750.

[93] Otterlei, M., Bruheim, P., Ahn, B., Bussen, W., Karmakar, P., Baynton, K., & Bohr, V. A. (2006). Werner syndrome protein participates in a complex with RAD51, RAD54, RAD54B and ATR in response to ICL-induced replication arrest. *J Cell Sci*, 119, 5137-5146.

[94] Machwe, A., Karale, R., Xu, X., Liu, Y., & Orren, D. K. (2011). The Werner and Bloom syndrome proteins help resolve replication blockage by converting (regressed) holliday junctions to functional replication forks. *Biochemistry*, 50(32), 6774-6788.

[95] Pichierri, P., & Rosselli, F. (2004). The DNA crosslink-induced S-phase checkpoint depends on ATR-CHK1 and ATR-NBS1-FANCD2 pathways. *EMBO J*, 23(5), 1178-1187.

[96] Bennetzen, M. V., Marino, G., Pultz, D., Morselli, E., Faergeman, N. J., Kroemer, G., & Andersen, J. S. (2012). Phosphoproteomic analysis of cells treated with longevity-related autophagy inducers. *Cell Cycle*, 11(9), 1827-1840.

[97] Patro, B. S., Frohlich, R., Bohr, V. A., & Stevnsner, T. (2011). WRN helicase regulates the ATR-CHK1-induced S-phase checkpoint pathway in response to topoisomerase-I-DNA covalent complexes. *J Cell Sci*, 124, 3967-3979.

[98] Pichierri, P., Nicolai, S., Cignolo, L., Bignami, M., & Franchitto, A. (2011). The RAD9-RAD1-HUS1 (9.1.1) complex interacts with WRN and is crucial to regulate its response to replication fork stalling. *Oncogene*, 31(23), 2809-2823.

[99] Svoboda, D.L., & Vos, J.M. (1995). Differential replication of a single, UV-induced lesion in the leading or lagging strand by a human cell extract: fork uncoupling or gap formation. *Proc Natl Acad Sci U S A*, 92(26), 11975-11979.

[100] Mc Inerney, P., & O'Donnell, M. (2004). Functional uncoupling of twin polymerases: mechanism of polymerase dissociation from a lagging-strand block. *J Biol Chem*, 279(20), 21543-21551.

[101] Heller, R.C., & Marians, K.J. (2006). Replisome assembly and the direct restart of stalled replication forks. *Nat Rev Mol Cell Biol*, 7(12), 932-943.

[102] Wu, L., & Hickson, I. D. (2003). The Bloom's syndrome helicase suppresses crossing over during homologous recombination. *Nature*, 426(6968), 870-874.

[103] Wu, L., Chan, K. L., Ralf, C., Bernstein, D. A., Garcia, P. L., Bohr, V. A., Vindigni, A., Janscak, P., Keck, J. L., & Hickson, I. D. (2005). The HRDC domain of BLM is required for the dissolution of double Holliday junctions. *EMBO J*, 24(14), 2679-2687.

[104] Opresko, P. L., Sowd, G., & Wang, H. (2009). The Werner syndrome helicase/exonuclease processes mobile D-loops through branch migration and degradation. *PLoS One*, 4(3), e4825.

[105] Kamath-Loeb, A. S., Johansson, E., Burgers, P. M., & Loeb, L. A. (2000). Functional interaction between the Werner Syndrome protein and DNA polymerase delta. *Proc Natl Acad Sci U S A*, 97(9), 4603-4608.

[106] Kunkel, T. A., Sabatino, R. D., & Bambara, R. A. (1987). Exonucleolytic proofreading by calf thymus DNA polymerase delta. *Proc Natl Acad Sci U S A*, 84(14), 4865-4869.

[107] Kamath-Loeb, A., Loeb, L. A., & Fry, M. (2012). The Werner syndrome protein is distinguished from the Bloom syndrome protein by its capacity to tightly bind diverse DNA structures. *PLoS One*, 7(1), e30189.

[108] Bukowy, Z., Harrigan, J. A., Ramsden, D. A., Tudek, B., Bohr, V. A., & Stevnsner, T. (2008). WRN Exonuclease activity is blocked by specific oxidatively induced base lesions positioned in either DNA strand. *Nucleic Acids Res*, 36(15), 4975-4987.

[109] Maser, R.S., & DePinho, R.A. (2004). Telomeres and the DNA damage response: why the fox is guarding the henhouse,. *DNA Repair (Amst)*, 3(8-9), 979-988.

[110] d'Adda, F., di Fagagna, S. H., & Jackson, S. P. (2004). Functional links between telomeres and proteins of the DNA-damage response. *Genes Dev*, 18(15), 1781-1799.

[111] Griffith, J. D., Comeau, L., Rosenfield, S., Stansel, R. M., Bianchi, A., Moss, H., & de Lange, T. (1999). Mammalian telomeres end in a large duplex loop. *Cell*, 97(4), 503-514.

[112] de Lange, T. (2005). Shelterin: the protein complex that shapes and safeguards human telomeres. *Genes Dev*, 19(18), 2100-2110.

[113] Broccoli, D., Smogorzewska, A., Chong, L., & de Lange, T. (1997). Human telomeres contain two distinct Myb-related proteins, TRF1 and TRF2. *Nat Genet*, 17(2), 231-235.

[114] Liu, D., O'Connor, M. S., Qin, J., & Songyang, Z. (2004). Telosome, a mammalian telomere-associated complex formed by multiple telomeric proteins. *J Biol Chem*, 279(49), 51338-51342.

[115] Baumann, P., & Cech, T. R. (2001). Pot1, the putative telomere end-binding protein in fission yeast and humans. *Science*, 292(5519), 1171-1175.

[116] O'Donovan, A., Tomiyama, A. J., Lin, J., Puterman, E., Adler, N. E., Kemeny, M., Wolkowitz, O. M., Blackburn, E. H., & Epel, E. S. (2012). Stress appraisals and cellular aging: a key role for anticipatory threat in the relationship between psychological stress and telomere length. *Brain Behav Immun*, 26(4), 573-579.

[117] Epel, E. S., Lin, J., Dhabhar, F. S., Wolkowitz, O. M., Puterman, E., Karan, L., & Blackburn, E. H. (2010). Dynamics of telomerase activity in response to acute psychological stress. *Brain Behav Immun*, 24(4), 531-539.

[118] Damjanovic, A. K., Yang, Y., Glaser, R., Kiecolt-Glaser, J. K., Nguyen, H., Laskowski, B., Zou, Y., Beversdorf, D. Q., & Weng, N. P. (2007). Accelerated telomere erosion is associated with a declining immune function of caregivers of Alzheimer's disease patients. *J Immunol*, 179(6), 4249-4254.

[119] Smogorzewska, A., & de Lange, T. (2004). Regulation of telomerase by telomeric proteins. *Annu Rev Biochem*, 73, 177-208.

[120] Harley, C. B., Futcher, A. B., & Greider, C. W. (1990). Telomeres shorten during ageing of human fibroblasts. *Nature*, 345(6274), 458-460.

[121] Harley, C. B., Kim, N. W., Prowse, K. R., Weinrich, S. L., Hirsch, K. S., West, M. D., Bacchetti, S., Hirte, H. W., Counter, C. M., Greider, C. W., et al. (1994). Telomerase, cell immortality, and cancer,. *Cold Spring Harb Symp Quant Biol*, 59, 307-315.

[122] Cesare, A.J., & Reddel, R.R. (2010). Alternative lengthening of telomeres: models, mechanisms and implications. *Nat Rev Genet*, 11(5), 319-330.

[123] Verdun, R. E., & Karlseder, J. (2007). Replication and protection of telomeres. *Nature*, 447(7147), 924-931.

[124] Schulz, V. P., Zakian, V. A., Ogburn, C. E., Mc Kay, J., Jarzebowicz, A. A., Edland, S. D., & Martin, G. M. (1996). Accelerated loss of telomeric repeats may not explain accelerated replicative decline of Werner syndrome cells. *Hum Genet*, 97(6), 750-754.

[125] Baird, D. M., Davis, T., Rowson, J., Jones, C. J., & Kipling, D. (2004). Normal telomere erosion rates at the single cell level in Werner syndrome fibroblast cells. *Hum Mol Genet*, 13(14), 1515-1524.

[126] Lombard, D. B., Beard, C., Johnson, B., Marciniak, R. A., Dausman, J., Bronson, R., Buhlmann, J. E., Lipman, R., Curry, R., Sharpe, A., Jaenisch, R., & Guarente, L. (2000). Mutations in the WRN gene in mice accelerate mortality in a p53-null background. *Mol Cell Biol*, 20(9), 3286-3291.

[127] Chang, S. (2005). A mouse model of Werner Syndrome: what can it tell us about aging and cancer? *Int J Biochem Cell Biol*, 37(5), 991-999.

[128] Du, X., Shen, J., Kugan, N., Furth, E. E., Lombard, D. B., Cheung, C., Pak, S., Luo, G., Pignolo, R. J., De Pinho, R. A., Guarente, L., & Johnson, F. B. (2004). Telomere shortening exposes functions for the mouse Werner and Bloom syndrome genes. *Mol Cell Biol*, 24(19), 8437-8446.

[129] Chang, S., Multani, A. S., Cabrera, N. G., Naylor, M. L., Laud, P., Lombard, D., Pathak, S., Guarente, L., & De Pinho, R. A. (2004). Essential role of limiting telomeres in the pathogenesis of Werner syndrome. *Nat Genet*, 36(8), 877-882.

[130] Stavropoulos, D. J., Bradshaw, P. S., Li, X., Pasic, I., Truong, K., Ikura, M., Ungrin, M., & Meyn, M. S. (2002). The Bloom syndrome helicase BLM interacts with TRF2 in ALT cells and promotes telomeric DNA synthesis. *Hum Mol Genet*, 11(25), 3135-3144.

[131] Lillard-Wetherell, K., Machwe, A., Langland, G. T., Combs, K. A., Behbehani, G. K., Schonberg, S. A., German, J., Turchi, J. J., Orren, D. K., & Groden, J. (2004). Association and regulation of the BLM helicase by the telomere proteins TRF1 and TRF2. *Hum Mol Genet*, 13(17), 1919-1932.

[132] Machwe, A., Xiao, L., & Orren, D. K. (2004). TRF2 recruits the Werner syndrome (WRN) exonuclease for processing of telomeric DNA. *Oncogene*, 23(1), 149-156.

[133] Opresko, P. L., von, C., Kobbe, J. P., Laine, J., Harrigan, I. D., & Hickson, V. A. (2002). Telomere-binding protein TRF2 binds to and stimulates the Werner and Bloom syndrome helicases. *J Biol Chem*, 277(43), 41110-41119.

[134] Opresko, P. L., Mason, P. A., Podell, E. R., Lei, M., Hickson, I. D., Cech, T. R., & Bohr, V. A. (2005). POT1 stimulates RecQ helicases WRN and BLM to unwind telomeric DNA substrates. *J Biol Chem*, 280(37), 32069-32080.

[135] Nora, G. J., Buncher, N. A., & Opresko, P. L. (2010). Telomeric protein TRF2 protects Holliday junctions with telomeric arms from displacement by the Werner syndrome helicase. *Nucleic Acids Res*, 38(12), 3984-3998.

[136] Lee, J. Y., Kozak, M., Martin, J. D., Pennock, E., & Johnson, F. B. (2007). Evidence that a RecQ helicase slows senescence by resolving recombining telomeres. *PLoS Biol*, 5(6), e160.

[137] Duquette, M. L., Handa, P., Vincent, J. A., Taylor, A. F., & Maizels, N. (2004). Intracellular transcription of G-rich DNAs induces formation of G-loops, novel structures containing G4 DNA. *Genes Dev*, 18(13), 1618-1629.

[138] Zaug, A. J., Podell, E. R., & Cech, T. R. (2005). Human POT1 disrupts telomeric G-quadruplexes allowing telomerase extension in vitro. *Proc Natl Acad Sci U S A*, 102(31), 10864-10869.

[139] Mohaghegh, P., & Hickson, I. D. (2001). DNA helicase deficiencies associated with cancer predisposition and premature ageing disorders. *Hum Mol Genet*, 10(7), 741-746.

[140] Opresko, P. L., Laine, J. P., Brosh, R. M., Seidman, M. M., & Bohr, V. A. (2001). Coordinate action of the helicase and 3' to 5' exonuclease of Werner syndrome protein. *J Biol Chem*, 276(48), 44677-44687.

[141] Kusumoto-Matsuo, R., Opresko, P. L., Ramsden, D., Tahara, H., & Bohr, V. A. (2010). Cooperation of DNA-PKcs and WRN helicase in the maintenance of telomeric D-loops. *Aging (Albany NY)*, 2(5), 274-284.

[142] Crabbe, L., Verdun, R. E., Haggblom, C. I., & Karlseder, J. (2004). Defective telomere lagging strand synthesis in cells lacking WRN helicase activity. *Science*, 306(5703), 1951-1953.

[143] Dewar, J. M., & Lydall, D. (2010). Pif1- and Exo1-dependent nucleases coordinate checkpoint activation following telomere uncapping. *EMBO J*, 29(23), 4020-4034.

[144] Barefield, C., & Karlseder, J. (2012). The BLM helicase contributes to telomere maintenance through processing of late-replicating intermediate structures. *Nucleic Acids Res*, 40(15), 7358-7367.

[145] Opresko, P. L. (2008). Telomere ResQue and preservation--roles for the Werner syndrome protein and other RecQ helicases. *Mech Ageing Dev*, 129(1-2), 79-90.

[146] Arnoult, N., Saintome, C., Ourliac-Garnier, I., Riou, J. F., & Londono-Vallejo, A. (2009). Human POT1 is required for efficient telomere C-rich strand replication in the absence of WRN. *Genes Dev*, 23(24), 2915-2924.

[147] Schulz, V.P., & Zakian, V.A. (1994). The saccharomyces PIF1 DNA helicase inhibits telomere elongation and de novo telomere formation. *Cell*, 76(1), 145-155.

[148] George, T., Wen, Q., Griffiths, R., Ganesh, A., Meuth, M., & Sanders, C. M. (2009). Human Pif1 helicase unwinds synthetic DNA structures resembling stalled DNA replication forks. *Nucleic Acids Res*, 37(19), 6491-6502.

[149] Paeschke, K., Capra, J. A., & Zakian, V. A. (2011). DNA replication through G-quadruplex motifs is promoted by the Saccharomyces cerevisiae Pif1 DNA helicase. *Cell*, 145(5), 678-691.

[150] Cheung, I., Schertzer, M., Rose, A., & Lansdorp, P. M. (2002). Disruption of dog-1 in Caenorhabditis elegans triggers deletions upstream of guanine-rich DNA. *Nat Genet*, 31(4), 405-409.

[151] Vallur, A. C., & Maizels, N. (2010). Distinct activities of exonuclease 1 and flap endonuclease 1 at telomeric g4 DNA. *PLoS One*, 5(1), e8908.

[152] Eller, M. S., Liao, X., Liu, S., Hanna, K., Backvall, H., Opresko, P. L., Bohr, V. A., & Gilchrest, B. A. (2006). A role for WRN in telomere-based DNA damage responses. *Proc Natl Acad Sci U S A*, 103(41), 15073-15078.

[153] Li, B., Jog, S. P., Reddy, S., & Comai, L. (2008). WRN controls formation of extrachromosomal telomeric circles and is required for TRF2DeltaB-mediated telomere shortening. *Mol Cell Biol*, 28(6), 1892-1904.

[154] Reddy, S., Li, B., & Comai, L. (2010). Processing of human telomeres by the Werner syndrome protein. *Cell Cycle*, 9(16), 3137-3138.

[155] Li, B., Reddy, S., & Comai, L. (2009). Sequence-specific processing of telomeric 3' overhangs by the Werner syndrome protein exonuclease activity. *Aging (Albany NY)*, 1(3), 289-302.

[156] Orren, D. K., Theodore, S., & Machwe, A. (2002). The Werner syndrome helicase/exonuclease (WRN) disrupts and degrades D-loops in vitro. *Biochemistry*, 41(46), 13483-13488.

[157] Johnson, F. B., Marciniak, R. A., Mc Vey, M., Stewart, S. A., Hahn, W. C., & Guarente, L. (2001). The Saccharomyces cerevisiae WRN homolog Sgs1p participates in telomere maintenance in cells lacking telomerase. *EMBO J*, 20(4), 905-913.

[158] Cohen, H., & Sinclair, D. A. (2001). Recombination-mediated lengthening of terminal telomeric repeats requires the Sgs1 DNA helicase. *Proc Natl Acad Sci U S A* [6], 3174-3179.

[159] Neumann, A.A., & Reddel, R.R. (2002). Telomere maintenance and cancer-- look, no telomerase. *Nat Rev Cancer*, 2(11), 879-884.

[160] Dunham, M. A., Neumann, A. A., Fasching, C. L., & Reddel, R. R. (2000). Telomere maintenance by recombination in human cells. *Nat Genet*, 26(4), 447-450.

[161] Hickson, I. D. (2003). RecQ helicases: caretakers of the genome. *Nat Rev Cancer*, 3(3), 169-178.

[162] Bailey, S. M., Cornforth, M. N., Ullrich, R. L., & Goodwin, E. H. (2004). Dysfunctional mammalian telomeres join with DNA double-strand breaks. *DNA Repair (Amst)*, 3(4), 349-357.

[163] Bailey, S. M., Brenneman, M. A., & Goodwin, E. H. (2004). Frequent recombination in telomeric DNA may extend the proliferative life of telomerase-negative cells. *Nucleic Acids Res*, 32(12), 3743-3751.

[164] Laud, P. R., Multani, A. S., Bailey, S. M., Wu, L., Kingsley, J. C., Lebel, M., Pathak, S., De Pinho, R. A., & Chang, S. (2005). Elevated telomere-telomere recombination in WRN-deficient, telomere dysfunctional cells promotes escape from senescence and engagement of the ALT pathway. *Genes Dev*, 19(21), 2560-2570.

[165] Celli, G. B., Denchi, E. L., & de Lange, T. (2006). Ku70 stimulates fusion of dysfunctional telomeres yet protects chromosome ends from homologous recombination. *Nat Cell Biol 8* [8], 885-890.

[166] Opresko, P. L., Fan, J., Danzy, S., Wilson, D. M., & Bohr, V. A. (2005). Oxidative damage in telomeric DNA disrupts recognition by TRF1 and TRF2. *Nucleic Acids Res*, 33(4), 1230-1239.

[167] Cooper, M. P., Machwe, A., Orren, D. K., Brosh, R. M., Ramsden, D., & Bohr, V. A. (2000). Ku complex interacts with and stimulates the Werner protein. *Genes Dev 14* [8], 907-912.

[168] Petersen, S., Saretzki, G., & von Zglinicki, T. (1998). Preferential accumulation of single-stranded regions in telomeres of human fibroblasts. *Exp Cell Res*, 239(1), 152-160.

[169] von, T., Zglinicki, G., Saretzki, J., Ladhoff, F., d'Adda di, Fagagna., & Jackson, S. P. (2005). Human cell senescence as a DNA damage response. *Mech Ageing Dev*, 126(1), 111-117.

[170] Passos, J. F., Saretzki, G., Ahmed, S., Nelson, G., Richter, T., Peters, H., Wappler, I., Birket, M. J., Harold, G., Schaeuble, K., Birch-Machin, M. A., Kirkwood, T. B., & von Zglinicki, T. (2007). Mitochondrial dysfunction accounts for the stochastic heterogeneity in telomere-dependent senescence. *PLoS Biol*, 5(5), e110.

[171] Passos, J. F., Saretzki, G., & von Zglinicki, T. (2007). DNA damage in telomeres and mitochondria during cellular senescence: is there a connection? *Nucleic Acids Res*, 35(22), 7505-7513.

[172] Karlseder, J., Smogorzewska, A., & de Lange, T. (2002). Senescence induced by altered telomere state, not telomere loss. *Science*, 295(5564), 2446-2449.

[173] Harrigan, J. A., Wilson, D. M., 3rd Prasad, R., Opresko, P. L., Beck, G., May, A., Wilson, S. H., & Bohr, V. A. (2006). The Werner syndrome protein operates in base excision repair and cooperates with DNA polymerase beta. *Nucleic Acids Res*, 34(2), 745-754.

[174] Harrigan, J. A., Opresko, P. L., von, C., Kobbe, P. S., Kedar, R., Prasad, S. H., Wilson, , & Bohr, V. A. (2003). The Werner syndrome protein stimulates DNA polymerase beta strand displacement synthesis via its helicase activity. *J Biol Chem*, 278(25), 22686-22695.

[175] Pagano, G., Zatterale, A., Degan, P., d'Ischia, M., Kelly, F. J., Pallardo, F. V., & Kodama, S. (2005). Multiple involvement of oxidative stress in Werner syndrome phenotype. *Biogerontology*, 6(4), 233-243.

[176] Labbe, A., Turaga, R. V., Paquet, E. R., Garand, C., & Lebel, M. (2010). Expression profiling of mouse embryonic fibroblasts with a deletion in the helicase domain of the Werner Syndrome gene homologue treated with hydrogen peroxide. *BMC Genomics*, 11, 127.

[177] Ghosh, A., Rossi, M. L., Aulds, J., Croteau, D., & Bohr, V. A. (2009). Telomeric D-loops containing 8-oxo-2'-deoxyguanosine are preferred substrates for Werner and Bloom syndrome helicases and are bound by POT1. *J Biol Chem*, 284(45), 31074-31084.

[178] Muftuoglu, M., Wong, H. K., Imam, S. Z., Wilson, D. M., 3rd Bohr, V. A., & Opresko, P. L. (2006). Telomere repeat binding factor 2 interacts with base excision repair proteins and stimulates DNA synthesis by DNA polymerase beta. *Cancer Res*, 66(1), 113-124.

[179] Miller, A., Balakrishnan, L., Buncher, N. A., Opresko, P. L., & Bambara, R. A. (2012). Telomere proteins POT1, TRF1 and TRF2 augment long-patch base excision repair in vitro. *Cell Cycle*.

Permissions

The contributors of this book come from diverse backgrounds, making this book a truly international effort. This book will bring forth new frontiers with its revolutionizing research information and detailed analysis of the nascent developments around the world.

We would like to thank David Stuart, for lending his expertise to make the book truly unique. He has played a crucial role in the development of this book. Without his invaluable contribution this book wouldn't have been possible. He has made vital efforts to compile up to date information on the varied aspects of this subject to make this book a valuable addition to the collection of many professionals and students.

This book was conceptualized with the vision of imparting up-to-date information and advanced data in this field. To ensure the same, a matchless editorial board was set up. Every individual on the board went through rigorous rounds of assessment to prove their worth. After which they invested a large part of their time researching and compiling the most relevant data for our readers. Conferences and sessions were held from time to time between the editorial board and the contributing authors to present the data in the most comprehensible form. The editorial team has worked tirelessly to provide valuable and valid information to help people across the globe.

Every chapter published in this book has been scrutinized by our experts. Their significance has been extensively debated. The topics covered herein carry significant findings which will fuel the growth of the discipline. They may even be implemented as practical applications or may be referred to as a beginning point for another development. Chapters in this book were first published by InTech; hereby published with permission under the Creative Commons Attribution License or equivalent.

The editorial board has been involved in producing this book since its inception. They have spent rigorous hours researching and exploring the diverse topics which have resulted in the successful publishing of this book. They have passed on their knowledge of decades through this book. To expedite this challenging task, the publisher supported the team at every step. A small team of assistant editors was also appointed to further simplify the editing procedure and attain best results for the readers.

Our editorial team has been hand-picked from every corner of the world. Their multi-ethnicity adds dynamic inputs to the discussions which result in innovative outcomes. These outcomes are then further discussed with the researchers and contributors who give their valuable feedback and opinion regarding the same. The feedback is then collaborated with the researches and they are edited in a comprehensive manner to aid the understanding of the subject.

Apart from the editorial board, the designing team has also invested a significant amount of their time in understanding the subject and creating the most relevant covers. They scrutinized every image to scout for the most suitable representation of the subject and create an appropriate cover for the book.

The publishing team has been involved in this book since its early stages. They were actively engaged in every process, be it collecting the data, connecting with the contributors or procuring relevant information. The team has been an ardent support to the editorial, designing and production team. Their endless efforts to recruit the best for this project, has resulted in the accomplishment of this book. They are a veteran in the field of academics and their pool of knowledge is as vast as their experience in printing. Their expertise and guidance has proved useful at every step. Their uncompromising quality standards have made this book an exceptional effort. Their encouragement from time to time has been an inspiration for everyone.

The publisher and the editorial board hope that this book will prove to be a valuable piece of knowledge for researchers, students, practitioners and scholars across the globe.

List of Contributors

Yoshizumi Ishino and Sonoko Ishino
Department of Bioscience and Biotechnology, Graduate School of Bioresource and Bio-environmental Sciences, Kyushu University, Japan

David Stuart
Department of Biochemistry, University of Alberta, Edmonton, Alberta, Canada

John C. Fisk, Michaelle D. Chojnacki and Thomas Melendy
Department of Microbiology & Immunology, University at Buffalo School of Medicine & Biomedical Sciences, Buffalo, NY, USA

Dianne C. Daniel, Ayuna V. Dagdanova and Edward M. Johnson
Department of Microbiology and Molecular Cell Biology, Eastern Virginia Medical School, Norfolk, Virginia, USA

Agustino Martinez-Antonio, Laura Espindola-Serna and Cesar Quiñones-Valles
Departamento de Ingeniería Genética, Cinvestav, Irapuato-León, Irapuato Gto, México

Apolonija Bedina Zavec
National Institute of Chemistry, Department for Molecular biology and Nanobiotechnology, Slovenia

Lindsay A. Matthews and Alba Guarné
Department of Biochemistry and Biomedical Sciences, McMaster University, Hamilton, ON, Canada

Sarah A. Sabatinos and Susan L. Forsburg
Department of Molecular and Computational Biology, University of Southern California, USA

Lynne S. Cox and Penelope A. Mason
Department of Biochemistry, University of Oxford, South Parks Road, Oxford

Printed in the USA
CPSIA information can be obtained
at www.ICGtesting.com
JSHW011437221024
72173JS00004B/843